Chemistry
Students' Book I
Topics 1 to 12

14 082651 3
Nuffield Advanced Science

Project team

E. H. Coulson, formerly of County High School, Braintree (organizer)
A. W. B. Aylmer-Kelly, formerly of Royal Grammar School, Worcester
Dr E. Glynn, formerly of Croydon Technical College
H. R. Jones, formerly of Carlett Park College of Further Education
A. J. Malpas, formerly of Highgate School
Dr A. L. Mansell, formerly of Hatfield College of Technology
J. C. Mathews, King Edward VII School, Lytham
Dr G. Van Praagh, formerly of Christ's Hospital
J. G. Raitt, formerly of Department of Education, University of Cambridge
B. J. Stokes, King's College School, Wimbledon
R. Tremlett, College of St Mark and St John
M. D. W. Vokins, Clifton College

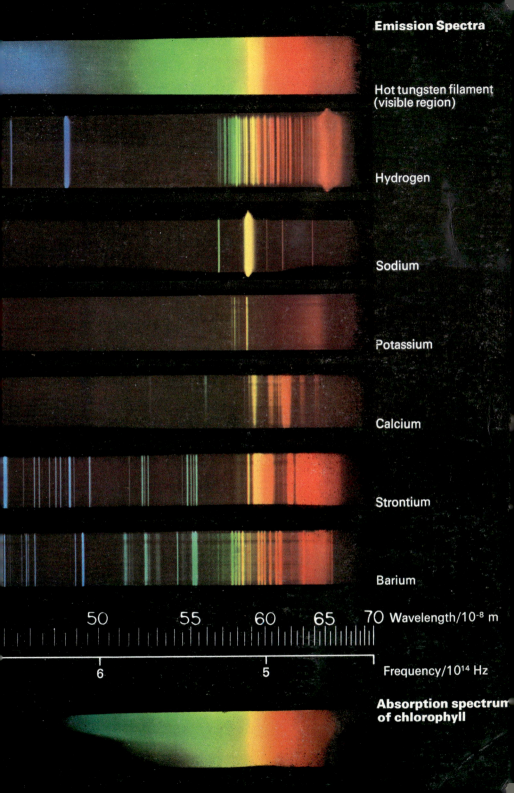

Emission Spectra

Hot tungsten filament
(visible region)

Hydrogen

Sodium

Potassium

Calcium

Strontium

Barium

50   55   60   65   70 Wavelength/10^{-8} m

6   5   Frequency/10^{14} Hz

**Absorption spectrum
of chlorophyll**

BOOK PRICE	s	d	BOOK NUMBER.	36.

COURSE	STUDENT'S NAME	DATE

Chemistry

Students' Book I

Topics 1 to 12

Nuffield Advanced Science
Published for the Nuffield Foundation by Penguin Books

Penguin Books Ltd, Harmondsworth, Middlesex, England
Penguin Books Inc., 7110 Ambassador Road, Baltimore, Md 21207, U.S.A.
Penguin Books Ltd, Ringwood, Victoria, Australia

Copyright © The Nuffield Foundation 1970
First published 1970
Reprinted with corrections 1971

Filmset in 'Monophoto' Times New Roman by J. W. Arrowsmith Ltd, Bristol
and made and printed in Great Britain by Gilmour and Dean Ltd, Hamilton.

Design and art direction by Ivan and Robin Dodd
Illustrations designed and produced by Penguin Education

Contents

Foreword

Sixth form courses in Britain have received more than their fair share of blessing and cursing in the last twenty years: blessing, because their demands, their compass, and their teachers are often of a standard which in other countries would be found in the first year of a longer university course than ours: cursing, because this same fact sets a heavy cloud of university expectation on their horizon (with awkward results for those who finish their education at the age of 18) and limits severely the number of subjects that can be studied in the sixth form.

So advanced work, suitable for students between the ages of 16 and 18, is at the centre of discussions on the curriculum. It need not, of course, be in a 'sixth form' at all, but in an educational institution other than a school. In any case, the emphasis on the requirements of those who will not go to a university or other institute of higher education is increasing, and will probably continue to do so; and the need is for courses which are satisfying and intellectually exciting in themselves – not for courses which are simply passports to further study.

Advanced science courses are therefore both an interesting and a difficult venture. Yet fresh work on advanced science teaching was obviously needed if new approaches to the subject (with all the implications that these have for pupils' interest in learning science and adults' interest in teaching it) were not to fail in their effect. The Trustees of the Nuffield Foundation therefore agreed to support teams, on the same model as had been followed in their other science projects, to produce advanced courses in Physical Science, in Physics, in Chemistry, and in Biological Science. It was realized that the task would be an immense one, partly because of the universities' special interest in the approach and content of these courses, partly because the growing size of sixth forms underlined the point that advanced work was not *solely* a preparation for a degree course, and partly because the blending of Physics and Chemistry in a single Advanced Physical Science course was bound to produce problems. Yet, in spite of these pressures, the emphasis here, as in the other Nuffield Science courses, is on learning rather than on being taught, on understanding rather than amassing information, on finding out rather than on being told: this emphasis is central to all worthwhile attempts at curriculum renewal.

If these advanced courses meet with the success and appreciation which I believe they deserve, then the credit will belong to a large number of people, in the teams and the consultative committees, in schools and universities, in authorities and councils, and associations and boards: once again it has been the

Foundation's privilege to provide a point at which the imaginative and helpful efforts of many could come together.

Brian Young
Director of the Nuffield Foundation

Contributors

Many people have contributed to this book. Final decisions on the content and treatment of the teaching scheme with which it is concerned were made by members of the Headquarters Team, who were also responsible for assembling and writing the material for the several draft versions that were used in school trials of the course. During this exercise much valuable help and advice was generously given by teachers in schools, and in universities and other institutions of higher education. In particular, the comments and suggestions of teachers taking part in the school trials have made a vital contribution to the final form of the published material. The editing of the final manuscript was carried out by B. J. Stokes.

At a time when the Headquarters Team was small in number and under heavy pressure, the following teachers undertook between them the task of developing experimental investigations and planning the theoretical treatment for the first drafts of Topics 12, 14, 15, 16, 17, 18, and 19: W. H. Francis and G. R. Grace (Apsley Grammar School, Hemel Hempstead), K. C. Horncastle (Exeter School), D. R. P. Jolly (Berkhamsted School), Dr R. Kempa (College of St Mark and St John), J. A. Kent (Southfield School, Oxford), P. Meredith (Exeter School), A. G. Moll (Plymouth College), A. B. Newall (The Grammar School for Boys, Cambridge), M. Pailthorpe (Harrow School), Miss K. Rennert (Cheney Girls' School, Oxford), T. A. G. Silk (Blundell's School), Dr R. C. Whitfield (Department of Education, University of Cambridge). It is fitting that their contribution should be acknowledged here. Thanks are also due to D. R. Browning, for advice and assistance with regard to safety measures in laboratory work.

E. H. Coulson

Introduction

This book

Not a textbook, not a background book, not a book of data – this book has been deliberately left in many ways incomplete. Like an outline map of an unexplored island, it will provide you with a series of starting points from which to explore the interior, to build up your own picture of chemistry at this level. To do so, you do not merely need to read the book: you need to do the course, using all the other resources available to you – textbooks, data books, films, discussion with others, and, above all, your own practical work.

This course

You are now beginning a new course in chemistry. What are you hoping to achieve by doing it? What is chemistry like at this level? Where will it get you? These are questions you can only answer at all fully after completing the course. But consider the essential nature of chemistry – what chemists do and how they think – and you will have some idea of what to expect.

Like all other sciences, chemistry is the study of the behaviour of materials. Behaviour implies change, and the type of change that interests you to a large extent influences the sort of science you do. Chemistry is the study of materials, but so is physics and so is engineering. What then makes chemistry distinctively chemical and different from physics or engineering?

Take a piece of aluminium as an example of a typical material. If you are interested, say, in the way in which the heat capacity or the electrical resistance of the aluminium changes when it is cooled to nearly the absolute zero of temperature, you are studying it as a physicist. If you are interested in the stresses which develop in a piece of aluminium when it forms part of a larger structure such as a supersonic aeroplane, you may be studying the aluminium as an engineer. If you are interested in the changes that take place when the aluminium is heated in an atmosphere of chlorine, what happens to the aluminium atoms, how they react with the chlorine molecules, what energy changes accompany the reaction, and to what uses the product can be put, you are studying the aluminium as a chemist.

Physicists, engineers, metallurgists, and chemists, in studying materials such as aluminium, all share many common objectives and methods of working. Indeed, as time goes on the boundaries between chemistry, physics, engineering, metallurgy, etc. may disappear and the subjects merge into one. At present, however, different scientists still have their own different emphases. Where does the emphasis lie in chemistry?

The entire physical world is composed of little more than one hundred different elements, but the atoms of these elements in linking to form compounds can combine in millions of different ways. Modern chemistry involves the study of the way the atoms are linked together by chemical bonds to form larger structures such as molecules. Much of chemistry is concerned with elucidating chemical structure and for this many powerful methods such as molecular spectroscopy or X-ray crystallography are available to the chemist. But chemists are not only concerned with structure; they also study the changes which take place and the patterns of change when the atoms of a structure become disengaged from one another (the 'chemical bonds are broken') and link up to form new structures. This is the essence of a chemical, as opposed to a physical or any other type of process – the breaking and making of chemical bonds. Chemistry is the study of this bond breaking and bond making process in all its aspects, including how rapidly it occurs and what energy changes accompany it.,

In understanding the underlying structure of a material we can often explain some of its macroscopic (large-scale) properties. For example, the hardness of diamond and the elasticity of rubber can be explained in terms of the ways in which the atoms and molecules, respectively, of these materials are arranged and interlinked. In understanding the bonding between atoms, and the material and energy changes that happen when bonds are made and broken, chemists can build structures to their own design to replace and often to supersede naturally occurring materials. This opens up exciting possibilities. Macromolecules for miniskirts, polymers for plastics, alloys for aero engines, drugs against disease – these are a few examples of the contribution chemistry can make to a more interesting, easier, faster, longer life for all.

Experiments and ideas: a simple example
For chemists, as for other workers in science, experiments and ideas go hand in hand. The two are very closely linked, as the following example shows.

Suppose you have been investigating the way in which hydrogen reacts with other chemical elements. After reading the reports of others who have worked in this field you have succeeded in making a small sample of lithium hydride by passing hydrogen gas over heated lithium metal using an apparatus similar to that shown opposite.

It called for considerable care and skill to cause an inflammable gas to combine safely with a vigorous reactive metal. The compound is a white crystalline material, which reacts rapidly with water, so it must be kept in the dry. What would you do next? You have a sample of an interesting and (to you) new compound, lithium hydride. You could examine how it reacts in the presence of

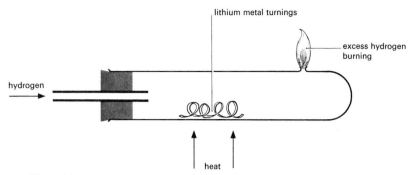

Figure 1.1a
Apparatus for making lithium hydride from lithium metal and hydrogen gas.

various laboratory reagents to see how your observations compare with those of others who have worked on lithium hydride. You could also follow up some of the many interesting questions which are raised as a result of the experiment you have just done. Here are a few.

A   Could other hydrides be prepared in this way? Are there other ways of preparing hydrides? Are all hydrides similar in their behaviour or are there differences? Is there any pattern in the behaviour of hydrogen gas with other elements (e.g. the other alkali metals) or in the behaviour of the hydrides?

B   What is happening at the level of the atoms and molecules during this reaction? What chemical bonds are broken and made and how much energy is required to do this? What is the structure of lithium hydride? Is it ionic or co-valent? How could this be settled?

C   How rapidly did the reaction between hydrogen and lithium go? What would affect how fast it goes and in what way?

D   Could we get the reaction to go backwards? What would be needed to do so? Was all the lithium converted to the hydride? Has this reaction 'gone to completion' or is it an equilibrium process?

E   Does the reaction need heat to keep it going as opposed to starting it off? Is there any evidence of heat being produced by the reaction? If so, how much?

F   Is lithium hydride likely to be of any use, for example in medicine or industry?

These groups of questions, which can apply to any chemical process you are studying, illustrate the main themes running through this course. They are summarized on the next page.

1   Particular chemical changes in materials
2   Patterns in the chemical behaviour of materials
3   Structure, including the structure of atoms and the structure of molecules and crystals
4   Rates of reactions
5   Equilibria in chemical systems
6   Energy changes accompanying chemical changes
7   Applications – industrial, medical, economic, and other social aspects

Starting with a particular chemical change in a particular material you are at
once led to questions of patterns of chemical change, energy, structure, rates of
reaction, equilibrium, and applications. These questions in turn will lead you
back to experiment as a means of answering them. You may decide next, for
example, to start a series of experiments to try to make other metal hydrides by
this method. Or you may start an investigation of the factors influencing the
rate of the chemical combination, and so on. Experiments lead to ideas and ideas
to further experiments. In a sense we traverse a helix of understanding, moving
periodically in and out of the realms of ideas and experiments (see figure 1.1b).

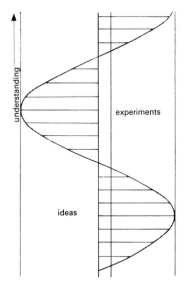

Figure 1.1b
A 'helix of understanding'.

The main point, then, about the activities of a chemist is that he deals in both
ideas and experiments which go hand in hand; just as ideas without experiments
are merely speculation, so experiments without ideas are merely trivial.

It is impossible to convey in a short section the great variety of interest and activity there is in modern chemistry. This you can find out only by *doing* chemistry, not by reading about it. Enough has been said, however, to make it clear that being 'good at' chemistry involves being able to report clearly and accurately what you do, what you observe, and what you think. This means developing your ability to write clear, unambiguous, readable English. Science is essentially an activity done by a community of people. This implies personal relationships. Good relationships depend on good communication.

Topic 1
Amount of substance

One fundamental quantity chemists need to know when measuring out materials is the amount of chemical substance.

In physics it is sometimes sufficient to measure quantities of material in units of mass (grammes or kilogrammes) for example in considering the motions of bodies. For chemical purposes this is not sufficient because proper chemical comparisons between different materials can only be made if comparable numbers of particles are considered in each case. The example of the experiment to make lithium hydride (see Introduction) can be used to make this point clearer. One question we were led to ask as a result of that experiment was whether other hydrides could be prepared in the same way. Let us take this one stage further.

Suppose we have one gramme of hydrogen available for combination and a supply of the metals lithium, Li, sodium, Na, and potassium, K. Assuming that all the hydrogen can be efficiently converted into hydride and that the formulae of all three hydrides are of the form MH, using which metal could we obtain the largest amount of hydride?

From one gramme of hydrogen the masses of the three hydrides obtainable are 8 g LiH, 24 g NaH, and 40 g KH, so that at first sight it may seem that with potassium the largest amount of hydride is obtained and with lithium the least. It is clear, however, that in the three equations,

$$Li(s) + \tfrac{1}{2}H_2(g) \longrightarrow LiH(s)$$

$$Na(s) + \tfrac{1}{2}H_2(g) \longrightarrow NaH(s)$$

$$K(s) + \tfrac{1}{2}H_2(g) \longrightarrow KH(s)$$

for every atom of hydrogen reacting, one atom of metal reacts and one formula unit of the hydride is formed, so there is a sense in which the same amount of metal reacts with one gramme of hydrogen in each case to give the same amount of hydride.

For chemists, equal amounts of different substances mean equal numbers of particles. So to compare equal amounts we have to weigh out, not equal masses, but masses of substances which are in the ratio of the masses of the particles present in them. The fact that the atoms of different elements have different masses is in a sense an inconvenience.

When chemists were first realizing, just over one hundred years ago, that what mattered was numbers of particles of substances, hydrogen, as the lightest element, was chosen as a reference element and the mass of the hydrogen atom was chosen as the atomic unit of mass. Masses of other atoms and molecules were expressed in terms of the number of times they were as heavy as one hydrogen atom.

Since the unit of mass was the gramme it was natural to choose one gramme of hydrogen as a unit amount of substance. This unit amount of substance later came to be called one mole of hydrogen atoms. The number of hydrogen atoms in one gramme of hydrogen was called first the Avogadro number, and later the Avogadro constant.

The carbon-twelve scale of atomic mass
In 1961 a new standard of atomic mass was adopted, based on the isotope of carbon of mass twelve (^{12}C). It was chosen for various reasons, most of them concerned with the existence of isotopes. The standard unit is defined as twelve grammes exactly of the carbon-twelve isotope. The masses of all other atoms are now expressed in terms of this. For example, the atomic mass of gold, Au, relative to ^{12}C is 196.967. This means that, on the average, the masses of gold atoms and the mass of atoms of the ^{12}C isotope are in the ratio

196.967 : 12.0000

Definition of the Avogadro constant, L
The Avogadro constant, represented by the symbol L, is now defined as the number of atoms of the ^{12}C isotope in exactly twelve grammes of ^{12}C.

Since carbon atoms have masses about twelve times as great as hydrogen atoms there is little difference between the definition of the Avogadro constant in this way and the original definition as the number of atoms in one gramme of hydrogen. In Topic 3 you will do an experiment to estimate approximately the numerical value of the Avogadro constant.

Definition of the mole
The mole (unit amount of substance) is now also defined in terms of the ^{12}C isotope: one mole of any substance is that amount of substance which contains as many particles (molecules, atoms, ions, etc.) as there are atoms of ^{12}C in twelve grammes (exactly) of ^{12}C.

This is equivalent to saying that one mole of any substance is that amount of substance which contains the Avogadro number of its particles as defined on the ^{12}C scale. It is for this reason that the Avogadro number is now referred to as Avogadro's constant (rather than number) – because it is a constant number of particles per mole of particles for all substances. It has units $mole^{-1}$.

You will notice that this definition of the mole refers to particles (molecules, atoms, ions, etc.). It is therefore vital in referring to amounts of substances to specify what particles, in this sense, are meant. For example, the phrase 'one mole of chlorine' is ambiguous; it could mean one mole of chlorine molecules (Cl_2), that is, a gramme-molecule of chlorine, 71.0 g, or it could mean one mole of chlorine atoms (Cl), that is 35.5 g. For this reason the *formula* of the substance being considered must always be stated, and the correct phrases should be 'one mole of chlorine molecules Cl_2' and 'one mole of chlorine atoms Cl'.

In substances consisting of giant lattices, whether they are covalent, such as silicon dioxide, SiO_2, or ionic, such as sodium chloride, Na^+Cl^-, the *gramme-formula* must be taken as the mole. In this latter example, one mole of sodium chloride, Na^+Cl^-, contains one mole of sodium ions and one mole of chloride ions, whereas one mole of calcium chloride, $Ca^{2+}Cl_2^-$, consists of one mole of calcium ions, Ca^{2+}, and two moles of chloride ions Cl^-. Difficulties which might arise, are dealt with by stating the formula being considered in every case. The mole is simply another way of referring to a gramme-formula of any substance.

Dissolving a mole of a substance in sufficient water to make 1 cubic decimetre (litre)* of solution gives a *molar solution* of that substance.

From the above considerations you can see that a solution may be molar (M) with respect to the substance dissolved, but two molar (2M) with respect to one of its ions. For example, molar calcium chloride solution is said to be M with respect to calcium ions, but 2M with respect to chloride ions. Because of this, solutions may sometimes be referred to as, for example, '2M with respect to Al^{3+}' if the nature of the other ion is irrelevant to the use to which the solution is to be put.

---

* The cubic decimetre, symbol dm^3, is now the preferred international name for the unit of volume otherwise known as the litre. In this book, when referring to volumes, we shall normally use cubic centimetres, cm^3, (and not millilitres, ml) and cubic decimetres, dm^3, (and not litres, l). You may find the alternative names in other books.

Problems

*1   Instructions for practical work often give the quantities of reactants in terms of moles. But for the actual measurement of these quantities they must be converted into the units of the measuring instrument; for example, grammes, cm^3, etc. The following questions require you to convert molar quantities in this manner with the aid of your *Book of Data*.

  i   What is the mass of 0.1 mole of zirconium atoms, Zr?
  ii   What is the mass of 0.02 mole of thorium atoms, Th?
  iii   What is the mass of 2 moles of nickel atoms, Ni?
  iv   What is the volume of 0.1 mole of mercury atoms, Hg?
  v   What is the volume of 1 g of red phosphorus?
  vi   What is the volume of 1 mole of phosphorus molecules, P_4?
  vii   What is the volume of 1 mole of sodium chloride, NaCl?

*2   Calculate the mass of each of the following.

  i   1 mole of hydrogen, H_2
  ii   1 mole of hydrogen, H
  iii   1 mole of silica, SiO_2
  iv   0.5 mole of carbon dioxide, CO_2
  v   0.5 mole of sodium chloride, NaCl
  vi   0.25 mole of hydrated sodium carbonate, $Na_2CO_3, 10H_2O$

*3   How many moles or part of a mole are each of the following?

  i   32 g of oxygen molecules, O_2
  ii   32 g of oxygen atoms, O
  iii   31 g of phosphorus molecules, P_4
  iv   32 g of sulphur molecules, S_8
  v   50 g of calcium carbonate, $CaCO_3$
  vi   24.5 g of sulphuric acid, H_2SO_4

*4   What mass of each of the following is dissolved in 250 cm^3 of 0.100M solution?

  i   Silver nitrate
  ii   Hydrochloric acid
  iii   Sulphuric acid
  iv   Sodium hydroxide
  v   Potassium permanganate $(KMnO_4)$
  vi   Sodium thiosulphate $(Na_2S_2O_3, 5H_2O)$

* Indicates that the *Book of Data* is needed. References to the *Book of Data* refer to the Nuffield Advanced Science *Book of Data* published jointly for the Chemistry, Physics, and Physical Science Projects. The Nuffield Chemistry *Book of Data*, used in connection with the O-level course, is also referred to in places.

*5   How many moles of each solute is contained in the following solutions? (Express your answer as a decimal, if necessary.)

  i $1000 \, cm^3$ of 0.100M silver nitrate
  ii $250 \, cm^3$ of 0.100M potassium dichromate
  iii $25 \, cm^3$ of 0.100M sodium chloride
  iv $10 \, cm^3$ of 2.00M sodium chloride
  v $12.2 \, cm^3$ of 1.56M nitric acid
  vi $12.2 \, cm^3$ of 1.56M sulphuric acid

*6   What is the molarity of each of the following solutions?

  i   1.70 g of silver nitrate in $1000 \, cm^3$ of solution
  ii   0.425 g of silver nitrate in $250 \, cm^3$ of solution
  iii 5.85 g of sodium chloride in $250 \, cm^3$ of solution
  iv   3.16 g of potassium permanganate $(KMnO_4)$ in $2 \, dm^3$ (litres) of solution
  v   6.20 g of sodium thiosulphate $(Na_2S_2O_3, 5H_2O)$ in $250 \, cm^3$ of solution
  vi   5.62 g of hydrated copper(II) sulphate $(CuSO_4, 5H_2O)$ in $250 \, cm^3$ of solution

*7   Calculate the molarity of:

  i   Ethanol (C_2H_6O), 23 g in $1 \, dm^3$ of solution
  ii   Hydrogen ion in $1 \, dm^3$ of solution containing 3.65 g of hydrogen chloride (assume that the hydrogen chloride is fully ionized)
  iii Hydroxide ion in $1 \, dm^3$ of solution containing 17.1 g of barium hydroxide, $Ba(OH)_2$ (assume that the barium hydroxide is fully ionized)
  iv   Sulphate ion in an 0.1M solution of aluminium sulphate, $Al_2(SO_4)_3$, $12H_2O$

Topic 2

Periodicity

If elements are arranged in order of increasing atomic weight, elements having similar properties recur at periodic intervals in this list. One of the first people to notice this periodicity of properties was Newlands, who presented a paper on the subject to the Chemical Society in 1866. (An extract from his paper is reproduced on page 21.)

This periodicity is particularly well seen if the elements are arranged in the form of a table. The first successful table to group elements according to their chemical behaviour in this way was devised by Mendeleev, the great Russian chemist, in 1869. Mendeleev based his table on the sixty-odd elements then known. Since that time the table has grown to accommodate over one hundred elements, and has been rearranged to take account of the electronic structures of the atoms, which were quite unknown to Mendeleev. That the original concept has proved capable of absorbing this new knowledge, however, shows the correctness of Mendeleev's original proposal.

Now known as the *Periodic Table*, this classification of elements is one of the great achievements of chemical science. The history of its development into its present form is given in the Nuffield Chemistry Background Book *The Periodic Table*.

In this topic the pattern of chemical behaviour across the Periodic Table will be considered, and in Topics 5 and 6 trends down the Periodic Table will be studied.

2.1 The Periodic Table

Have a look at a modern Periodic Table. A horizontal row across the Table is known as a *period*. Periods are numbered from the top downwards. Period 1 thus consists of hydrogen and helium; the elements lithium (Li) to neon (Ne) form period 2, and so on.

A vertical column is called a *group*. Groups having elements in periods 2 and 3 are numbered from left to right, with the exception of the group headed by neon, which is called group 0. In addition to numbers, several groups have names, and the following are in general use.

Group number	Name
I	alkali metals
II	alkaline earth metals
VII	halogens
0	inert gases

Three horizontal regions of the Table have names, and these are the transition elements, the lanthanides, and the actinides. The *transition elements* are those in the groups headed by titanium (Ti) to copper (Cu) inclusive. The *lanthanides* are the elements cerium (Ce) to lutetium (Lu) inclusive. Lanthanum and the lanthanides are sometimes known as the rare earth elements. The *actinides* are the elements thorium (Th) to lawrencium (Lw) inclusive.

These names are indicated on the outline Periodic Table (figure 2.1a).

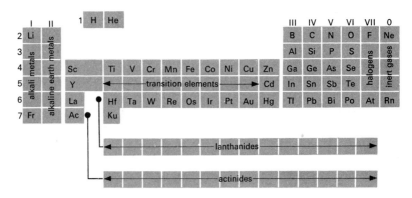

Figure 2.1a
Periodic Table of the elements.

If the elements are numbered along each period from left to right, starting at period 1, then period 2, and so on, the number given to each element is called its *atomic number*. This number has a greater significance for the Periodic Table than has atomic weight, as will be seen later.

Periodicity of physical properties

Some interesting information can be obtained by plotting graphs of various physical quantities against atomic number. If you look at figures 2.1b to 2.1f you will see that the units of the physical quantities are per mole of atoms of the element (mol^{-1}). Comparisons of substances must be based on appropriate quantities, which might be equal weights, equal volumes, or equal cost; but for chemists a valid comparison can only be based on equal numbers of particles.

Try to answer the following questions about the graphs.

Questions on figure 2.1b: *atomic volume plotted against atomic number*
1   What are the elements (a) at the peaks and (b) at the troughs of the graph?
Do the peaks occur at regular intervals on the graph?
2   In which group of the Periodic Table are these elements found?
3   Whereabouts on the graph are the elements of group II to be found?
4   If elements are found at or near a peak on the graph, what does this tell
you about their atoms?

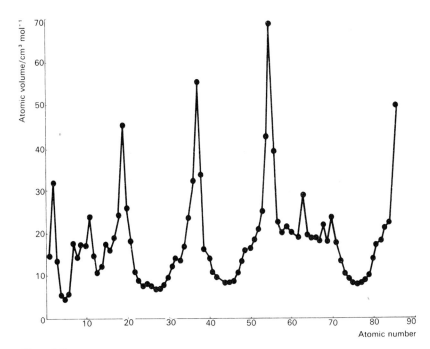

Figure 2.1b
Atomic volumes of the elements.

Questions on figures 2.1c *to* 2.1f *inclusive*
1   Which elements appear at the highest points in the various graphs?
2   Where do the elements of (a) group II and (b) group VII appear on the graphs?
3   Do these graphs give any evidence of periodicity of physical properties of the elements?

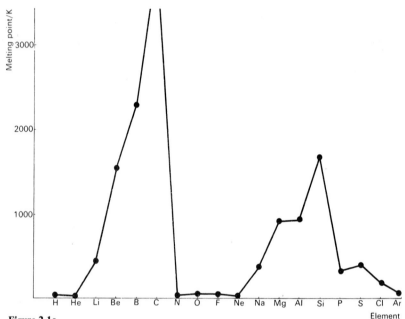

Figure 2.1c
Melting points of the elements 1–18.

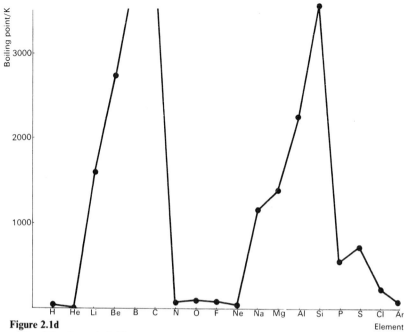

Figure 2.1d
Boiling points of the elements 1–18.

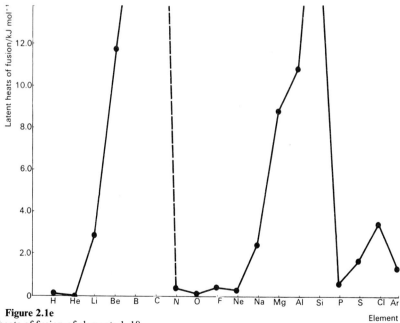

Figure 2.1e
Latent heats of fusion of elements 1–18.

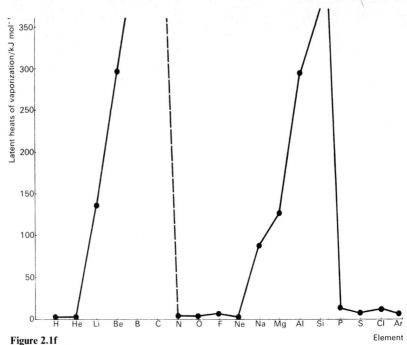

Figure 2.1f
Latent heats of vaporization of elements 1–18.

The physical properties plotted in figures 2.1c to 2.1f reflect a change in structure on moving across the Periodic Table. Structural investigations have shown that the elements on the lefthand side of the Periodic Table exist as giant lattices, whereas those on the right consist of molecules. The graphs reflect this change by the steep drop which occurs after carbon, and again after silicon.

The data on melting points and latent heats of fusion are quoted in table 2.1. Boiling points and latent heats of vaporization follow a similar pattern, as may be seen from the values given in the *Book of Data*.

Element	Li	Be	B	C	N	O	F	Ne
Structural type	← giant lattices →				← molecules →			
Melting point/°C	180	1300	2000	3700	−210	−220	−220	−250
Latent heat of fusion/kJ mol⁻¹	3.0	11.7	22.2		0.36	0.22	0.26	0.33

Element	Na	Mg	Al	Si	P(white)	S	Cl	Ar
Structural type	← giant lattices →				← molecules →			
Melting point/°C	98	650	660	1400	44	120	−100	−190
Latent heat of fusion/kJ mol⁻¹	2.6	8.9	11	46	0.63	1.4	3.2	1.2

Table 2.1

Look at some models of the molecules of the elements. From these you will see that neon and argon exist as single atoms; nitrogen, oxygen, fluorine, and chlorine have two atoms in each molecule; white phosphorus has four; and sulphur has eight.

Background reading
Early ideas of periodicity
The following is an extract from the report of a meeting of the Chemical Society on 1 March 1866, at which Newlands announced his observations on the periodicity of properties of the elements. It will be seen from the report that his ideas met with a good deal of scepticism.

> Mr John A. R. Newlands read a paper entitled 'The Law of Octaves, and the Causes of Numerical Relations among the Atomic Weights'. The author claims the discovery of a law according to which the elements analogous in their properties exhibit peculiar relationships, similar to those subsisting in music between a note and its octave. Starting from the atomic weights on Cannizzaro's system, the author arranges the known elements in order of succession, beginning with the lowest atomic weight (hydrogen) and ending with thorium (= 231.5); placing, however, nickel and cobalt, platinum and iridium, cerium and lanthanum, etc., in positions of absolute equality or in the same line. The fifty-six elements so arranged are said to form the compass of eight octaves, and the author finds that chlorine, bromine, iodine, and fluorine are thus brought into the same line, or occupy corresponding places in his scale. Nitrogen and phosphorus, oxygen and sulphur, etc., are also considered as forming true octaves.

> The author's supposition will be exemplified in Table II, shown to the meeting, and here subjoined.

No.		No.		No.		No.		No.		No.		No.		No.		No.	
H	1	F	8	Cl	15	Co and Ni	22	Br	29	Pd	36	I	42	Pt and Ir	50
Li	2	Na	9	K	16	Cu	23	Rb	30	Ag	37	Ca	44	Os	51
G	3	Mg	10	Ca	17	Zn	24	Sr	31	Cd	38	Ba and V	45	Hg	52
Bo	4	Al	11	Cr	19	Y	25	Ce and La	33	U	40	Ta	46	Tl	53
C	5	Si	12	Ti	18	In	26	Zr	32	Sn	39	W	47	Pb	54
N	6	P	13	Mn	20	As	27	Di and Mo	34	Sb	41	Nb	48	Bi	55
O	7	S	14	Fe	21	Se	28	Ro and Ru	35	Te	43	Au	49	Th	56

Table II
Elements arranged in octaves

Dr Gladstone made objection on the score of its having been assumed that no elements remain to be discovered. The last few years had brought forth thallium, indium, caesium, and rubidium, and now the finding of one more would throw out the whole system. The speaker believed there was as close an analogy subsisting between the metals named in the last vertical column as in any of the elements standing on the same horizontal line.

Professor G. F. Foster humorously inquired of Mr Newlands whether he had ever examined the elements according to the order of their initial letters? For he believed that any arrangement would present occasional coincidences, but he condemned one which placed so far apart manganese and chromium, or iron from nickel and cobalt.

Mr Newlands said that he had tried several other schemes before arriving at that now proposed. One founded upon the specific gravity of the elements had altogether failed, and no relation could be worked out of the atomic weights under any other system than that of Cannizzaro.

2.2   Trends in chemical properties

Having seen evidence for a periodic recurrence of physical properties in the elements when arranged in atomic number order, it would be interesting to find out if chemical properties show this periodicity as well.

A start can be made by comparing the behaviour of the elements across a period towards another element with which they all react. This can then be extended by investigating the nature of the products formed. Reaction with chlorine, and investigation of the chlorides that are formed, is suitable here, and the period sodium to argon is a good one with which to start.

Experiment 2.2a

The preparation of some chlorides of elements in the third period of the Periodic Table

The chlorides of many elements can be prepared by heating the element in a current of dry chlorine gas. Often the products boil or sublime at a relatively low temperature and can be collected by cooling the vapours which leave the reaction vessel. As chlorine is a poisonous gas care should be exercised in carrying out the procedures described below. You will probably be asked to make one of these chlorides, and other members of the class will make the others.

Record in your notebook the observations that you make, and then record the principal observations that are made by the class as a whole. These can be collected together most easily in the form of a table, and suggestions for doing this are given as table 2.2. The last column, formula of product, should be filled in when the results of the analysis are available from experiment 2.2b. One example is given to illustrate how the table should be completed.

Element examined	Vigour of reaction	State of product	Appearance of product	Formula of product
Sodium	great	solid	white powder	NaCl
Magnesium etc.				

Table 2.2
Reaction of elements with chlorine

1   The reaction between aluminium and chlorine

In this experiment aluminium (in the form of foil or turnings) is heated in a slow stream of dry chlorine. The product, aluminium chloride, vaporizes at the temperature of the reaction, and condenses as a solid in the cooler parts of the apparatus. A suitable form of apparatus for small scale work is shown in figure 2.2a.

First fill the combustion tube. Using forceps or tongs and a glass rod (*glass wool must not be handled* as small pieces may penetrate the skin and cause painful irritation) put a plug of glass wool about 3 cm from the narrow end of the combustion tube. From a folded piece of paper add a 2 cm length of granular anhydrous calcium chloride (loosely packed to allow the chlorine to pass easily), and then insert a second plug of glass wool. What is the purpose of the calcium chloride?

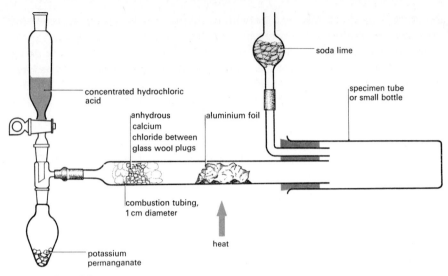

Figure 2.2a
Preparation of aluminium chloride.

Weigh out approximately 0.25 g of aluminium foil or turnings. If turnings are used, they should be freed from grease by washing them with tetrachloromethane (about $5\ cm^3$ of solvent in a test-tube) and dried with filter paper. Insert the aluminium into the combustion tube so that it forms a layer about 3 cm long, starting about 2 cm from the calcium chloride layer. If foil is used, crumple it so that as much air space as possible is left, otherwise attack by chlorine will be slow. Attach the receiver (a 75×25 mm specimen tube) to the combustion tube, and fit a bulb tube containing soda lime granules to the outlet.

Put about 5 g of potassium permanganate crystals into the flask. Attach this to the combustion tube and the dropping funnel as shown in figure 2.2a. Pour into the dropping funnel about $10\ cm^3$ of concentrated hydrochloric acid (measured roughly using a measuring cylinder).

Heat the aluminium foil gently at the end nearest the calcium chloride. Carefully open the tap of the dropping funnel so that one or two drops of hydrochloric acid fall onto the potassium permanganate and then close the tap. The chlorine which is generated will attack the aluminium and it will glow brightly. When the glow begins to die down add a further drop or two of acid to maintain the re-action. Continue in this way, using the burner flame to heat successive parts of the aluminium foil. You may find it necessary to heat the top of the combustion tube occasionally during this process. With care most of the aluminium can be converted to aluminium chloride, which will condense to a solid in the receiver.

When no further action appears to be taking place, stop the flow of chlorine and allow the combustion tube to cool. Taking care not to inhale any chlorine (keep your face away) remove the receiver from the apparatus and immediately stopper it with a well-fitting cork. Keep some of the product for later experiments. If your product is to be analysed as in experiment 2.2b, then the necessary amount will have to be treated at once according to the procedure described for that experiment.

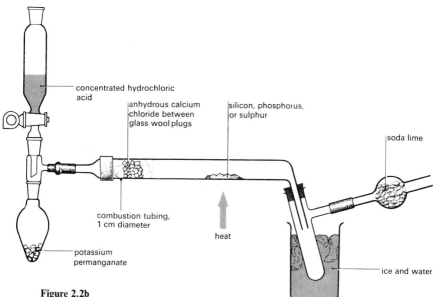

concentrated hydrochloric acid

anhydrous calcium chloride between glass wool plugs

silicon, phosphorus, or sulphur

soda lime

combustion tubing, 1 cm diameter

heat

potassium permanganate

ice and water

Figure 2.2b
Preparation of silicon, phosphorus, or sulphur chlorides.

2 Reaction between silicon and chlorine

Assemble the apparatus shown in figure 2.2b and follow the same general procedure as for aluminium chloride. Use about 0.5 g of silicon powder, retained between glass wool plugs. Fairly strong heating is needed to maintain the re-action in the combustion tube. The product condenses as a liquid in the cooled receiver. It is rather impure and needs to be redistilled using the apparatus shown in figure 2.2c.

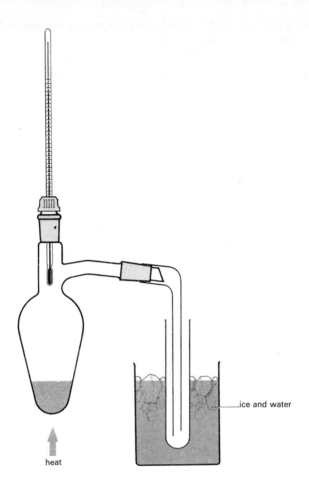

Figure 2.2c
Apparatus for the redistillation of silicon, phosphorus, and sulphur chlorides.

Set up the apparatus as shown in figure 2.2c using a dry test-tube as receiver. Collect the fraction of distillate boiling between 56 and 58 °C. Cork the receiver securely as soon as distillation is finished, and keep the product for later experiments. For analysis according to experiment 2.2b a portion must be treated at once, following the procedure described for that experiment.

3   Reaction between phosphorus and chlorine
The apparatus of figures 2.2b and 2.2c and the same general procedure can be used to examine the reaction between phosphorus and chlorine.

Use about 0.5 g of dry *red* phosphorus. A slow stream of chlorine is essential. Heat the phosphorus gently to start the reaction which will then proceed without further heating. When redistilling, collect the fraction boiling at 72–74 °C, stopper it securely and keep it for later experiments. If experiment 2.2b is to be done, treat a portion at once according to the procedure described for that experiment.

4 Reaction between sulphur and chlorine

The apparatus of figures 2.2b and 2.2c and the same general procedure may be used to examine the reaction between sulphur and chlorine.

Use about 0.8 g of powdered roll sulphur. Heat until the sulphur is just melted and maintain a good supply of chlorine. Use a 0–200 °C thermometer for the redistillation. Collect the fraction boiling at 133–138 °C. Stopper the product securely and keep it for later experiments. If experiment 2.2b is to be done, treat a portion at once according to the procedure described for that experiment.

Experiment 2.2b
Analysis of the chlorides sodium to sulphur

Having prepared some chlorides, the next step is to analyse them, that is, to find out how many moles of chlorine atoms are combined with one mole of atoms of the other element. This can be done using a titration technique, but first of all conditions must be found in which chloride ions can be determined quantitatively.

Silver ions and chloride ions in solution react together to give a precipitate of silver chloride:

$$Ag^+(aq) + Cl^-(aq) \longrightarrow Ag^+Cl^-(s)$$

To 5 cm^3 of 0.05M sodium chloride solution in a test-tube add a few drops of potassium chromate solution followed by 0.05M silver nitrate solution in portions of 1 cm^3. Shake carefully after each addition and note the volume required to produce a permanent red precipitate. What is this red precipitate?

Now repeat the experiment using 5 cm^3 of 0.05M hydrochloric acid. Add the same total volume of silver nitrate solution. Is a red precipitate formed? Add powdered calcium carbonate in small amounts so as to neutralize the acid. Does a red precipitate appear?

You can now see that if silver nitrate solution is added to a solution containing both chloride ions and chromate ions only silver chloride is precipitated as long as there is an appreciable amount of the chloride ions present in the solution. When all (or almost all) of the chloride ions have reacted with silver ions, precipitation of red silver chromate begins.

The first appearance of a permanent red tinge to the precipitate marks the *end-point* of the titration of a chloride solution with silver nitrate solution. Silver chromate is soluble in acids so titrations using a *chromate indicator* must be carried out in neutral solution.

An analysis of sodium chloride

There is no doubt that you will know the formula for sodium chloride already. This does not matter since it is the *method* that is important in this experiment. You will be using the same method for some less familiar chlorides in the next experiment.

You are provided with a solution of sodium chloride of known concentration (expressed in grammes per cubic decimetre (litre) or $g\,dm^{-3}$) and also a standard solution of silver nitrate of concentration 0.05 mole of $AgNO_3$ per cubic decimetre, or 0.05M.

Using a pipette, measure 10.0 cm^3 of the sodium chloride solution into a conical flask. Add 10 drops of chromate indicator. There is no need to add calcium carbonate for this titration since sodium chloride solution is not acidic. Fill a burette with 0.05M silver nitrate solution. Make sure that the jet is filled with solution and the level of the liquid is on the scale. Note the burette reading.

Run the silver nitrate solution from the burette into the solution in the conical flask, 1 cm^3 at a time, shaking the mixture in the flask between each addition. Stop the additions when silver chromate begins to be precipitated, i.e. the precipitate has a reddish tint. You will now have a rough idea of how much silver nitrate solution is required to react with 10 cm^3 of the sodium chloride solution. (For example, if the precipitate becomes red after 15 cm^3 have been added, the volume of silver nitrate solution required to react with the sodium chloride is between 14 and 15 cm^3.)

Wash out the conical flask with distilled water and repeat the titration with a fresh 10.0 cm^3 sample of sodium chloride solution. Run in silver nitrate solution from the burette until within 1 cm^3 of your rough end-point.

Now add one drop of solution at a time, shaking the mixture between each addition, until the end-point is reached. Record the exact volume (to the nearest 0.05 cm^3) of 0.05M silver nitrate solution which reacts with the volume of sodium chloride taken.

You may find it difficult to decide exactly when the end-point has been reached. At this point the colour of the precipitate changes from bright white to off-white; a definite reddish tone does not appear until 0.1 cm^3 beyond the end-point. (If the solution is shaken the change is from bright yellow to dirty yellow.) It may be helpful to have a conical flask alongside for comparison, containing the complete titration mixture but 1 cm^3 of silver nitrate short of the end-point.

The titration should be repeated to obtain a second accurate end-point. The two results should agree to within 0.1 cm^3 of each other. If they differ by more than this repeat the titration until you obtain two results within this limit. Use the average of these two for calculation purposes.

Record your results as follows:

Titration of 10 cm^3 portions of sodium chloride solution $(\dots \text{g dm}^{-3})$ with 0.05M silver nitrate solution

	1st titration	2nd titration etc.
2nd burette reading	....	....
1st burette reading	....	....
Volume delivered	....	....

Analysis of some other chlorides (Al, Si, P, S)

Because of the reactive nature of these chlorides it is as well to weigh and dissolve a sample as soon as it has been prepared. If you weigh out 0.5–1.0 g of chloride, a volumetric flask of 100 cm^3 capacity would be a suitable size in which to make the solution.

When investigating the properties of these chlorides, you will notice that hydrogen chloride is often given off when they react with water. Obviously when making up a solution of known concentration any loss of hydrogen chloride must be avoided. It is therefore best to make the initial solution in a closed container.

Furthermore you may have observed that many chlorides fume in moist air, and this also involves loss of chlorine. For these reasons a special weighing technique will be necessary if the sample you weigh is to be *pure* chloride rather than partially decomposed material.

A small weighing bottle should be cleaned, dried, accurately weighed, and then left in a desiccator. Collect a sample of your chloride (e.g. during redistillation) directly into the weighing bottle, stopper *at once*, and weigh accurately.

Have ready a wide-mouthed 250 cm^3 stoppered bottle containing about 40 cm^3 of distilled water, or in the case of sulphur chloride about 40 cm^3 of 2M nitric acid. Open the weighing bottle, drop it gently into the water, and *at once* stopper the wide-mouthed bottle. Hold the stopper in place (to prevent it from blowing out) and swirl the bottle and contents gently. Then leave them for ten minutes until the reaction is complete.

Transfer the solution to a volumetric flask, using a funnel so that no liquid is lost. To ensure that all the solute is transferred, the wide-mouthed bottle should be washed out at least three times with small portions of distilled water. Finally make up the volumetric flask to the calibration mark, stopper, and mix well by repeated inversions.

Record your results so far as follows:

Weight of container and chloride   .... g

Weight of container before use   .... g

Weight of chloride taken   .... g

This was dissolved in water and made up to 100 cm^3.

The titration procedure is the same as described for sodium chloride except that the solutions are acidic and must therefore be neutralized before being titrated.

After measuring out each 10.0 cm^3 portion of solution add small portions of pure powdered calcium carbonate until there is no further effervescence and a small excess of unreacted powder remains. Then add chromate indicator and titrate with silver nitrate to the end-point.

Calculate the number of moles of chlorine atoms that were combined with one mole of atoms of the other element. If your result is not a whole number would you suggest that it should have been a whole number but

1   The difference is due to experimental error in the analysis, or
2   The sample of chloride was impure, or
3   The result obtained is the true formula?

Calculations

The calculation can be carried out in the following stages:

1   Work out the number of moles of silver ion used in each titration (0.05M solution will have 0.05 moles in 1000 cm^3). This will be the same as the number of moles of chloride ion taken in each pipette volume.

2   From the number of moles of chloride ions in a pipette volume calculate the grammes weight of chloride ion present in a pipette volume.

3   From the grammes weight of the chloride compound put in the volumetric flask work out the grammes weight taken in each pipette volume.

4   Subtracting (2) from (3) will give the grammes weight of the element combined with chlorine.

5   Convert the weights of the element and the chlorine into the numbers of moles combined together, by dividing the grammes weight of each by their atomic weights.

6   Finally, work out the number of moles of chlorine atoms that would combine with one mole of the element [Al = 27, Si = 28, P = 31, S = 32, Cl = 35.5].

When you have determined the formula of your sample of chloride, compare it with the results obtained by other members of your class.

2.3   **Periodicity of chemical properties in chlorides and oxides**

You have seen something of the preparation of chlorides, and may have come to some conclusions about their nature. In the next experiment you will carry the investigation further by comparing the properties of some chlorides.

Experiment 2.3a

Properties of some chlorides

Begin by comparing the properties of two very different chlorides, such as lithium chloride and tetrachloromethane.

1   Note their appearance.

2   Note what happens when they are mixed with water.

3   Compare the electrical conductivity of molten lithium chloride with that of tetrachloromethane.

If there is time, repeat the comparisons using sodium chloride and silicon tetrachloride.

Now try to widen the field by extending the investigation to other chlorides. Use the rest of the sample of chloride that you prepared, or a sample from a bottle. The following tests can be carried out on a drop or two of the liquid chlorides and similar volumes of the solid chlorides.

If there is time you should also examine the chlorides of the other elements of the third period of the Periodic Table, including sodium and magnesium, most of which may have been prepared by other members of the class. Collect your results in the form of a table.

1 Examine the effect of heat on the chlorides. Which ones are volatile?

2 Place a little of the chloride in about $1–2 \text{ cm}^3$ of water and note what happens. Does the chloride dissolve or does it react as well? What is the pH of the resulting solution?

3 Add a little of the chloride to hexane. Does it react or dissolve? Is the result different from that obtained in (2)?

What conclusions can you draw about the structures and reactions with water of the various chlorides? Note that water dissolves ionic compounds while hexane is better at dissolving molecular compounds. Note also that volatile compounds are generally molecular, whereas involatile ones usually have giant structures, either ionic or with giant lattice-like molecules.

Principal halide	Formula and state at room temperature	Structure	Standard heat of formation at 298 K/kJ mol^{-1}	Other known chlorides (n = whole number)
Hydrogen chloride	$HCl(g)$	HCl molecules	-92.8	–
Helium	no chloride	–	–	–
Lithium chloride	$LiCl(s)$	ionic lattice Li^+ and Cl^-	-410	–
Beryllium chloride	$BeCl_2(s)$	infinite chains giant molecules	-470	–
Boron trichloride	$BCl_3(g)$	BCl_3 molecules	-396	$B_2Cl_4, B_4Cl_4, B_8Cl_8$
Carbon tetrachloride	$CCl_4(l)$	CCl_4 molecules	-140	C_nCl_{2n+2}
Nitrogen trichloride	$NCl_3(l)$	NCl_3 molecules	–	–
Oxygen: dichlorine monoxide	$Cl_2O(g)$	Cl_2O molecules	$+76.4$	ClO_2, Cl_2O_7
Fluorine: chlorine fluoride	$ClF(g)$	ClF molecules	-87.9	ClF_3
Neon	no chloride	–	–	–
Sodium chloride	$NaCl(s)$	ionic lattice Na^+ and Cl^-	-414	–
Magnesium chloride	$MgCl_2(s)$	layer lattice	-643	–
Aluminium chloride	$AlCl_3(s)$	layer lattice	-697	–
Silicon tetrachloride	$SiCl_4(l)$	$SiCl_4$ molecules	-613	Si_nCl_{2n+2}
Phosphorus trichloride	$PCl_3(l)$	PCl_3 molecules	-280	PCl_5, P_2Cl_4
Disulphur dichloride	$S_2Cl_2(l)$	S_2Cl_2 molecules	-60.5	SCl_2
Chlorine	$Cl_2(g)$	Cl_2 molecules	0	–
Argon	no chloride	–	–	–

Table 2.3a

Is there a pattern of behaviour in the chlorides of elements of the third period, and are there general similarities between sodium and lithium chlorides and silicon and carbon chlorides? This comparison of chlorides can be taken further by looking at the information given in table 2.3a and answering the following questions:

1 Do you notice any pattern about the formulae of the principal chlorides across the Periodic Table, and, if so, what is it?

2 Consider the chlorides having the greatest number of moles of chlorine atoms for every mole of atoms of the other element. Does this give a pattern different from any you have noted in answer to question 1?

3 Whereabouts in the Periodic Table are the chlorides with giant lattice structures, and where are those which are composed of molecules?

4 Plot a graph of heat of formation, in kilojoules per mole of chlorine atoms, against atomic number and stick it into your notebook. Is there a pattern in this graph?

5 Try to complete additional columns for this table in your notebook. These should be headed 'electrical conductivity of melt', 'pH of solution', 'solubility in hexane', and any other comparative properties that you can investigate.

Now that some fairly detailed comparisons of chlorides have been made, a similar comparison can be made for the oxides.

This can be begun by carrying out the following experiment.

Experiment 2.3b
Properties of some oxides
Oxides of the elements are usually prepared easily by burning the element in oxygen. Prepare small samples of the oxides of the following elements by heating a 2 mm cube of the element in a combustion spoon and lowering it into a 150 × 25 mm test-tube of oxygen.

Note.
1 The relative vigour of the reactions,
2 The effect of water on the oxides, and
3 The pH of the water after mixing with the oxide.

Suitable elements to examine are lithium, boron, sodium, magnesium (use 3 cm of ribbon), aluminium, silicon, sulphur, and phosphorus (red). The reaction with phosphorus must be carried out in a fume cupboard.

Having carried out the experiments, try to compile a table in your notebook for the oxides similar to the ones already given for chlorides.

Draw a graph of heats of formation against atomic number. Formulaê of the principal oxides, and values of heats of formation, are given in the *Book of Data*. When plotting the graph, values of heat of formation should be calculated per mole of oxygen atoms rather than per mole of oxide formed, if a fair comparison is to be made. Structures of the compounds are given in table 2.3b.

Giant lattices			Molecules			
Li_2O	BeO	B_2O_3	CO_2	NO_2	O_2	
Na_2O	MgO	Al_2O_3	SiO_2	P_2O_5	SO_2	Cl_2O_7

Table 2.3b
Structures of some oxides

Background reading
'From Periodic Table to production'
In 1937 the Perkin medal for valuable work in applied chemistry was awarded to the American chemist Thomas Midgley, Jr. The lecture he gave at the time was entitled 'From Periodic Table to production', and the extracts below have been adapted from that lecture and also from an account of Midgley's work given by Robert E. Wilson.

In 1916, Midgley got a job with the Delco Light Company. It so happened that his first task was that of trying to get more power out of the small Delco light units when they were operated on kerosine. It had been found that, operating on gasoline, the compression on these engines could be raised to a point where they gave fairly good power output and efficiency, but when an attempt was made to use these compressions with kerosine as fuel, severe knocking and even cracked cylinder heads resulted.

Midgley found that the knock was due to a rapid rise in pressure after ignition and near top dead centre. In attempting to theorize as to why kerosine knocked and gasoline did not, he first seized upon the most obvious difference between the two products – that of volatility – and thought that possibly the kerosine vaporized rather slowly until after combustion started and then vaporized very suddenly with a resultant too-rapid explosion. If this explanation were correct, he reasoned that it might be possible to make the droplets absorb radiant heat from the incipient flame and hence vaporize sooner. Had Midgley been a good physicist he could have doubtless found by calculation that this theory was untenable, but being a mechanical engineer he fortunately decided that it was much easier to try it out than do the calculations. He accordingly went to the stockroom in search of some oil-soluble dye, and as usual the stockroom was just out of the desired product. However, someone suggested that iodine was oil-soluble and would tend to dye the kerosine, so Midgley dissolved a substantial quantity of iodine in the kerosine, tested it in a moderately high-compression engine, and found to his delight that the knocking was eliminated. Midgley immediately sent out to scour Dayton for all available samples of oil-soluble dyes and that afternoon tested out a dozen different ones in rapid succession without getting the slightest result from any of them. To clinch the matter, he added colourless ethyl iodide to the gasoline and found that this stopped the knock. Thus the first theory of detonation, based on differences of volatility of the fuels, was shown to be false.

Thomas Midgley continues his story:

In the course of my education, I had occasion to learn about the Periodic Table and to have it impressed upon my memory as a very useful tool in research work.

For example, on the search for a material with which to control knocking in an internal combustion engine, the following determinations were arrived at by the method of trial and error:

1 Elemental iodine, dissolved in motor fuel in very small quantities, greatly enhanced the antiknock characteristics of the fuel (the basic discovery described above).
2 Oil-soluble iodine compounds had a similar, though modified, effect.
3 Aniline, its homologues and some other nitrogenous compounds were effective, though their effectiveness varied over a wide range depending upon the hydrocarbon radicals attached to the nitrogen.
4 Bromine, carbon tetrachloride, nitric acid, hydrochloric acid, nitrites, and nitro-compounds in general increased knocking when added to the fuel and air mixture.
5 Selenium oxychloride was extremely effective as an antiknock material.
6 A large number of compounds of other elements had shown no effect.

With these facts before us, we profitably abandoned the method of trial and error in favour of a correlational procedure based on the Periodic Table. What had seemed at times a hopeless quest, covering many years and costing a considerable amount of money, rapidly turned into a 'fox hunt'. Predictions began fulfilling themselves instead of fizzling out. Diethyl selenide was prepared and worked as expected; diethyl telluride next fulfilled our predictions, and seemingly our wildest dreams of success had been realized. There are, however, good reasons for not using tellurium compounds. A systematic survey was then conducted using a periodic arrangement suggested by R. E. Wilson (figure 2.3a). The ethyl and phenyl derivatives of the elements on the right of the arrangement were prepared and their effects measured in motor operation. The results, plotted as antiknock effect against atomic number, are shown in figure 2.3b. Tin was the first element investigated from the group immediately to the left of those previously reported. Its ethyl derivative was studied. This compound exhibited a much more powerful effect than had been expected. We thereupon predicted that tetraethyl lead would solve the problem. The record of the past decade has borne out that prediction.

However, as Robert Wilson recalls, there were still many problems to solve before the basic discovery could be successfully marketed.

Diethyl telluride was the best antiknock agent found up to the time of discovery of tetraethyl lead. Its cost and efficiency were such that it could have found a definite place in the commercial field had it not been for one serious drawback – the fact that even minute quantities of diethyl telluride, or almost any form of tellurium, when inhaled or even absorbed through the skin gives an individual tellurium poisoning, the outstanding symptom of which is a strong garlic-like odour given off apparently from every pore of the body and affecting everything and everyone with whom the victim comes into close contact. During the period of active experimentation on this compound, Midgley had to travel considerably and found that the only feasible way was to travel by day, go into the smoking car, and sit down beside the swarthiest individual in the car in hope that the other passengers would blame this hapless individual for the odour which soon pervaded the car.

6	7	8	9
C	N	O	F
14	15	16	17
Si	P	S	Cl
32	33	34	35
Ge	As	Se	Br
50	51	52	53
Sn	Sb	Te	I
82	83	84	85
Pb	Bi	Po	At

Figure 2.3a

There were also major difficulties to be encountered in putting tetraethyl lead on the market. In the first place Midgley was up against the fundamental difficulty that tetraethyl lead was not of much advantage except in motors of higher compression than were on the market; on the other hand, motors of higher compression could not be put on the market until a better fuel was available. In the second place the oil industry could foresee an investment of many millions of dollars in storage facilities, distributing facilities, and extra pumps and tanks at service stations if this new and more expensive fuel were to be put on the market.

Standard Oil of Indiana was the first major company to take up the marketing of tetraethyl lead. Before this was well under way, however, a serious outbreak of poisoning cases resulting from the manufacture of tetraethyl lead caused concern that it might be a threat to public health. Ethyl gasoline was therefore temporarily withdrawn from the market, pending a thorough investigation by a committee. Its sale was resumed only after they approved its manufacture and general distribution under regulations directed particularly toward reducing the hazard in manufacturing tetraethyl lead and blending the concentrated product into the gasoline.

Another recurrent difficulty was the fact that the lead oxide formed in the combustion of the minute amount of tetraethyl lead added to the fuel in ethyl gasoline tended to deposit on spark plugs, exhaust valves, etc., and in some cases to flux with the iron oxides, thus accelerating corrosion. This difficulty led to a tremendous amount of study and the gradual evolution of the present composition of ethyl fluid in which chlorine and bromine compounds are added to convert the products of combustion to a harmless form (as volatile lead salts).

As a result, a serious shortage of bromine was soon created and Midgley and his associates accordingly developed two or three alternative methods of recovering bromine from sea water. This operation is now performed on a tremendous scale. The total consumption of bromine in ethyl fluid is about ten times as great as this country's consumption of bromine prior to the advent of tetraethyl lead – another result of the impact of Midgley's basic discovery.

Reprinted from *Industrial and Engineering Chemistry*, **29**, 1937, 241–244. Copyright 1937 by the American Chemical Society.

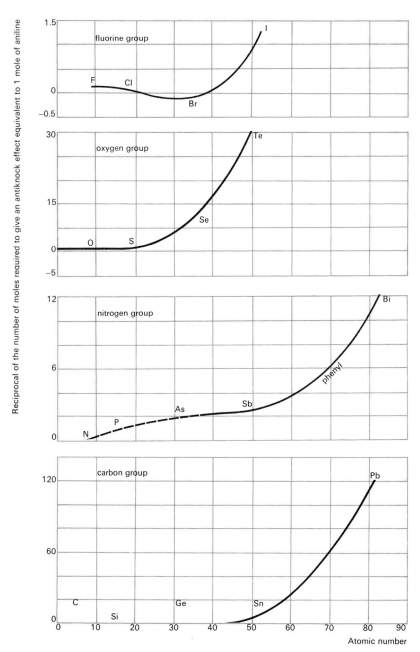

Figure 2.3b
Detonation-influencing effects of the ethyl compounds of four groups of elements.

Problems

* Indicates that the *Book of Data* is needed.

1   1.19 g of an anhydrous chloride of an element in group II of the Periodic Table was weighed and dissolved in demineralized water. The solution was made up to 250 cm^3 in a volumetric flask. 25.0 cm^3 of the resulting solution reacted exactly with 15.0 cm^3 of 0.100M silver nitrate solution.

Calculate:
i   The formula weight of the chloride
ii   The atomic weight of the element
iii   The volume of silver nitrate solution which would have been required to react with 25 cm^3 of the chloride solution, if the silver nitrate solution had been 0.096M.

2   The elements X, Y, and Z (these letters are not their symbols) are in the same period of the Periodic Table; use the information about their chlorides given below to answer the questions which follow.

X   The chloride of X has a melting point of 776 °C; it is soluble in water and both the aqueous solution and the molten chloride are good conductors of electricity. 50 cm^3 of 0.2M AgNO$_3$ are required to convert 0.01 mole of the chloride into silver chloride.

Y   The chloride of Y has a melting point of 772 °C. It is soluble in water and both the aqueous solution and the molten chloride are good conductors of electricity. 100 cm^3 of 0.2M AgNO$_3$ are required to convert 0.01 mole of the chloride into silver chloride.

Z   The chloride of Z has a melting point of -50 °C, and a boiling point of 87 °C. It is not a good conductor of electricity at room temperature. It reacts with water to form an acidic solution which is a good conductor of electricity. 200 cm^3 of 0.2M AgNO$_3$ are required to convert 0.01 mole of the chloride into silver chloride.

i   Determine the formula of each of the three chlorides.
ii   In which group of the Periodic Table should each element be placed?
iii   Into which of the following structure categories would you place each chloride at room temperature?
    A   A giant structure of ions
    B   A giant structure of atoms
    C   Molecular
    D   Widely separated ions
    E   Widely separated molecules
iv   Use your knowledge of the Periodic Table to write a brief description of the properties you would expect of the chloride of the element the atomic number

of which is one less than that of Z. Confine yourself to the formula of the chloride, its appearance, the effect of heat on it, and the effect of water on it.

3   Figure 2.5 shows a graph of the atomic volume of some elements plotted against their atomic number at 298 K and 1 atmosphere pressure. The actual atomic numbers have been omitted. Each letter on the horizontal axis represents an element, but the letter is not the chemical symbol for the element. Each element differs from the next by one atomic number. (The volume of 1 mole of atoms of element J is 24.3 dm^3 at 298 K and 1 atmosphere pressure.)

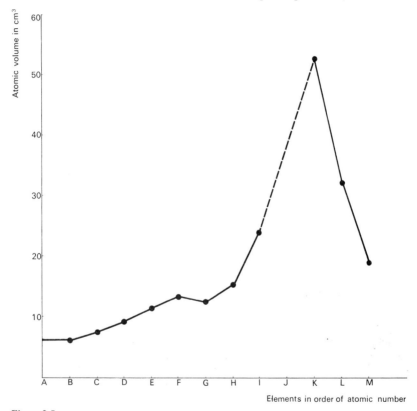

Figure 2.5

i   Which of these elements would you expect to be a gas at room temperature and pressure?

ii   Which of these elements would you expect to be an alkali metal?

iii   Which of these elements will contain the greatest number of atoms per cubic centimetre of the element?

iv   The density of element G is 5.73 g cm^{-3}; calculate the mass of 1 mole of atoms of the element.

v   Is it possible, from the information given in the graph, to determine which of these elements has the highest density in grammes per cubic centimetre? If so, state which one has the highest density and give your reasons. If not, explain why not.

4   In the blank Periodic Table shown below certain elements are indicated by letters which are not their usual symbols. In the questions which follow you should study the position of these elements in the table and you should refer to the elements by means of the letters which have been given to them.

I	II												III	IV	V	VI	VII
														Q			
A													R		L	P	
		E															Y
Z				F													C
	B			H													D
	M																

i   State two of the elements which will react with water to form an alkaline solution.

ii   State two elements the characteristic ions of which have two positive charges each.

iii   State two of the elements which will be inert gases.

iv   State two of the elements which are radioactive. (Do not include those that contain traces of radioactive isotopes or which can be artificially converted into radioactive isotopes.)

v   How many moles of atoms of the element P would you expect to react with 1 mole of atoms of the element B to form a stable compound?

vi   State two of the elements – other than the inert gases – which would have a molecular structure.

vii   What name is given to the family of elements of which E, F, and H are members? State three ways in which this family of elements differs from the group of elements of which A and Z are members.

Here is some information about four of the elements. Give the letter of each of the elements to which the information could refer.

Element	Melting point/°C	Electrical conductance	Density /g cm^{-3}	Other information
viii	97	good	0.97	Forms a chloride in which the element is in the form of a cation with a single charge.
ix	−112	poor	0.0059	Forms a fluoride in which 4 moles of fluorine atoms are combined with 1 mole of atoms of the element.
x	113	poor	2.1	Forms a stable hydride in which 2 moles of hydrogen atoms are combined with 1 mole of atoms of the element.
xi	660	good	2.7	Forms an amphoteric oxide.

xii Suppose you were given a sample of an element and asked to determine in which of the groups of the Periodic Table it would be most suitably classified. Describe the tests you would perform on the element and show how your observations would guide you to an answer to this problem. You should give some indication of the apparatus you would use but do not give a detailed account of practical techniques. (You are advised to spend not more than about twenty minutes on this part of the question.)

*5 Suppose you were about to do some work on the element gallium and its chloride and oxide. Use your Periodic Table and your *Book of Data* to compile a summary of information about the element and the two compounds. You should indicate which information you *obtain directly* from the *Book of Data* and which information you *deduce* from your general knowledge of the Periodic Table. The information should be given briefly and concisely and in a form which would be clear to others.

Topic 3
The masses of molecules and atoms; the Avogadro constant

In this topic we shall discuss how to find out the relative masses of molecules and atoms. We shall also consider ways of counting the number of particles (atoms, molecules, ions, etc.) in a mole of particles.

3.1 Gay-Lussac's law
Having studied the results of a large number of reactions involving gases, the French scientist Gay-Lussac in the years just after 1800 saw that the volume measurements were very simply related. For example, when hydrogen and oxygen combine,

2 volumes of hydrogen + 1 volume of oxygen → 2 volumes of steam,

and for the hydrogen and chlorine reaction,

1 volume of hydrogen + 1 volume of chlorine → 2 volumes of hydrogen chloride.

He expressed this in his *Law of Gaseous Combining Volumes*, which he published in 1808:

When gases react, the volumes in which they do so bear a simple ratio to one another, and to the volume of the products if gaseous, all volumes being measured under the same conditions of temperature and pressure.

3.2 Avogadro's theory
An explanation of the law in terms of atoms and molecules was sought, and in 1811 the Italian scientist Avogadro put forward his theory:

Equal volumes of gases, under the same conditions of temperature and pressure, contain equal numbers of molecules.

How this theory applies to the results of experiments can be seen by taking an example. Suppose one volume contains x molecules. Then the first example quoted above would be interpreted as follows:

$2x$ molecules of hydrogen + x molecules of oxygen → $2x$ molecules of steam.

Dividing all through by x,

2 molecules of hydrogen + 1 molecule of oxygen → 2 molecules of steam.

Relative masses of molecules ('molecular weights')
Assuming the correctness of Avogadro's theory, it is possible to compare the masses of different molecules, by comparing the masses of equal volumes of different gases (measured under the same conditions of temperature and pressure). In this way a scale of relative molecular masses (also known as molecular weights) is obtained.

In the early part of the nineteenth century the atom of hydrogen was chosen as the standard for this scale, the mass of other atoms and molecules being compared with the mass of one hydrogen atom. Chemical evidence shows that hydrogen *molecules* consist of two atoms, and so the mass of a hydrogen molecule is therefore two units on this scale.

This chemical evidence is of the following type:

Experiment shows that

one volume + one volume → two volumes
of hydrogen   of chlorine   of hydrogen chloride

Avogadro's theory therefore suggests that

one molecule + one molecule → two molecules of
of hydrogen   of chlorine   hydrogen chloride

When hydrogen chloride reacts with metals only one type of salt can be made, and this contains no hydrogen. There are no salts corresponding to sodium hydrogen sulphate, for example. Each hydrogen chloride molecule therefore contains only one hydrogen atom. Two hydrogen chloride molecules are made from every hydrogen molecule, and so hydrogen molecules must contain two hydrogen atoms.

If we find by experiment that 1 cubic decimetre (say) of a certain gas weighs 14 times as much as 1 cubic decimetre of hydrogen at the same temperature and pressure, then we know that its molecules must be 14 times as heavy as a hydrogen molecule, and therefore 28 times as heavy as a hydrogen atom. Its molecular mass, relative to that of a hydrogen atom, is therefore 28.

This argument can be summarized as follows:

$$\frac{\text{mass of one volume of gas A}}{\text{mass of one volume of hydrogen}}$$

$$= \frac{\text{mass of } x \text{ molecules of gas A}}{\text{mass of } x \text{ molecules of hydrogen}} \quad \text{(Avogadro's theory)}$$

$$= \frac{\text{mass of 1 molecule of gas A}}{\text{mass of 1 molecule of hydrogen}} \quad \text{(dividing by } x\text{)}$$

$$= \frac{\text{mass of 1 molecule of gas A}}{\text{mass of 2 atoms of hydrogen}} \quad \begin{array}{l}\text{(since hydrogen molecules} \\ \text{contain 2 atoms)}\end{array}$$

$$= \tfrac{1}{2} \text{ (relative molecular mass of gas A)}$$

Determining the masses of known volumes of gases, or of vapours obtained by boiling volatile liquids, is quite easily done by weighing. Some instructions for doing so are given in experiment 3.4a.

3.3   The effect of temperature and pressure changes on the volume of a gas

It is important to remember that all measurements of gas volumes are dependent upon temperature and pressure. The volume of a given mass of gas is inversely proportional to its pressure, p, if its temperature remains unchanged (Boyle's law) and directly proportional to its absolute temperature, T, if its pressure remains unchanged (Charles's law).

Suppose one mole of gas has a volume V_0. Boyle's law and Charles's law can be expressed mathematically, and combined as follows:

$$V_0 \propto \frac{1}{p} \quad \text{(Boyle)}$$

$$V_0 \propto T \quad \text{(Charles)}$$

$$\therefore V_0 \propto \frac{T}{p}$$

$$\therefore V_0 = \frac{RT}{p}$$

or, multiplying both sides by p,

$$pV_0 = RT$$

p being the pressure, T the temperature, and R a constant known as the *gas constant*.

For n moles of gas, volume V, the equation becomes

$$pV = nRT$$

and is known as the *ideal gas equation*.

If p is measured in atmospheres, V in cubic decimetres, and T in kelvins, the gas constant has units of atm dm^3 K^{-1} mol^{-1}.

The most convenient procedure when calculating the results of experimental work is to 'correct' observations to standard temperature and pressure (s.t.p., 273 K or 0 °C and 1 atm or 760 mmHg*) using the gas laws, in order to obtain a proper comparison between one set of results and another.

Correction to s.t.p. – a simple example

The volume of a certain mass of gas was measured at 40 °C and 750 mmHg pressure and found to be 65 cm^3. What is its volume at s.t.p.?

At 313 K and 750 mmHg, volume of gas is 65 cm^3

$\therefore$ at 273 K and 760 mmHg, volume is $65 \times \dfrac{273}{313} \times \dfrac{750}{760}$ cm^3

i.e. *volume of gas at s.t.p. is 54.5 cm^3.*

Notice that if the pressure on the gas is increased its volume decreases, i.e. is multiplied by $\dfrac{750}{760}$, not by $\dfrac{760}{750}$.

Correction to s.t.p. can also be made by direct substitution in the ideal gas equation.

At s.t.p., one mole of hydrogen, H$_2$, (that is, 2 g of hydrogen) occupies 22.4 cubic decimetres. By Avogadro's theory it therefore follows that 22.4 cubic decimetres (at s.t.p.) of any other gas contains the same number of molecules as one mole of hydrogen molecules. In other words, one mole of any gas occupies 22.4 cubic decimetres at s.t.p.

---

* The internationally accepted unit of pressure is the pascal, Pa. One pascal is defined as a pressure of one newton per square metre, N m^{-2}.

For many practical purposes, gas pressures are normally recorded as a height of a column of mercury in a barometer, and one 'atmosphere' is the pressure which will support a column of mercury 760 mm high. It is related to the international unit in the following way:

1 atmosphere = 101 325 Pa

In this book we shall normally be concerned with the comparison of gas pressures, and will use the practical unit of measurement, the mmHg.

3.4 Finding the relative masses of molecules by weighing known volumes of gases

This section describes some experiments you can do to find the relative masses of the molecules of some gases and volatile liquids.

Experiment 3.4a

To find the relative masses of molecules by weighing known volumes of gases

In these experiments we shall assume that 1 mole of molecules of a gas occupies a volume of 22.4 cubic decimetres at standard temperature and pressure (273 K and 760 mmHg). So that if we find the mass of a known volume of any gas, at a known temperature and pressure, we can find its relative molecular mass by calculating the mass of 22.4 cubic decimetres of the gas at standard temperature and pressure.

Method 1   Measuring gas densities using a graduated syringe

A syringe is first weighed with a known volume of air in it. An equal volume of the gas under investigation is then put into the syringe, which is weighed again. The weight of air in the syringe during the first weighing must be taken into account when calculating the molecular weight of the gas.

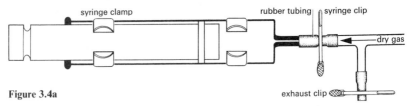

Figure 3.4a

Figure 3.4a shows the apparatus used for filling the syringe. Fit a drying tube containing anhydrous calcium chloride to a rubber tube on the righthand end of the T-piece. Open the syringe and exhaust clips (if you are using a glass syringe you can do without the exhaust clip by closing the rubber tube on this limb of the T-piece with your finger and thumb when required). Push the plunger of the syringe in as far as it will go. Close the exhaust clip and pull out the syringe plunger slowly until 100 cm³ (or 50 cm³ if you are using a plastic syringe) of dry air have been drawn into the syringe. Close the syringe clip. Remove the T-piece from the end of the rubber tube attached to the syringe. Weigh the syringe + dry air + rubber tube and clip, on an accurate balance. Open the clip and expel the air from the syringe.

Attach the T-piece to the end of the rubber tube on the syringe again and connect the righthand end of the T-piece to a source of the dry gas whose relative molecular mass you are going to determine.

Note.

If you are using poisonous or unpleasant-smelling gases the filling must be done in a fume cupboard.

Filling the syringe

1   Open the exhaust clip and allow the gas to displace the air from the T-piece for a few seconds. Close the exhaust clip and allow about 30 cm^3 of gas to enter the syringe. If you are using a plastic syringe twist the plunger once or twice to make sure that it is running freely. Expel this sample of gas from the apparatus by opening the exhaust clip and pushing in the plunger.

2   Close the exhaust clip and allow another 30 cm^3 of gas to enter the syringe. Expel this as before. What is the purpose of these two operations?

3   Close the exhaust clip again and allow dry gas to enter the syringe until it is filled to just beyond the highest graduation mark (100 cm^3 or 50 cm^3).

4   Open the exhaust clip and stop the flow of gas by disconnecting the T-piece from the gas generator. Expel the gas beyond the highest graduation mark by pushing in the syringe plunger until it is exactly on the 100 cm^3 or 50 cm^3 mark.

5   Close the syringe clip and disconnect the T-piece from the syringe.

6   Weigh dry gas + rubber tube + syringe clip.

7   Record the laboratory temperature (°C) and atmospheric pressure (mmHg).

8   Make a table of your results.

Method 2   Measuring gas densities using a flask

The principle is the same here as in the method described above but the procedure for weighing a known volume of the gas is slightly different.

1   Weigh accurately a stoppered flask of approximately 100 cm^3 capacity. It will, of course, be filled with air.

2   Remove the stopper and insert a glass delivery tube connected to a suitable gas generator into the flask so that the open end of the tube is very near the bottom of the flask. Pass gas into the flask until you judge that all the air is displaced. This should take 2–3 minutes. Remove the delivery tube slowly and stopper the flask immediately. Why should the delivery tube be removed slowly?

3   Weigh flask plus gas. To check that all the air was displaced pass gas through the flask again for 2–3 minutes and reweigh. If the two weights agree to within 0.001 g it can be assumed that little or no air was in the flask when the first weighing of flask plus gas was done. If the weights do not agree repeat the process until agreement between successive weighings is obtained.

4   To find the capacity of the flask mark the position of the bottom of the stopper on the neck of the flask (use a grease pencil or felt-tipped marker). Fill

the flask with water to the level of this mark and find the volume of water added by pouring it into a measuring cylinder. A volumetric flask is convenient for this experiment. If a flask of this type is used, its capacity can be found more accurately by filling to the graduation mark with water and then running in more water from a burette until the level reaches the mark indicating the bottom of the stopper.

5   Record the laboratory temperature (°C) and atmospheric pressure (mmHg).

6   Make a table of your results.

Calculations

These steps apply to both methods.

1   Find the volume which the air (and the gas used) would have at s.t.p.

2   Taking the density of dry air at s.t.p. as 1.293 g dm^{-3}, calculate the mass of air in the vessel used.

3   Allowing for the mass of air present during the first weighing, calculate the mass of gas taken.

4   Calculate the mass of 22.4 cubic decimetres of the gas at s.t.p. This is the mass of one mole of gas molecules.

5   Divide this molar mass by the mass (in grammes) of one mole of hydrogen atoms. This gives the relative molecular mass of the gas.

Additional exercise

Estimate the accuracy which you could attach to your result. How do you think the method you used could be improved to give more accurate results?

Experiment 3.4b

To determine the relative molecular mass of a liquid of fairly low boiling point

By finding the volume of the vapour formed from a known mass of a liquid, at a known temperature and pressure, the relative molecular mass of the liquid can be calculated. The simplest way to do this is to allow a known mass of liquid to vaporize in a heated graduated syringe, the temperature of which is at least 20° above the boiling point of the liquid. If steam is used to heat the syringe, as in the method described below, liquids of boiling point lower than 80 °C must be used. A glass syringe is essential for this experiment; plastic syringes soften at steam temperature.

The apparatus used is shown in figure 3.4b. A metal can (or a glass flask) fitted with an outlet tube and a safety tube can be used as a steam generator. Place a beaker under the outlet from the steam jacket to catch drops of water from the condensing steam.

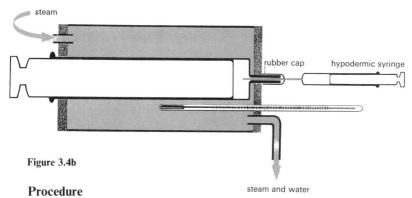

steam

rubber cap   hypodermic syringe

Figure 3.4b

steam and water

Procedure

1   Assemble the apparatus without the self-sealing rubber cap and the hypodermic syringe. Draw about 5 cm^3 of air into the syringe and fit the self-sealing cap over the syringe nozzle. Pass steam through the jacket until the thermometer reading and the volume of air in the syringe reach steady values. Record the temperature and air volume. Continue to pass steam through the jacket while you prepare the hypodermic syringe.

2   Fit a hollow needle to the hypodermic syringe. Push in the plunger, put the needle into the liquid that is to be investigated, and draw about 1 cm^3 of this into the syringe. Reject this liquid into the sink by pushing the plunger home, thus washing the syringe barrel and needle. Recharge the syringe with about 1 cm^3 of liquid, hold it vertically with the needle pointing upwards and expel air from the needle by pushing in the plunger gently until a few drops of the liquid are driven out. Dry the outside of the needle with a piece of filter paper. Close off the orifice of the needle by pushing the needle into a small cap of self-sealing rubber. Weigh syringe, cap, and contents.

3   Push the needle of the hypodermic syringe through its own self-sealing jacket, and through that of the large graduated syringe in the steam jacket, so that the needle tip is well clear of the narrow nozzle (see diagram). Inject about 0.2 cm^3 of liquid into the air space in the large syringe.

4   Withdraw the hypodermic syringe into the needle cap, and weigh the syringe and cap *immediately*. Between the two weighings handle the hypodermic syringe as little as possible to reduce loss of liquid by expansion or evaporation. Whilst you are weighing the hypodermic syringe for the second time, the liquid in the graduated syringe will evaporate.

5   Record the volume of air plus vapour in the graduated syringe when the reading has become steady.

6   Whilst steam is passing through the apparatus, remove the large self-sealing cap and push the plunger of the syringe in and out several times to expel all vapour. The apparatus is then ready for another determination of molecular weight. If it is to be used by you or somebody else immediately, keep the steam

passing through the jacket. If not, remove the rubber delivery tube from the steam generator (use a cloth – it will be hot!) and only then turn out the Bunsen burner flame under the generator.

Record the atmospheric pressure (mmHg).

Calculations
1   Subtract the first reading of the graduated syringe from the second to find the volume of vapour produced.
2   Calculate the volume which the vapour would occupy at s.t.p. if it still remained in the vapour state.
3   Calculate the mass of liquid which would yield 22.4 cubic decimetres of vapour at s.t.p. This is the mass of 1 mole of the liquid.

Questions
1   What assumptions do you make in regarding the result that you obtained as the relative molecular mass of the liquid?
2   How accurate is your result (that is, how well does it agree with estimates by other workers of the relative molecular mass of the liquid you used)? What do you think are the main sources of error in this experiment?
3   How would you measure the relative molecular mass of a liquid which boils at 250 °C?

Note.
There are a number of other methods which have been used to find the relative molecular masses of volatile liquids; for example, those devised by Victor Meyer, Dumas, and Regnault. Experimental details for these methods can be found in standard textbooks of practical chemistry. Relative molecular masses are often referred to as 'molecular weights', a point which should be borne in mind when consulting indexes of other books.

3.5 Relative atomic masses ('atomic weights')
Once we know the relative molecular masses of some compounds of an element, we can find the relative atomic mass of the element itself. This section describes the method for doing this developed by Cannizzaro and first used by him in 1858.

Cannizzaro found the relative molecular masses of a number of compounds of a given element, and then found the weight in grammes of this element present in 1 mole of each of these compounds. He saw that there would be a good chance that one of these compounds would contain only one atom of the element per molecule, and so the smallest weight of element in one mole of these compounds would be that of one mole of atoms of that element.

An example of the working of this method may make it clearer. Table 3.5a shows the relative molecular masses, and weights of carbon present in one mole, of a number of carbon compounds.

Compound	Relative molecular mass	Weight of carbon in one mole of compound/g
Methane	16	12
Ethane	30	24
Acetylene	26	24
Benzene	78	72
Carbon monoxide	28	12
Carbon dioxide	44	12

Table 3.5a

The smallest weight of carbon present in one mole of any compound is 12 g, and the other figures are all multiples of 12. It seems likely, therefore, that 12 g is the weight of one mole of carbon atoms. In this way Cannizzaro was able to work out a number of relative atomic masses, the only limitation to the method being the need to have gaseous, or volatile liquid, compounds for analysis.

The impact of Cannizzaro's method can be judged by looking at table 3.5b which shows the values of relative atomic masses for the first 18 elements as accepted at various times over the last 140 years.

Name	Symbol	Atomic number	Relative atomic masses			
			in 1831 ($H = 1$)	in 1871 ($H = 1$)	in 1951 ($O = 16$)	in 1961 ($^{12}C = 12$)
Hydrogen	H	1	1	1	1.0080	1.00797
Helium	He	2	unknown	unknown	4.003	4.0026
Lithium	Li	3	6	7	6.94	6.939
Beryllium	Be	4	18	14	9.013	9.0122
Boron	B	5	8	11	10.82	10.811
Carbon	C	6	6	12	12.011	12.01115
Nitrogen	N	7	14	14	14.008	14.0067
Oxygen	O	8	8	16	16.0000 (defined)	15.9994
Fluorine	F	9	18	19	19.00	18.9984
Neon	Ne	10	unknown	unknown	20.183	20.183
Sodium	Na	11	24	23.0	22.991	22.9898
Magnesium	Mg	12	12	24	24.32	24.312
Aluminium	Al	13	10	27.5	26.98	26.9815
Silicon	Si	14	8	28.5	28.09	28.086
Phosphorus	P	15	16	31.0	30.975	30.9738
Sulphur	S	16	16	32.0	32.066	32.064
Chlorine	Cl	17	36	35.5	35.457	35.453
Argon	Ar	18	unknown	unknown	39.994	39.948

Table 3.5b

Relative atomic masses give the mass of atoms relative to the mass of one kind of atom chosen as a standard. The changes, over more than a century, in the accepted values illustrate three points:

1   The difficulty in early years (1805–1861) in agreeing on the values on any scale.

2   The changes between 1871 and 1951 both in the number of elements known and the precision with which relative atomic masses have been measured.

3   The change in opinion as to the most useful scale. At first it was assumed that all atoms of an element were alike in mass, just as they were alike in chemical properties. An obvious standard was the lightest atom – that of hydrogen (H = 1.00). On this scale the relative atomic mass of oxygen is very nearly, but not quite, 16. By the end of the nineteenth century it was felt that, because of the way relative atomic masses were then measured, it was more reasonable to accept the standard of the relative atomic mass of oxygen as exactly 16. By that time, however, it was realized that some elements at least were mixtures of *isotopes* (see Topic 4) and that not all atoms of an element had the same mass. By 1933 it was known that most elements, including oxygen, were mixtures of isotopes, and that most relative atomic masses were average values: one particular kind of atom would have to be chosen. In 1961 it was internationally decided that $^{12}C = 12$ exactly should be the standard.

A similar table for all the elements, in alphabetical order, is given in the Nuffield Chemistry *Book of Data*. Entries such as the following

			1831	1871	1951	1961
Polonium	Po	84	unknown	unknown	210.0	—
Astatine	At	85	unknown	unknown	(210)	—
Radon	Rn	86	unknown	unknown	222.0	—

illustrate a further point, namely the change over the years in attitude towards relative atomic masses. In the nineteenth century the relative atomic mass was regarded as the fundamental characteristic of an element. In the twentieth century, it is the atomic number which is regarded as the fundamental characteristic of an element, and by 1961 the relative atomic masses of certain elements which were quoted in 1951 were no longer given in the International Table. Moreover, because of the precision with which relative atomic masses can now be measured, those of some elements must be given as variable, the value depending on the source of the element.

However, the average relative atomic mass of an element is still of great practical importance. A chemist needs to work with quantities of an element which, *within the limits of experimental accuracy*, contain the *same number of atoms*. This is done by weighing.

3.6   The Avogadro constant

The number of molecules per mole of a substance is called the Avogadro constant (sometimes also the Avogadro number). It is usually referred to by the symbol L and has units $mole^{-1}$. Not all substances, of course, have a molecular constitution: many are composed of atoms or ions arranged in various kinds of crystal structure. For such substances it is still true that the Avogadro constant is the number of particles per mole of particles, whether these particles are molecules, atoms, ions, or electrons.

A rough value for the Avogadro constant can be found for a molecular liquid by means of a simple laboratory experiment, details of which are now given.

Experiment 3.6

To make an estimate of the Avogadro constant

This experiment is an adaptation of the experiment you may have seen in your O-level course, in which a monomolecular layer was made from a drop of liquid. There it was used to obtain an idea of the size of molecules. This time it will be used to find a rough value of the number of molecules in a mole. The liquid is a solution of stearic (or palmitic or oleic) acid in petroleum ether containing 0.1 g per cubic decimetre of the acid. When dropped on to water this liquid spreads out rapidly, the solvent evaporates, and a layer of acid which we assume is monomolecular remains. If the surface of the water has first been dusted with talc, or with flowers of sulphur, this will be pushed aside by the spreading drop and thus the area covered by the drop will be made more easily visible. The area covered by the drop is measured and, since the quantity of liquid added is known, estimates can be made of the size of the acid molecules and of the number of molecules present in a known volume of liquid.

Experimental details

1   A good way of obtaining a clean water surface, and of renewing it quickly for a repeat experiment, is as follows. Hold a large funnel in a ring-stand over the sink and connect the stem of the funnel to the water tap by a piece of rubber tubing. Fill the funnel. A fresh water surface can be obtained by turning the tap on again for a few moments. Alternatively use a large flat dish or plastic tray, or even the sink if it can be made scrupulously clean.

2   Cover the clean surface of the water with powdered talc or flowers of sulphur from a muslin bag, chalk from the blackboard rubber, or by shaking a little lycopodium powder over it.

3   With the aid of a teat pipette drop one drop of the solution of acid on to the centre of the water surface. Note that spreading occurs. If the size of the spread-out drop is small compared with the area of water surface available, add a further drop or two, but remember to count how many have been added.

Now make the following measurements.

4 Estimate the area of surface film formed.

5 Obtain the volume of one drop of the solution of acid by allowing 1 cm³ to drop from the pipette into a 10 cm³ measuring cylinder, counting the number of drops.

Calculations
Thickness of the acid layer

1 Knowing the volume of one drop of the solution and that the concentration of stearic acid in the solution is 0.1 g dm^{-3}, calculate the mass of acid in one drop of solution.

2 Using the fact that stearic acid has a density of 0.941 g cm^{-3} and the result from (1), calculate the volume of stearic acid in one drop of solution. (If you have used another acid instead of stearic obtain its density from tables.)

3 Using the volume, calculated in (2), of stearic acid which was spread over the water surface and your estimate of the surface area, calculate the thickness of the acid layer. The molecules of which it is composed must be of this order of size or smaller.

Avogadro constant

4 Assume that the molecules are cubes with sides of length equal to the thickness of the monomolecular layer. Calculate the volume of one molecule.

5 Look up the relative molecular mass of the acid you have used and, using its density, work out the volume occupied by one mole of the acid molecules.

6 Divide the volume occupied by one mole by the volume occupied by one molecule to obtain the number of molecules in a mole of acid.
These steps are illustrated in figure 3.6.

Alternative method for the 'monomolecular layer' experiment
Another way of doing the experiment is to float a loop of cotton or unspun silk on the water and to count the number of drops of acid solution required to fill the loop with a monolayer of acid molecules.

Make a loop from a piece of cotton or unspun silk thread about 30 cm long by tying the ends together. Cut off the loose ends near the knot. Slightly grease the thread by drawing it through fingers greased with a trace of Vaseline or candle wax. Float the loop on the water. Add drops of the petroleum ether solution from the teat pipette one at a time until the loop is taut. When the loop is filled by a solid film of acid molecules, it can be pushed across the surface of the water by a glass rod, so after each drop has been added, push gently against the cotton loop with a glass rod. It will be found that the loop can be dented at first, but

One drop of known volume, dropped onto a water surface spreads out to give a monomolecular layer. Its area can be measured and then its thickness calculated.

Suppose molecules are cubes of side equal to the thickness of the layer.

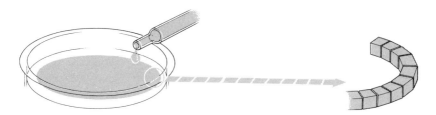

The volume of one such cube could be calculated and divided into the volume of one mole of the liquid forming the original drop. The result is the number of molecules in a mole.

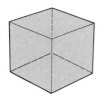

Figure 3.6

after a certain number of drops have been added the loop will not dent but will move across the surface as a solid raft.

At this point it may be assumed that a monomolecular layer fills the area enclosed by the loop, and the thickness of the layer can be calculated. This gives an upper limit to the size of the particles. The calculation is similar to the one given for the first method.

Other ways of estimating the Avogadro constant

There are several other more accurate ways of finding the Avogadro constant. These include an electrical method, based upon the determination of the charge on the electron, together with the charge required to deposit a mole of a substance in electrolysis; a radioactivity method, which is described below; and a method, the most accurate so far known, which depends upon the X-ray measurement of the internuclear distances of atoms or ions in crystals. Details about the X-ray method are to be found in Topic 8.

Radioactivity method

When radium decays each atom emits an α-particle. These particles can be collected in a suitable container, where they pick up stray electrons and become helium atoms. The volume of the trapped helium can be measured. It is very small, but measureable if the collection is continued over a long period. A typical figure is 0.043 cm^3 per gramme of radium per year.

The number of α-particles emitted can be counted by means of a Geiger counter. The fraction emitted in a small solid angle is measured, and then multiplied to find the total number emitted; this is very large, a typical value being 11.6×10^{17} per gramme of radium per year.

These figures tell us that 0.043 cm^3 of helium consists of 11.6×10^{17} atoms. One mole of helium, which occupies 22.4 cubic decimetres at a standard temperature and pressure, therefore contains

$$\frac{11.6 \times 10^{17} \times 22\,400}{0.043} = 6.04 \times 10^{23} \text{ atoms.}$$

The best available estimate of L

Determinations of L using many methods, for instance the X-ray method and measurements on the viscosity of gases, in addition to the ones already described, lead to values which are very close to each other; we can therefore feel confident that these values can be relied upon.

The most recent values lie between 6.02224×10^{23} and 6.02280×10^{23} (National Bureau of Standards, Miscellaneous Publications No. 253, 1963).

Background reading

There now follow some extracts from papers by Gay-Lussac, Avogadro, and Cannizzaro. These three scientists have already been mentioned as key figures in the development of the ideas discussed in this topic.

1 Gay-Lussac

Joseph Louis Gay-Lussac (1778–1850) made many important discoveries in physics and chemistry. Most of his early work was done on gases. In 1802 he put forward, independently of Dalton, the law governing the expansion of gases by heat. In 1805, with Alexander von Humboldt (1769–1854), he observed the already known fact that the volumes of the gases hydrogen and oxygen which combine together are in the ratio of two to one. This led him to study the reactions of other gases to see whether their reacting volumes were also in simple ratios, and in 1808 he announced the law of gaseous combining volumes which is now known as Gay-Lussac's law.

Figure 3.7
Joseph Louis Gay-Lussac. *Photo, Ronan Picture Library*

The following extract is from a paper by Gay-Lussac which was originally published in *Mémoires de la Société d'Arcueil*. This translation is that of the Alembic Club (Alembic Club Reprints No. 4, Edinburgh 1899), and is entitled *Memoir on the combination of gaseous substances with each other*.

The paper begins by pointing out that gases alone amongst the various forms of matter appear to obey simple and regular laws. It then announces some new properties of gases that the author had observed.

Suspecting, from the exact ratio of 100 of oxygen to 200 of hydrogen, which Mr Humboldt and I had determined for the proportions of water, that other gases might also combine in simple ratios, I have made the following experiments. I prepared fluoboric, muriatic, and carbonic gases,* and made them combine successively with ammonia gas. 100 parts of muriatic gas saturate precisely 100 parts of ammonia gas, and the salt which is formed from them is perfectly neutral, whether one or other of the gases is in excess. Fluoboric gas, on the contrary, unites in two proportions with ammonia gas. When the acid gas is first put into the graduated tube, and the other gas is then passed in, it is found that equal volumes of the two condense, and that the salt formed is neutral. But, if we begin by first putting the ammonia gas into the tube, and then admitting the fluoboric gas in single bubbles, the first gas will then be in excess with regard to the second, and there will result a salt with excess of base, composed of 100 of fluoboric gas and 200 of ammonia gas. If carbonic gas is brought into contact with ammonia gas, by passing it sometimes first, sometimes second into the tube, there is always formed a sub-carbonate (ammonium carbamate) composed of 100 parts of carbonic gas and 200 of ammonia gas. It may, however, be proved that neutral carbonate of ammonia would be composed of equal volumes of each of these components.

Thus we may conclude that muriatic, fluoboric, and carbonic acids take exactly their own volume of ammonia gas to form neutral salts, and that the last two take twice as much to form *sub-salts*. It is very remarkable to see acids so different from one another neutralize a volume of ammonia gas equal to their own. . . .

Davy, from the analysis of various compounds of nitrogen with oxygen, has found the following proportions by weight:

	Nitrogen	Oxygen
Nitrous oxide (dinitrogen monoxide)	63.30	36.70
Nitrous gas (nitrogen monoxide)	44.05	55.95
Nitric acid (nitrogen dioxide)	29.50	70.50

Reducing these proportions to volumes we find:

	Nitrogen	Oxygen
Nitrous oxide	100	49.5
Nitrous gas	100	108.9
Nitric acid	100	204.7

The first and last of these proportions differ only slightly from 100 to 50, and 100 to 200; it is only the second which diverges somewhat from 100 to 100. The difference, however, is not very great, and is such as we might expect in experiments of this sort; and I have assured myself that it is actually nil. On burning the new combustible substance† from

* Boron trifluoride, hydrogen chloride, and carbon dioxide.
† Potassium, discovered in 1807 by Humphry Davy.

potash in 100 parts by volume of nitrous gas, there remained over exactly 50 parts of nitrogen, the weight of which, deducted from that of the nitrous gas (determined with great care by M. Bérard at Arceuil), yields as result that this gas is composed of equal parts by volume of nitrogen and oxygen.

We may then admit the following numbers for the proportions by volume of the compounds of nitrogen with oxygen:

	Nitrogen	Oxygen
Nitrous oxide	100	50
Nitrous gas	100	100
Nitric acid	100	200

After giving several other examples, the paper concluded in the following way.

Thus it appears evident to me that gases always combine in the simplest proportions when they act on one another; and we have seen in reality in all the preceding examples that the ratio of combination is 1 to 1, 1 to 2, or 1 to 3. It is very important to observe that in considering weights there is no simple and finite relation between the elements of any one compound; it is only when there is a second compound between the same elements that the new proportion of the element that has been added is a multiple of the first quantity. Gases, on the contrary, in whatever proportions they may combine, always give rise to compounds whose elements by volume are multiples of each other.

2 Avogadro

Lorenzo Romano Amedeo Carlo Avogadro (1776–1856), in an attempt to explain Gay-Lussac's findings, suggested that equal volumes of gases under the same physical conditions contain equal numbers of ultimate particles. This suggestion had been made by others before 1811, when Avogadro published his paper; for example, Dalton* considered the idea for a time, but because he did not make any distinction between atoms and molecules, he had to reject it as unworkable. He also refused to believe in Gay-Lussac's law. Avogadro succeeded where Dalton failed because he was one of the first to see clearly the need to distinguish between atoms and molecules. Even so, his ideas were ignored for almost half a century after they were published, until Cannizzaro took them up again in 1858.

The following extract is from the translation of Alembic Club Reprint No. 4.

---

* John Dalton (1766–1844) put forward in 1808 an atomic theory of matter to explain the laws of constant and multiple proportions by weight of substances in chemical combination. See Nuffield Chemistry Background Book *Dalton and the Atomic Theory*.

Figure 3.8
Amedeo Avogadro. *Photo, Università degli Studi di Torino*

Essay on a manner of determining the relative masses of the elementary molecules of bodies, and the proportions in which they enter into these compounds

1

M. Gay-Lussac has shown in an interesting Memoir that gases always unite in a very simple proportion by volume, and that when the result of the union is a gas, its volume also is very simply related to those of its components. But the quantitative proportions of substances in compounds seem only to depend on the relative number of molecules which combine, and on the number of composite molecules which result. It must then be admitted that very simple relations also exist between the volumes of gaseous substances and the numbers of simple or compound molecules which form them. The first hypothesis to present itself in this connection, and apparently even the only admissible one, is the supposition that the number of integral molecules* in any gases is always the same for equal volumes, or always proportional to the volumes. Indeed, if we were to suppose that the number of molecules contained in a given volume were different for different gases, it would scarcely be possible to conceive that the law regulating the distance of molecules could give in all cases relations so simple as those which the facts just detailed compel us to acknowledge between the volume and the number of molecules.

Setting out from this hypothesis, it is apparent that we have the means of determining very easily the relative masses of the molecules of substances obtainable in the gaseous state, and the relative number of these molecules in compounds; for the ratios of the masses of the molecules are then the same as those of the densities of the different gases at equal temperature and pressure, and the relative number of molecules in a compound is given at once by the ratio of the volumes of the gases that form it. For example, since the numbers 1.10359 and 0.07321 express the densities of the two gases oxygen and hydrogen compared to that of atmospheric air as unity, and the ratio of the two numbers consequently represents the ratio between the masses of equal volumes of these two gases, it will also represent on our hypothesis the ratio of the masses of their molecules. Thus the mass of the molecule of oxygen will be about 15 times that of the molecule of hydrogen, or more exactly as 15.074 to 1. In the same way the mass of the molecule of nitrogen will be to that of hydrogen as 0.96913 to 0.07321, that is, as 13, or more exactly 13.238, to 1. On the other hand, since we know that the ratio of the volumes of hydrogen and oxygen in the formation of water is 2 to 1, it follows that water results from the union of each molecule of oxygen with two molecules of hydrogen. Similarly, according to the proportions by volume established by M. Gay-Lussac for the elements of ammonia, nitrous oxide, nitrous gas, and nitric acid, ammonia will result from the union of one molecule of nitrogen with three of hydrogen, nitrous oxide from one molecule of oxygen with two of nitrogen, nitrous gas† from one molecule of nitrogen with one of oxygen, and nitric acid‡ from one of nitrogen with two of oxygen.

2

There is a consideration which appears at first sight to be opposed to the admission of our hypothesis with respect to compound substances. It seems that a molecule composed of two or more elementary molecules should have its mass equal to the sum of the masses of those molecules; and that in particular, if in a compound one molecule of one substance unites with two or more molecules of another substance, the number of compound molecules should remain the same as the number of molecules of the first substance. Accordingly, on our hypothesis when a gas combines with two or more times its volume of another gas, the resulting compound, if gaseous, must have a volume equal to that of the first of these gases. Now, in general, this is not actually the case. For instance, the volume of water in the gaseous state is, as M. Gay-Lussac has shown, twice as great as the volume of oxygen which enters into it, or, what comes to the same thing, equal to that of the hydrogen instead of being equal to that of the oxygen. But a means of explaining facts of this type in conformity with our hypothesis presents itself naturally enough: we suppose, namely, that the constituent molecules of any simple gas whatever (i.e. the molecules which are at such a distance from each other that they cannot exercise their mutual action) are not formed of a solitary elementary molecule, but are made up of a certain number of these molecules united by attraction to form a single one; and further, that when molecules of another

* The words 'atom' and 'molecule' had not yet been confirmed in their modern meanings. By 'integral molecule' Avogadro denoted a molecule of a compound; by 'constituent molecule' a molecule of an element; and by 'elementary molecule' an atom.
† Nitrogen monoxide.
‡ Nitrogen dioxide.

substance unite with the former to form a compound molecule, the integral molecule which should result splits up into two or more parts (or integral molecules) composed of half, quarter, etc., the number of elementary molecules going to form the constituent molecule of the first substance, combined with half, quarter, etc., the number of constituent molecules of the second substance that ought to enter into combination with one constituent molecule of the first substance (or, what comes to the same thing, combined with a number equal to this last of half-molecules, quarter-molecules, etc., of the second substance); so that the number of integral molecules of the compound becomes double, quadruple, etc., what it would have been if there had been no splitting-up, and exactly what is necessary to satisfy the volume of the resulting gas. Thus, for example, the integral molecule of water will be composed of a half-molecule of oxygen with one molecule, or, what is the same thing, two half-molecules of hydrogen.

3   Cannizzaro

Between the publication of Avogadro's paper in 1811 and the circulation of a pamphlet in 1858 by Cannizzaro, there occurred a period of chaos in chemistry. There was no agreement about atomic weights, nor about the correct formulae of compounds. There was no understanding of the distinction between atoms and molecules. Most chemists either did not understand, or refused to accept, the work of Gay-Lussac and Avogadro.

It was left to Stanislao Cannizzaro (1826–1910), in the pamphlet from which the following extract is taken, to clear up the muddle and set chemical theory on its modern course. He stated, even more clearly and explicitly than Avogadro had done, the distinction between molecules and atoms; he revived Avogadro's method of determining the relative masses of molecules of substances obtainable in the gaseous state; and he employed a common standard (H = 1) for atomic weights. It was this system which Newlands used to develop his Law of Octaves ten years later, as described in Topic 2.

In this paper Cannizzaro set out the details of the various lectures that constituted his chemistry course at Genoa. The lectures began with Gay-Lussac's law and Avogadro's theory, and then went on to show that it was failure to distinguish between atoms and molecules that delayed the acceptance of these ideas by the majority of chemists. It may also be noted, from the reference to 'youths not well accustomed to the comparison of quantities,' that there were problems in the teaching of chemistry even in the early days! We begin the extract at the fifth lecture.

I begin the fifth lecture by applying the hypothesis of Avogadro and Ampère to determine the weights of molecules even before their composition is known.

On the basis of the hypothesis cited above, the weights of the molecules are proportional to the densities of the substances in the gaseous state. If we wish the densities of vapours to express the weights of the molecules, it is expedient to refer them all to the density of a simple gas taken as unity, rather than to the weight of a mixture of two gases such as air.

Figure 3.9
Stanislao Cannizzaro. *Photo, Ronan Picture Library*

Hydrogen being the lightest gas, we may take it as the unit to which we refer the densities of other gaseous bodies, which in such a case express the weights of the molecules compared to the weight of the molecule of hydrogen = 1.

Since I prefer to take as a common unit for the weights of the molecules and for their fractions, the weight of a half and not of a whole molecule of hydrogen, I therefore refer the densities of the various gaseous bodies to that of hydrogen = 2. If the densities are referred to air = 1, it is sufficient to multiply by 14.438 to change them to those referred to that of hydrogen = 1; and by 28.87 to refer them to the density of hydrogen = 2.

I write the two series of numbers, expressing these weights in the following manner:

Name of substances	Densities or weights of one volume, the volume of hydrogen being made = 1, i.e. weights of the molecules referred to the weight of a whole molecule of hydrogen taken as unity	Densities referred to that of hydrogen = 2, i.e. weights of the molecules referred to the weight of a half a molecule of hydrogen taken as unity
Hydrogen	1	2
Oxygen, ordinary	16	32
Oxygen, electrized	64	128
Sulphur below 1000°	96	192
Sulphur* above 1000°	32	64
Chlorine	35.5	71
Bromine	80	160
Arsenic	150	300
Mercury	100	200
Water	9	18
Hydrochloric acid	18.25	36.50†
Acetic acid	30	60

Whoever wishes to refer the densities to hydrogen = 1 and the weights of the molecules to the weight of half a molecule of hydrogen, can say that the weights of the molecules are all represented by the weight of two volumes.

I myself, however, for simplicity of exposition, prefer to refer the densities to that of hydrogen = 2, and so the weights of the molecules are all represented by the weight of one volume.

From the few examples contained in the table, I show that the same substance in its different allotropic states can have different molecular weights, without concealing the fact that the experimental data on which this conclusion is founded still require confirmation.

I assume that the study of the various compounds has been begun by determining the weights of the molecules, i.e. their densities in the gaseous state, without enquiring if they are simple or compound.

I then come to the examination of the composition of these molecules. If the substance is undecomposable, we are forced to admit that its molecule is entirely made up by the weight of one and the same kind of matter. If the body is composite, its elementary analysis is made, and thus we discover the constant relations between the weights of its components; then the weight of the molecule is divided into parts proportional to the numbers expressing the relative weights of the components, and thus we obtain the quantities of these components contained in the molecule of the compound, referred to the same unit as that to which we refer the weights of all the molecules. . . .

All the numbers contained in the preceding table are comparable amongst themselves, being referred to the same unit. And to fix this well in the minds of my pupils, I have recourse

*This determination was made by Bineau, but I believe it requires confirmation.
†The numbers expressing the densities are approximate: we arrive at a closer approximation by comparing them with those derived from chemical data, and bringing the two into harmony.

to a very simple artifice: I say to them, namely: 'Suppose it to be shown that the half molecule of hydrogen weighs a millionth of a milligram, then all the numbers of the preceding table become concrete numbers, expressing in millionths of a milligram the concrete weights of the molecules and of their components: the same thing would follow if the common unit had any other concrete value', and so I lead them to gain a clear conception of the comparability of these numbers, whatever to be the concrete value of the common unit.

Once this artifice has served its purpose, I hasten to destroy it by explaining how it is not possible in reality to know the concrete value of this unit; but the clear ideas remain in the minds of my pupils whatever may be their degree of mathematical knowledge. I proceed pretty much as engineers do when they destroy the wooden scaffolding which has served them to construct their bridges, as soon as these can support themselves. But I fear that you will say, 'Is it worth the trouble and the waste of time and ink to tell me of this very common artifice?' I am, however, constrained to tell you that I have paused to do so because I have become attached to this pedagogic expedient, having had such great success with it amongst my pupils, and thus I recommend it to all those who, like myself, must teach chemistry to youths not well accustomed to the comparison of quantities.

Once my students have become familiar with the importance of the numbers as they are exhibited in the preceding table, it is easy to lead them to discover the law which results from their comparison. 'Compare', I say to them, 'the various quantities of the same element contained in the molecule of the free substance and in those of all its different compounds, and you will not be able to escape the following law: *the different quantities of the same element contained in different molecules are all whole multiples of one and the same quantity, which, always being entire, has the right to be called an atom'*.

Thus:

One molecule of free hydrogen	contains 2 of hydrogen = 2×1
One molecule of hydrochloric acid	contains 1 of hydrogen = 1×1
One molecule of hydrobromic acid	contains 1 of hydrogen = 1×1
One molecule of hydriodic acid	contains 1 of hydrogen = 1×1
One molecule of hydrocyanic acid	contains 1 of hydrogen = 1×1
One molecule of water	contains 2 of hydrogen = 2×1
One molecule of sulphuretted hydrogen	contains 2 of hydrogen = 2×1
One molecule of formic acid	contains 2 of hydrogen = 2×1
One molecule of ammonia	contains 3 of hydrogen = 3×1
One molecule of gaseous phosphoretted hydrogen	contains 3 of hydrogen = 3×1
One molecule of acetic acid	contains 4 of hydrogen = 4×1
One molecule of ethylene	contains 4 of hydrogen = 4×1
One molecule of alcohol	contains 6 of hydrogen = 6×1
One molecule of ether	contains 10 of hydrogen = 10×1

Thus all the various weights of hydrogen contained in the different molecules are integral multiples of the weight contained in the molecule of hydrochloric acid, which justifies our having taken it as common unit of the weights of the atoms and of the molecules. The atom of hydrogen is contained twice in the molecule of free hydrogen.

In a similar way may be found the smallest quantity of each element which enters as a whole into the molecules which contain it, and to which may be given with reason the name of atom. In order, then, to find the atomic weight of each element it is necessary first of all to know the weights of all or of the greater part of the molecules in which it is contained and their composition . . .

After this I easily succeed in explaining how, expressing by symbols the different atomic weights of the various elements, it is possible to express by means of formulae the composition of their molecules and of those of their compounds, and I pause a little to make my pupils familiar with the passage from gaseous volume to molecule, the first directly expressing the fact and the second interpreting it. Above all, I study to implant in their minds thoroughly the difference between molecule and atom . . .

Problems

1 *i*   Calculate the maximum volume (at 25 °C and 1 atm) of carbon dioxide which could be obtained by burning 240 g of carbon. 1 cubic decimetre of carbon dioxide weighs 1.81 g at 25 °C and 1 atm.

ii   1 gramme-molecule of an ideal gas is said to have a volume of 22.4 cubic decimetres at s.t.p. Calculate the error you would have incurred in (*i*) if you had assumed that carbon dioxide is an ideal gas.

2   Arrange the following in order of increasing volume (s.t.p.) putting the greatest volume last (atomic weights: $C = 12$, $O = 16$, $H = 1$, $Cl = 35.5$, $He = 4$.)

    A   1 g of carbon dioxide
    B   1 g of hydrogen
    C   1 g of chlorine
    D   1 g of methane (CH_4)
    E   1 g of helium

3   The gas phosphine contains phosphorus and hydrogen only. At 25 °C and 1 atm, 34 g of phosphine has a volume of 24.30 cubic decimetres, and this volume yields 36.0 cubic decimetres (at 25 °C and 1 atm) of hydrogen on decomposition. Deduce the formula of phosphine. Explain how you obtained your answer.

4   Devise an alternative form of the ideal gas law using the symbols: p, V, R, T, w, M; where $w = $ the weight of gas and $M = $ the molecular mass of the gas.

5   Devise an alternative form of the gas law using the symbols: p, M, ρ, R, T; where $M = $ the molecular mass of the gas and $\rho = $ the density of the gas in grammes per cubic decimetre.

6   A volatile liquid compound was found to be composed of carbon and hydrogen in the ratio of 1 mole of carbon atoms to 2 moles of hydrogen atoms. 0.124 g of the liquid on evaporation at 100 °C and 1 atm gave rise to 45 cm³ of vapour.

    i   What is the molecular mass of the liquid?
    ii   What is the molecular formula of the liquid?

7 The following experiment is to determine the molecular mass of an acidic gas X.

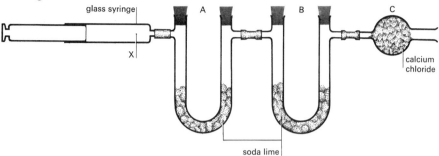

Some of X was drawn into a glass syringe and its volume noted. The syringe was then connected to two U-tubes, A and B, as shown in figure 3.10, and these to the tube C. The gas X was slowly forced out of the syringe. The combined weight of A and B was determined before and after the experiment.

> Volume of X at 16 °C and 1 atm = 100 cm³
> Combined weight of A and B before experiment = 72.640 g
> Combined weight of A and B after experiment = 72.912 g

i Calculate the molecular mass of X (you may assume that X behaves like an ideal gas, i.e., it has a gramme-molecular volume of 22.4 cubic decimetres at s.t.p.)

ii Suggest a reason why *two* soda lime tubes were used.

iii Suggest a modification to the procedure which would find out whether the tube B was necessary.

iv What is the purpose of C?

v What modification would you adopt to determine the molecular mass of ammonia?

8 Use the following information about five gaseous compounds of a certain element (Y) to deduce the approximate atomic mass of the element. Explain how you arrived at your answer.

Compound	Percentage of Y by weight in each compound	Approx. weight of 24.5 dm³ of compound at 25 °C and 1 atm/g
A	87	32
B	90	62
C	25	111
D	42	66.6
E	91	92

9 $10 \, cm^3$ of hydrogen fluoride gas react with $5 \, cm^3$ of dinitrogen difluoride gas (N_2F_2) to form $10 \, cm^3$ of a single gas. Which of the following is the most likely equation for the reaction? Show how you reach your decision.

A $HF + N_2F_2 \longrightarrow N_2HF_3$
B $2HF + N_2F_2 \longrightarrow 2NHF_2$
C $2HF + N_2F_2 \longrightarrow N_2H_2F_4$
D $HF + 2N_2F_2 \longrightarrow N_4HF_5$
E $2HF + 2N_2F_2 \longrightarrow 2N_2HF_3$

10   1 volume of the gaseous element X combined with 1 volume of the gaseous element Y to form 2 volumes of a gaseous compound Z. Z is the only product. Which of the following statements conflicts with this evidence? Show how you reach your decision.

A   When 1 molecule of X reacts with 1 molecule of Y an even number of molecules of Z are formed.

B   1 molecule of Z could contain an even number of atoms.

C   1 g-molecule of X reacts with 1 g-molecule of Y.

D   Both 1 molecule of X and 1 molecule of Y could contain an odd number of atoms.

E   Both 1 molecule of X and 1 molecule of Y could contain an even number of atoms.

11   Which of the following solutions contains the greatest number of ions (assume that each solute is fully ionized)? Show how you reach your decision.

A   1 cubic decimetre of 0.2M Na_2SO_4
B   1 cubic decimetre of 0.2M $CaCl_2$
C   1 cubic decimetre of 0.1M $BaCl_2$
D   1 cubic decimetre of 0.2M $Cr_2(SO_4)_3$
E   1 cubic decimetre of 0.3M $NaCl$

12   In a certain electrolysis experiment using silver nitrate solution as electrolyte, $0.216 \, g$ of silver was deposited by a steady current of 0.200 ampere flowing for 960 seconds. The atomic mass of silver is 108. The charge on the electron is 1.60×10^{-19} coulomb. What value for the Avogadro constant do these figures give?

13   The number of alpha particles emitted from a sample of radium was measured by a Geiger counter and found to be $8.20 \times 10^{10} \, second^{-1}$. The same sample produced $0.0790 \, cm^3$ of helium in 300 days (at s.t.p.). What value do these figures give for the Avogadro constant?

14   When 10 cm^3 of 2×10^{-4}M potassium permanganate (KMnO$_4$) are diluted to 1 dm^3, the solution still appears pink to normal eyesight. Calculate how many permanganate ions there are in 1 drop of the diluted solution (assume there are approximately 20 drops to each cm^3, and that the Avogadro constant is 6×10^{23}).

15   Refer to Part 2 of the Background reading at the end of this topic. Translate into modern scientific language the passage which begins: 'There is a consideration . . .' and ends: '. . . volume of the resulting gas.'

Topic 4
Atomic structure

4.1 Introduction

There is no simple model of atoms which enables us to understand all their observed properties. This is partly because the real nature of the atom is very complex, and partly because our knowledge is still very incomplete in spite of modern discoveries. But in order to discuss the behaviour of atoms it is necessary to suggest one or more models by which they may be pictured. This topic indicates the experimental evidence upon which one useful model is based.

For convenience, we may consider an atom in two principal parts: the nucleus and the electrons.

4.2 The nucleus: part 1

In 1897 J. J. Thomson discovered the electron, and in 1899 he put forward a model of the atom consisting of rings of negatively-charged electrons embedded in a sphere of positive charge so that a neutral atom resulted. As the mass of the atom was considered to be due only to the electrons, there had to be 1800 electrons in the hydrogen atom.

A few years later Dr H. Geiger, working in Manchester under the guidance of Professor Rutherford, discovered that when α-particles are fired at a thin metal foil in an evacuated container, most of the particles which penetrate the foil do so in an undeviated or an only very slightly deviated course. In this experiment, a narrow pencil of α-particles from a source fell on a zinc sulphide screen, and the distribution of the scintillations on the screen was observed when different metal foils were placed in the path of the particles.

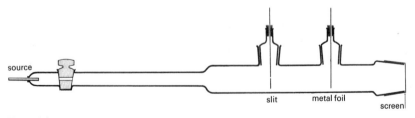

Figure 4.2a
The main components of Geiger's apparatus.

That most of the α-particles should penetrate undeviated or little deviated was to be expected on the Thomson model. Professor Rutherford takes up the story from here.

One day Geiger came to me and said, 'Don't you think that young Marsden, whom I am training in radioactive methods, ought to begin a small research?' Now I had thought that, too, so I said, 'Why not let him see if any α-particles can be scattered through a large angle?' I may tell you in confidence that I did not believe that they would be, since we knew that the α-particle was a very fast, massive particle, with a great deal of energy, and you could show that if the scattering was due to the accumulated effect of a number of small scatterings the chance of an α-particle's being scattered backwards was very small. Then I remember two or three days later Geiger coming to me in great excitement and saying, 'We have been able to get some of the α-particles coming backwards....' It was quite the most incredible event that has ever happened to me in my life. It was almost as incredible as if you fired a 15-inch shell at a piece of tissue paper and it came back and hit you. On consideration, I realized that this scattering backwards must be the result of a single collision, and when I made calculations I saw that it was impossible to get anything of that order of magnitude unless you took a system in which the greater part of the mass of the atom was concentrated in a minute nucleus. It was then that I had the idea of an atom with a minute massive centre carrying a charge. I worked out mathematically what laws the scattering should obey, and I found that the number of particles scattered through a given angle should be proportional to the thickness of the scattering foil, the square of the nuclear charge, and inversely proportional to the fourth power of the velocity. These deductions were later verified by Geiger and Marsden in a series of beautiful experiments...

This extract is taken from the 1936 essay by Ernest Rutherford, 'The Development of the Theory of Atomic Structure', in *Background to Modern Science*, J. Needham and W. Pagel (eds.), The MacMillan Company, New York, 1938.

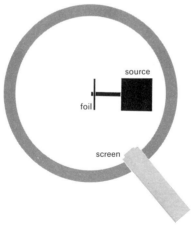

Figure 4.2b
A view from above showing the main components of Geiger and Marsden's apparatus.

The apparatus used by Geiger and Marsden, and shown in figure 4.2b, was similar to that used by Geiger in the earlier experiments, except that deviations of the α-particles through large angles could be observed by means of the moveable zinc sulphide screen.

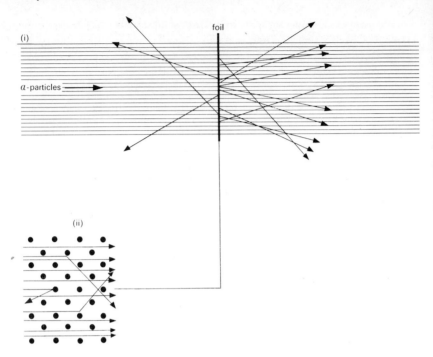

Figure 4.2c

The diagram illustrates (*i*) Geiger and Marsden's results, and (*ii*) Rutherford's interpretation of these results.

Figure 4.2c illustrates diagrammatically Rutherford's interpretation of this experimental result.

It was possible to calculate from these results that for a nucleus of atomic mass, A, the number of positive charges has a value of approximately $\frac{1}{2}A$. (The numerical magnitude of 'a charge' is equal to the value of the negative charge on an electron, which is 1.602×10^{-19} coulombs.) This meant that the number of electrons in the space around the nucleus in the atom must also be $\frac{1}{2}A$ since the atom is neutral. Since atomic masses were known to be non-integral, this was not a very satisfactory result. Van den Broek noticed that $\frac{1}{2}A$ was approximately equal to the atom's numbered position in the Periodic Table, and suggested that the number of positive charges on the nucleus (and thus also the number of electrons) was equal to this number, the atomic number, Z. But Z was merely a position in the Periodic Table: it could not be measured directly as could atomic masses. Did this number have any fundamental significance?

In 1914, H. G. J. Moseley published the results of the experiments which he had been conducting in Oxford. He found that when a metallic element was bombarded with an electron beam, X-rays were produced, and the wavelength or frequency of the X-rays depended only on which element was used. Whatever the conditions of the experiment, a particular element always produced X-rays of the same frequency, and the frequency of the X-rays was found to be given by a formula *which involved the atomic number of the element*:

$$\sqrt{v} = a(Z - b)$$

(*v* is the frequency of the X-rays, and a and b are constants.) Here was a method whereby the atomic number of an element could be determined directly. The atomic number of an element is thus a fundamental characteristic of that element.

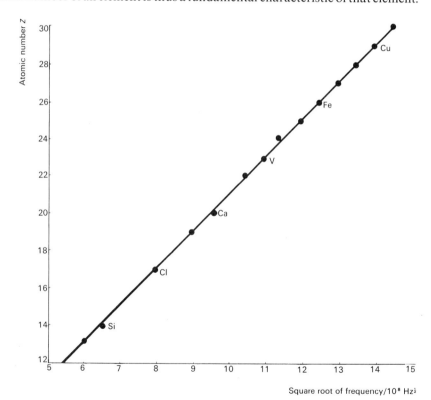

Figure 4.2d
X-ray spectra of the elements.

4.3 The nucleus: part 2
What particles are present in the nucleus?
In 1913 J. J. Thomson discovered that a given element can possess atoms with different masses, and the principle of his experiment was developed from 1919 onwards by Aston and others for use in the accurate comparison of these masses. This apparatus, the mass spectrometer, will be described later. As the atoms of a given element must all have the same atomic number (otherwise they would be atoms of a different element), they must have the same number of electrons and the same charge on the nucleus. To account for the different weights, Rutherford suggested that the nuclei contained the same number of positively-charged particles, protons; and he further suggested the existence of a particle with no charge but with the same mass as the proton which he named the neutron. Neutrons were first observed experimentally in 1932 by Chadwick.

Three different types of hydrogen atom are now known, with masses in the ratio $1:2:3$. The nuclei are represented below.

$$\left(+\right) \qquad \left(+\right)\left(n\right) \qquad \begin{array}{c}\left(n\right)\\\left(+\right)\left(n\right)\end{array} \qquad \begin{array}{l}+ = \text{proton}\\ n = \text{neutron}\end{array}$$

$$_1^1\text{H} \qquad\qquad _1^2\text{H} \qquad\qquad _1^3\text{H}$$

Each of these atoms contains one proton, and is therefore hydrogen. Each differs from the others by having a different number of neutrons. Atoms of an element which differ only in the number of neutrons that they contain, are known as *isotopes.*

A convention is adopted by which the number of protons and neutrons in atoms may be readily shown on paper. The convention is used in the diagram of the hydrogen isotopes. The number of protons (that is, the atomic number) is placed at the bottom left hand of the symbol for the element, and the total number of protons and neutrons (that is, the mass number) is placed at the top left hand of the symbol. The number of neutrons is obtained by subtraction.

Two of the isotopes of lithium are shown below.

Each of these atoms would have three electrons outside the nucleus. Why have they not been represented on this diagram, assuming that the same scale was maintained? In addition to the reason which you should have suggested, it is very difficult to say precisely where an electron is at any given instant.

The accurate determination of atomic masses

The most accurate method of determining atomic masses is by use of the mass spectrometer. The principle is to determine the relative abundance of the iso-topes of the element, and their isotopic masses; the weighted mean of these (that is, taking into account the percentage abundances) is then the atomic mass.

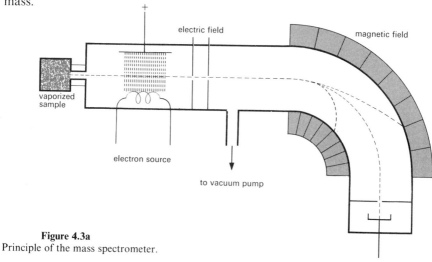

Figure 4.3a
Principle of the mass spectrometer.

Five main operations are performed by the spectrometer:
1 The sample of the element is vaporized
2 Positive ions are produced from the vapour
3 The positive ions are accelerated by a known electric field
4 The ions are then deflected by a known magnetic field
5 The ions are then detected.

The manner in which these stages are achieved may be seen from figure 4.3a. A stream of the vaporized element enters the main apparatus which is main-tained under high vacuum. The atoms of the element are bombarded by a stream of high-energy electrons, which on collision with the atoms knock electrons out of them and produce positive ions. The positive ion stream passes through holes in two parallel plates to which a known electric field is applied, and the ions are accelerated by this field. They then enter a region to which a magnetic field is applied, and they are deflected by it.

For given electric and magnetic fields only ions with one particular mass will reach the detector at the end of the apparatus, all other ions having hit the walls of the instrument. By gradually increasing the strength of the magnetic field, ions of different masses may be brought successively to the detector. Their masses are calculated from the known applied fields, and their relative abundance is found from the relative magnitudes of the current produced in the detector.

Figure 4.3b shows the type of trace which is obtained.

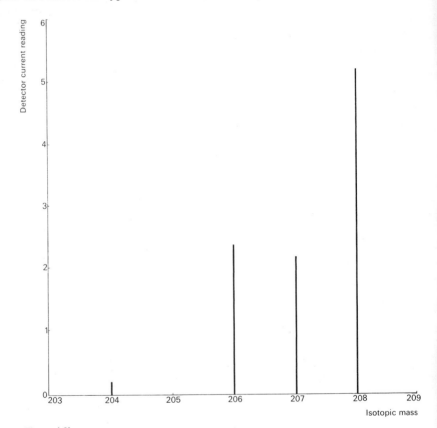

Figure 4.3b
Mass spectrometer trace for naturally occurring lead.

From it the relative abundances can be seen to be:

Isotopic mass	Relative abundance	% relative abundance
204.0	0.2	2
206.0	2.4	24
207.0	2.2	22
208.0	5.2	52
	10.0	100

From these values the atomic mass of naturally-occurring lead can be worked out as follows.

In 100 atoms of naturally-occurring lead there will be, on average, 2 atoms of isotopic mass 204.0, 24 of 206.0, 22 of 207.0, and 52 of 208.0. If we find the total mass of all of these 100 atoms we may find the average mass by dividing by 100.

Isotopic mass	Number of atoms in 100 atoms of mixture	Mass of isotopes in 100 atoms of mixture
204.0	2	408
206.0	24	4944
207.0	22	4554
208.0	52	10 816
		20 722

$$\text{Average mass of 1 atom} = \frac{20\,722}{100} = 207.2 \text{ atomic mass units}$$

(This is commonly known as the *atomic mass*, the reference standard for which is the mass of one atom of the $^{12}_{6}C$ isotope which is taken as exactly 12 units of atomic mass.)

Use the table of isotopic abundances in the *Book of Data* to work out the atomic mass of naturally-occurring chlorine, and of magnesium.
Examine the abundance of the isotopes of $_{52}Te$ and the atomic mass of tellurium. Compare this with the atomic mass of $_{53}I$, and comment on their relative positions in the Periodic Table and their atomic masses.

Do the same for $_{18}Ar$ and $_{19}K$.

How dense is the nucleus of the atom?
If the nucleus is so small and yet contains most of the mass of an atom, it must be very dense.

Calculate the density of the nucleus of a fluorine atom in g cm^{-3} and in tonnes cm^{-3} from the following data.

The nucleus of a fluorine atom has a radius of approximately 5×10^{-13} cm and a mass of approximately 3.15×10^{-23} grammes. (1 tonne or metric ton $= 1$ Mg $= 1000$ kg.)

What does the result tell us about the strength of the forces holding the nucleus together?

Do you think that these forces penetrate very far outside the nucleus?

4.4 The arrangement of electrons in atoms

It is the electrons of an atom which are involved in chemical changes, not the protons and neutrons; this is to be expected because the electrons constitute the outer part of an atom. When one atom combines with another, one or more electrons are transferred from one atom to the other, or electrons may be shared between the two atoms. The final arrangement need not concern us at the moment, but it will be clear that the amount of energy which is needed to remove an electron from the atom is an important quantity to be known if the energy changes involved in chemical bonding are to be understood. This energy is known as the ionization energy.

The first ionization energy of an element is the energy required to remove one mole of electrons from the element in the gaseous state to form ions.

$$M(g) \longrightarrow M^+(g) + e^- \qquad \Delta H = \text{ionization energy}$$

The ionization energy can be measured in electron volts, kilocalories per mole, or kilojoules per mole; in this book we shall use the last of these three units (kJ mol^{-1}). The reasons for using this unit are discussed in Topic 7.

Several different methods are available for determining the ionization energy of an element, and two will be considered here. One is for elements in the gaseous form and involves bombardment of the gas atoms by electrons; the other can be used for elements in either the gaseous or solid form and involves a study of their spectra.

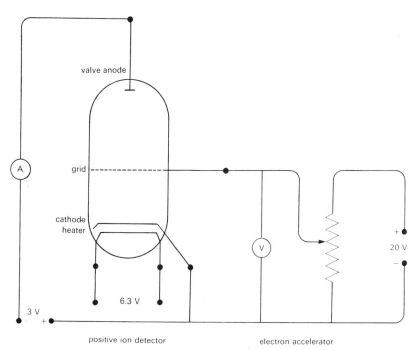

Figure 4.4a
Circuit for determining the ionization energy of an inert gas.

Determining the ionization energy of a gas by electron bombardment

This experiment involves the use of a radio valve containing gas at very low pressure. The valve is incorporated in a circuit as shown in figure 4.4a. When the cathode is heated by means of an electric current, electrons are emitted and stream into the body of the valve. This is illustrated in figure 4.4b.

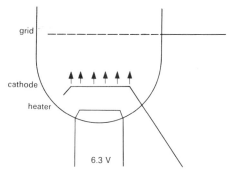

Figure 4.4b
The arrows represent electrons, and the lengths of the arrows represent the energy of the electrons.

The effect of applying a potential between the grid and the cathode of the valve is shown in figure 4.4c. The potential difference accelerates the electrons towards the grid.

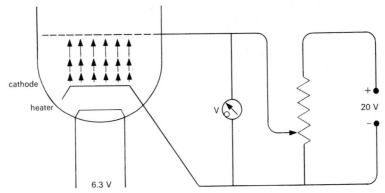

Figure 4.4c

The anode is made negative by a small voltage so that the electrons are repelled from it. This is shown in figure 4.4d. *Note :* the word *anode* refers to that part of the valve normally called the anode, even though in this experiment it is negatively charged.

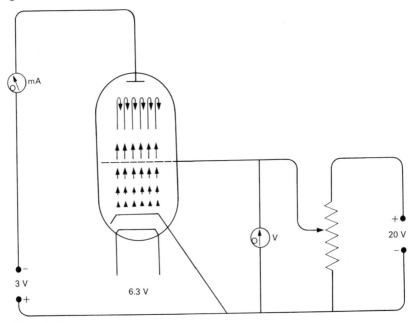

Figure 4.4d

Under these circumstances no current flows between the anode and the cathode and this is indicated on the milliammeter.

While the electrons travel through the space between the cathode and the anode they collide with the gas atoms (or molecules). If the electrons are travelling slowly the collisions will be gentle and the electrons will merely bounce off the atoms; but if the electrons are travelling sufficiently fast they will have enough energy to knock electrons forming part of the atoms of the gas right out of those atoms. This results in the formation of a positive ion, which is then attracted towards the anode. Current will then flow between the cathode and the anode, and this can be detected by the milliammeter. The anode therefore acts as a detector of positive ions. The formation of one positive ion is shown in figure 4.4e.

The ionization process may be represented in equation form. For argon the equation is:

$$\text{Ar(g)} + e^- \longrightarrow \text{Ar}^+\text{(g)} + 2e^-$$

(fast   (ionized   (slow
electron)   argon)   electrons)

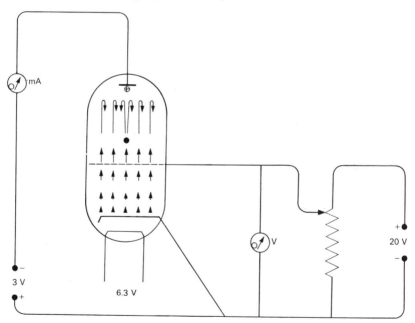

⊕ = positive ion

● = gas atom

Figure 4.4e

The speed at which the electrons travel can be controlled by the magnitude of the potential difference which is applied between the cathode and the grid.

Can you suggest an experiment by which the potential difference, in volts, at which ionization of the gas takes place could be determined? What would you do? What quantities would you measure? You will have an opportunity to see an experiment in which this potential is measured.

The first ionization energy of an element such as argon can be found from the potential difference in volts at which ionization of the gas takes place by multiplying this potential difference by 96.3. This figure arises in the following way.

As 1 joule of energy has to be expended when 1 coulomb of charge passes through a potential difference of 1 volt,

$$1 V = 1 J C^{-1}.$$

When one electron (charge $= 1.6 \times 10^{-19}$ C) is accelerated through a potential difference of 1 V, it acquires $1.6 \times 10^{-19} \times 1$ joule of energy. When 1 mole of electrons is accelerated through a potential difference of 1 V, it acquires

$$1.6 \times 10^{-19} \times 6.02 \times 10^{23} \text{ J}$$

$$= 9.63 \times 10^4 \text{ J}$$

$$= 96.3 \text{ kJ}$$

The amount of energy possessed by electrons when accelerated by a potential difference of 1 volt is therefore 96.3 kJ mol^{-1}. So the energy that 1 mole of electrons had when it ionized 1 mole of argon can be calculated from the accelerating voltage by multiplying the voltage by 96.3.

Emission spectra of elements

When atoms of an element are supplied with sufficient energy they will emit light. This energy may be provided in several ways. If the element is a gas it may be placed in an electric discharge tube at low pressure; neon signs work on this principle. Certain easily vaporized metals also emit light under these conditions; examples are the blueish-white street lamps which are mercury discharge tubes, and the yellow street lamps which are sodium discharge tubes.

The energy may also be supplied by a flame; many metals and their salts when vaporized in a flame emit light. A further method which is particularly appropriate for metals and alloys is to make the element part of one electrode in an electric arc discharge.

When the light which is emitted is examined through a spectroscope, it is found not to consist of a continuous range of colours like part of a rainbow, but to be made up of discrete lines of colour. This type of spectrum is known as a *line emission spectrum*.

You should examine the line emission spectra of some elements by means of a spectroscope or a diffraction grating. Each element has its own characteristic set of lines, and these enable elements to be identified by examination of their spectra. Indeed, spectroscopic examination of the sun revealed the existence there of an element which at that time had not been discovered on earth; it was named helium, from the Greek word *helios*, meaning the sun.

The photograph in figure 4.4f shows the emission spectrum of atomic hydrogen, in the visible region. (A coloured photograph of the hydrogen spectrum is given inside the front cover of this book.) Each line corresponds to a given frequency, and the lines fit into a series, known as the *Balmer series*.

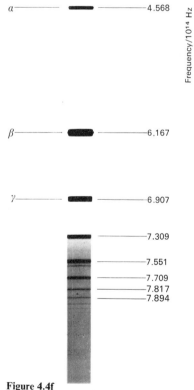

Figure 4.4f
Spectrum of atomic hydrogen, visible region.

Balmer series of lines in the spectrum of atomic hydrogen

Frequency, $v/10^{14}$ Hz
$(1 \text{ Hz} = 1 \text{ sec}^{-1})$

Red α 4.568

       β 6.167

Violet γ 6.907

       7.309

       7.551

       7.709

       7.817

       7.894

What do you notice about the spacing of the lines?

Plot a graph of v against $1/n^2$ where
   $n = (2+1)$ for the first line in the visible region of the spectrum, (α)
      $(2+2)$ for the second line, (β)
      $(2+3)$ for the third line, (γ)
      etc.

What do you think has happened to the lines, and also to the hydrogen when $1/n^2 = 0$?

What is the interpretation of the two main features of the spectrum of hydrogen, first the discrete lines, and second that the lines come closer together until they coalesce?

In order to explain these observations it is necessary to assume that an electron in an atom can exist only in certain energy levels; it cannot possess energy of intermediate magnitude. Energy is required to promote an electron from a low energy level to a high one (excitation); and if an electron falls from a high level to a lower one energy is released. This energy is not released over a continuous wide band of frequencies, but at a unique frequency, and we say that a quantum of radiation (light) has been emitted. For each electron transition, therefore, an associated quantum of radiation is emitted.

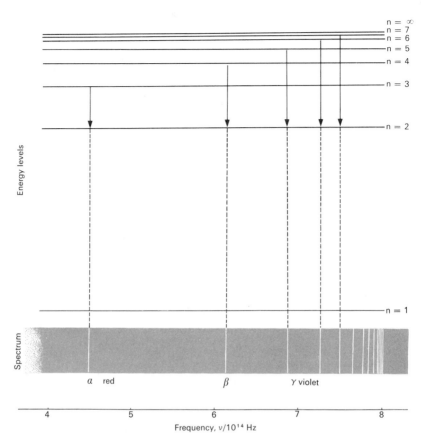

Figure 4.4g

Energy level diagram, showing the origin of the lines in the visible portion of the hydrogen spectrum (not to scale).

Suppose in figure 4.4g that energy level $n = 3$ is associated with an energy E_3, and the level $n = 2$ with an energy E_2; then if an electron is transferred from n_3 to n_2 an amount of energy

$$E_3 - E_2 = \Delta E$$

is released. We must also assume that there is a relationship between a quantum of energy and its frequency, if it appears as radiation, such that

$$\Delta E = \text{constant} \times v.$$

This relationship has been determined experimentally and the constant is called Planck's constant, h, and it has the value 3.99×10^{-13} kJ sec mol^{-1}. v is the frequency of the radiation.

Since ΔE for the change from E_3 to E_2 is always the same in a given atom, and h is a constant, v must also be a constant; that is, the radiation always has the same energy and is always of the same frequency for this particular electron transition. Thus a discrete line appears in the spectrum.

Since the spectral lines converge and finally come together it is assumed that the electronic energy levels in an atom also converge and finally come together. This is shown in figure 4.4g.

Transitions of an electron from various energy levels in hydrogen to energy level n = 2 involve energy changes such that the radiation emitted appears in the visible part of the spectrum. If, however, the transitions are to the n = 1 level, more energy is released, and the lines appear in the higher energy range of the spectrum, that is, in the ultraviolet region. Similarly, if transitions occur from high levels to the n = 3 level, much less energy is released and the lines appear in the low-energy region of the spectrum, that is, in the infra-red region.

Copy the energy levels of figure 4.4g and draw in transitions from high levels to either the n = 1 level or the n = 3 level. Then draw in schematically the general position of the spectral lines that would be produced.

If sufficient energy is supplied to an atom to promote an electron from one energy level to the highest possible one and just beyond it, then the electron is able to escape, and the atom becomes an ion. It should therefore be possible to determine the ionization energy of an element from its spectrum. This can be done if the frequency can be determined at which the converging spectral lines actually come together. This frequency is known as the 'convergence limit'.

The table below gives the frequency of lines in the ultraviolet spectrum of atomic hydrogen, which form a series known as the *Lyman series*.

Lyman series of lines in the spectrum of atomic hydrogen

Frequency, $v/10^{15}$ Hz

2.466
2.923
3.083
3.157
3.197
3.221
3.237
3.248

You will notice that the values of the frequencies come closer together, and once again converge to a limit. Since the ultraviolet spectrum represents electrons making transitions to the lowest energy level, $n = 1$, the convergence limit represents the energy required to ionize a hydrogen atom with its electron in the lowest level, and hence may be used to find the ionization energy of hydrogen. Find the frequency of the convergence limit as follows. Work out the difference in frequency, Δv, between successive lines, and plot a graph of Δv against the frequency v. Then extrapolate the curve to the point where the difference in frequency, Δv, becomes equal to 0 and read off the frequency. It does not matter whether you use the value of the higher or the lower frequencies for plotting v, as long as you are consistent in your choice. If two curves are plotted, one using the higher values of the frequencies and the other using the lower values, both on the same graph, it will be easier to estimate the value of the frequency when $\Delta v = 0$.

Convert this frequency into kJ mol^{-1} using Planck's constant given above. Compare the value for the ionization energy of hydrogen that you obtain with that given in the *Book of Data*.

4.5   What can ionization energies tell us about the arrangement of electrons in atoms?

The discussion of the ionization of an atom has so far considered the removal of one electron only; but if an atom containing several electrons is treated with sufficient vigour, then more than one electron may be removed from it. A succession of ionization energies is therefore possible. These may be determined, principally from spectroscopic measurements; a table of successive ionization energies for a number of elements is given in the *Book of Data*.

1   Using this table, attempt to plot for sodium a graph of number of electrons removed against the appropriate ionization energy. What do you notice, first about the general trend in values, and second about their magnitude? Why does the general trend (increase or decrease in values) occur?
Now plot a graph of the logarithm of the ionization energy, on the vertical axis, against number of electrons removed, on the horizontal axis.
Does this give any information about groups of electrons which can be removed more readily than others? How many electrons are there in each group?

2   Using the same table, study the change in the first ionization energy of the elements. For the first twenty elements plot the value of their first ionization energy, on the vertical axis, against their atomic number, on the horizontal axis. When you have plotted the points, draw lines between them to show the pattern, and label each point with the symbol for the element.

Where do the alkali metals lithium, sodium, and potassium appear?
Where do the inert gases occur?
Do you notice any groups of points in the pattern? How many elements are
there in each group? Do the numbers bear any relation to the numbers of
electrons in any pattern you may have found for the successive ionization
energies for sodium?

   3 What interpretations can be placed on these results?

In the successive ionizations of sodium, one electron needs much less energy
for its removal than did the others, and it must therefore have been in a high
energy level; eight electrons required much more energy, and must have been
in a lower energy level; two electrons must have been in a still lower energy
level. The lowest energy level is called the n = 1 level.

Thus, for sodium there would appear to be:
      2 electrons in the n = 1 energy level
      8 electrons in the n = 2 energy level
      and 1 electron in the n = 3 energy level.

We can represent this on an energy level diagram.

Energy level   Number of electrons
and quantum
shell

n = 3   Higher energy levels:
less energy required to
remove electrons
from these.

n = 2    Lower energy levels:
more energy required to
remove electrons
from these.

n = 1

Figure 4.5a
Energy levels of electrons in a sodium atom.

The two electrons in the n = 1 energy level are situated *most* of the time closer to the nucleus than the other electrons, and they are said to be in the first *quantum shell*. The eight electrons in the n = 2 energy levels spend much of their time further from the nucleus, and are said to be in the second quantum shell. The single electron in sodium spends much of its time further still from the nucleus and is said to be in the third quantum shell.

Thus there are two ways of looking at electrons in atoms: from the point of view of their energy level, n = 1, 2, 3, 4, etc., and from the point of view of how far from the nucleus they are on average, that is, in the first second, third, or fourth, etc., quantum shell.

The electrons in figures 4.5a, 4.5b, and 4.5c have been represented by arrows. When an energy level is half full the next electrons pair up with existing ones. Electrons behave as though they had the property of spin, and paired electrons must have their spins in opposite directions; this is represented by up and down arrows. The reasons why we believe that electrons behave as though they were spinning, and that the spins of paired electrons are opposed, are complicated, and it is not necessary to discuss them here. The evidence for this comes from a more detailed examination of line spectra.

Had you plotted the successive ionization energies for potassium, the pattern would have been 1 electron most easily removed, followed by 8 more difficult, followed by 8 even more difficult, followed by 2 extremely difficult to remove. How many electrons do the n = 1, n = 2, n = 3, and n = 4 energy levels and quantum shells hold in potassium?

Does a similar pattern show in the graph of first ionization energies which you plotted for 20 elements?

You will notice from this latter graph that the groups of eight are made up of groups of (2, 3, and 3) points on the curve. This indicates that the eight electrons are not all exactly the same as far as their energies are concerned. From this type of evidence, and also from studies of spectral lines, it has been concluded that the energy levels are split so that the n = 2 level has two electrons in a sub-level known as 2s (slightly more difficult to remove) and six electrons in a sub-level known as 2p (slightly less difficult to remove).

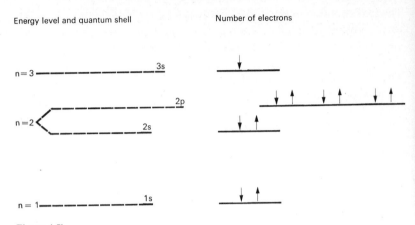

Figure 4.5b
Energy level of electrons in a sodium atom. showing sub-levels.

From similar evidence, it has been concluded that the n = 3 level is split into s, p, and d, and the n = 4 and n = 5 levels into s, p, d, and f sub-levels. All the s sub-levels can contain up to two electrons, the p sub-levels six, the d sub-levels ten, and the f sub-levels fourteen. The arrangement of these energy levels is shown in figure 4.5c.

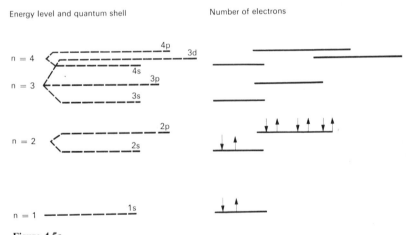

Figure 4.5c
Energy levels of electrons in atoms. The electronic structure of neon is shown (lower energy levels only are illustrated).

Figure 4.5c shows an electron representation for neon. The atomic number of neon is 10, and its atom therefore contains 10 electrons.

2 are in the n = 1 level, and are in the s sub-level $1s^2$
2 are in the n = 2 level, and are in the s sub-level $2s^2$
6 are in the n = 2 level, and are in the p sub-level $2p^6$

This electronic structure is written thus: $1s^2 2s^2 2p^6$.

The next element, sodium, atomic number 11, would have one electron in the n = 3 level, and it would be in an s sub-level, since this is the lowest n = 3 level. The structure of sodium is therefore $1s^2 2s^2 2p^6 3s^1$.

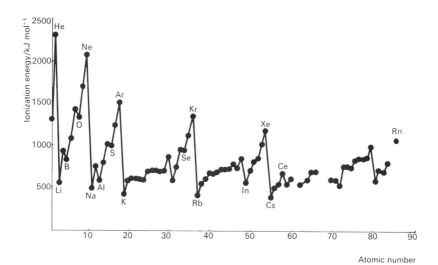

Figure 4.5d
First ionization energies of the elements.

From figure 4.5d you will see that after calcium, atomic number 20, the (2, 3, 3) grouping is broken. How many elements produce the break? What might this indicate in terms of electrons? What is the name of this grouping of elements in the Periodic Table?

After calcium, a new energy level belonging to the n = 3 quantum shell becomes occupied; the electrons in it are known as d electrons. A d sub-level can hold 10 electrons when full. You will notice from figure 4.5c that the 3d level is just above the 4s level and is just below the 4p level; this is of great importance in the chemistry of the transition elements.

Each major peak in the ionization energy curve in figure 4.5d represents an element with a completed quantum shell; the element concerned is an inert gas.

Where and what are the electrons?
We have seen that the electrons are in certain energy levels and that a particular level corresponds to the electrons in that level *having a certain mean radius.*

An electron can be considered to be distributed as a diffuse negative charge-cloud. The charge-cloud for an s electron is spherical in shape, with more charge near the centre of the sphere than near the periphery.

A hydrogen atom showing the density distribution of its 1s electron is shown in figure 4.5e. It can be seen that there is a certain distance from the nucleus that can be taken to be the average distance of the charge of the electrons.

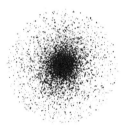

Figure 4.5e
Density of the electron charge-cloud in a cross section of a hydrogen atom (1s electron).

The distributions of p electron charge-clouds and d electron charge-clouds are not spherical. Our beliefs concerning their shapes are derived from quantum mechanical calculations, and it is not necessary in the present context to discuss them.

Under certain circumstances electrons behave as though they possess wave properties. Thus at least two models are required to represent the behaviour of electrons: a charge-cloud model and a wave model.

4.6   Some relationships between the atomic structure and periodic properties

1   Trends in ionization energies
As a broad trend how does the first ionization energy change across a period of the Periodic Table?
Why is this?
How does the first ionization energy change on going down a group in the Periodic Table?
Why is this?

2 Electron affinity

In moving across the Periodic Table the atom of each element contains one more positive charge in the nucleus than did the previous one. With the halogens the attraction of the nucleus for electrons has become very considerable.

Partly because of this, neutral atoms of the halogens are able to accept an electron, and a negative ion results. The process is often accompanied by a substantial release of energy, as the following figures indicate.

$$Cl(g) + e^- \longrightarrow Cl^-(g) \qquad \Delta H = -364 \text{ kJ mol}^{-1}$$

$$Br(g) + e^- \longrightarrow Br^-(g) \qquad \Delta H = -342 \text{ kJ mol}^{-1}$$

Values for oxygen are as follows:

$$O(g) + e^- \longrightarrow O^-(g) \qquad \Delta H = -141 \text{ kJ mol}^{-1}$$

$$O^-(g) + e^- \longrightarrow O^{2-}(g) \qquad \Delta H = +791 \text{ kJ mol}^{-1}$$

3 Atomic radius

The atomic radius of an atom in an element is taken as half of the distance between the centres of two adjacent atoms in a close-packed structure or in a molecule.

The *Book of Data* gives values of atomic radii for some elements. These values are given in nanometres, nm; one nanometre is 10^{-9} metre. Using a scale of 1 cm : 0.1 nm, draw circles representing atoms of the elements up to $_{22}$Ti.

Draw the circles in horizontal rows in Periodic Table order.

How does atomic radius change in the Periodic Table? Why is this?

4 Chemical similarities

The chemical similarities existing among members of a group of elements arise because of the similar configurations of the outer electron shells of their atoms.

The lack of reactivity of the inert gases is largely due to their very high ionization energies and the stability associated with a completed shell of electrons. They have no detectable electron affinity since, presumably, an incoming electron would have to occupy the next outer quantum shell on its own. This would be too far from the nucleus and the nucleus would be too well screened by inner shells for the electron to be held.

Background reading

There now follow extracts from papers by Geiger and Marsden, Moseley, and Rutherford concerning discoveries described earlier in this topic. After these, an account of atomic emission spectroscopy is given, as an example of the type of practical use to which these discoveries have been put.

The laws of deflexion of α-particles through large angles
by H. Geiger and E. Marsden
Extracts from a paper published in the *Philosophical Magazine* in 1913 (*Phil. Mag.* (1913), 25, 604–613)

In a former paper one of us has shown that in the passage of α-particles through matter the deflexions are, on the average, small and of the order of a few degrees only. In the experiments a narrow pencil of α-particles fell on a zinc-sulphide screen in vacuum, and the distribution of the scintillations on the screen was observed when different metal foils were placed in the path of the α-particles. From the distribution obtained, the most probable angle of scattering could be deduced, and it was shown that the results could be explained on the assumption that the deflexion of a single α-particle is the resultant of a large number of very small deflexions caused by the passage of the α-particle through the successive individual atoms of the scattering substance.

In an earlier paper, however, we pointed out that α-particles are sometimes turned out through very large angles. This was made evident by the fact that when α-particles fall on a metal plate, a small fraction of them, about 1/8000 in the case of platinum, appears to be diffusely reflected. This amount of reflexion, although small, is, however, too large to be explained on the above simple theory of scattering. It is easy to calculate from the experimental data that the probability of a deflexion through an angle of 90° is vanishingly small, and of a different order to the value found experimentally.

Professor Rutherford has recently developed a theory to account for the scattering of α-particles through these large angles, the assumption being that the deflexions are the result of an intimate encounter of an α-particle with a single atom of the matter traversed. In this theory an atom is supposed to consist of a strong positive or negative central charge concentrated within a sphere of less than about 3×10^{-12} cm radius, and surrounded by electricity of the opposite sign distributed throughout the remainder of the atom of about 10^{-8} cm radius. In considering the deflexion of an α-particle directed against such an atom, the main deflexion-effect can be supposed to be due to the central concentrated charge which will cause the α-particle to describe an hyperbola with the centre of the atom as one focus.

Assuming a narrow pencil of α-particles directed against a thin sheet of matter containing atoms distributed at random throughout its volume, if the scattered particles are counted by the scintillations they produce on a zinc-sulphide screen distance r from the point of incidence of the pencil in a direction making an angle ϕ with it, the number of α-particles falling on unit area of the screen per second is deduced to be equal to

$$\frac{Qntb^2 \cosec^4 \phi/2}{16r^2}$$

where Q is the number of α-particles per second in the original pencil, n the number of atoms in unit volume of the material, and t the thickness of the foil. The quantity

$$b = \frac{2ZE}{mu^2}$$

where Z is the central charge of the atom, and m, E, and u are the respective mass, charge, and velocity of the α-particle.

The number of deflected α-particles is thus proportional to (1) $\text{cosec}^4 \, \phi/2$, (2) thickness of scattering material t if the thickness is small, (3) the square of the central charge Z of the atoms of the particular matter employed to scatter the particles, (4) the inverse fourth power of the velocity u of the incident α-particles.

At the suggestion of Professor Rutherford, we have carried out experiments to test the main conclusions of the above theory. The following points were investigated:

1 Variation with angle.
2 Variation with thickness of scattering material.
3 Variation with atomic weight of scattering material.
4 Variation with velocity of incident α-particles.
5 The fraction of particles scattered through a definite angle.

The main difficulty of the experiments has arisen from the necessity of using a very intense and narrow source of α-particles owing to the smallness of the scattering effect. All the measurements have been carried out by observing the scintillations due to the scattered α-particles on a zinc-sulphide screen, and during the course of the experiments over 100 000 scintillations have been counted. It may be mentioned in anticipation that all the results of our investigations are in good agreement with the theoretical deductions of Professor Rutherford, and afford strong evidence of the correctness of the underlying assumption that an atom contains a strong charge at the centre of dimensions small compared with the diameter of the atom.

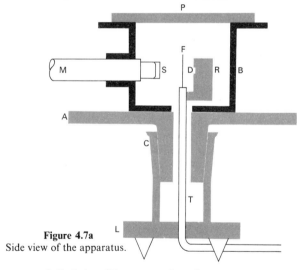

Figure 4.7a
Side view of the apparatus.

1 *Variation of Scattering with Angle*
We have already pointed out that to obtain measurable effects an intense pencil of α-particles is required. It is further necessary that the path of the α-particles should be in an evacuated chamber to avoid complications due to the absorption and scattering of the air. The apparatus used is shown in figure 4.7a and mainly consisted of a strong cylindrical metal box B, which contained the source of α-particles, R, the scattering foil F, and a microscope M to which the zinc-sulphide screen S was rigidly attached. The box was fastened down to a graduated circular platform A, which could be rotated by means of a

conical air-tight joint C. By rotating the platform the box and microscope moved with it, while the scattering foil and radiating source remained in position, being attached to the tube T, which was fastened to the standard L. The box B was closed by the ground-glass plate P, and could be exhausted through the tube T.

The source of α-particles employed was the gas radon.

By means of a diaphragm placed at D, a pencil of α-particles was directed normally on to the scattering foil F. By rotating the microscope the α-particles scattered in different directions could be observed on the screen S.

Table 4.7 gives the collected results for two series of experiments with foils of silver and gold.

Angle of deflexion ϕ	Cosec$^4\phi/2$	Silver Number of scintillations N	N/ cosec^4 $\phi/2$	Gold Number of scintillations N	N/ cosec^4 $\phi/2$
150°	1.15	22.2	19.3	33.1	28.8
135	1.38	27.4	19.8	43.0	31.2
120	1.79	33.0	18.4	51.9	29.0
105	2.53	47.3	18.7	69.5	27.5
75	7.25	136	18.8	211	29.1
60	16.0	320	20.0	477	29.8
45	46.6	989	21.2	1435	30.8
37.5	93.7	1760	18.8	3300	35.3
30	223	5260	23.6	7800	35.0
22.5	690	20 300	29.4	27 300	39.6
15	3445	105 400	30.6	132 000	38.4

Table 4.7a
Variation of scattering with angle (collected results)

These experiments, therefore, prove that the number of α-particles scattered in a definite direction varies as cosec^4 $\phi/2$.

Summary
The experiments described in the foregoing paper were carried out to test a theory of the atom proposed by Professor Rutherford, the main feature of which is that there exists at the centre of the atom an intense highly concentrated electrical charge. The verification is based on the laws of scattering which were deduced from this theory.

The high frequency spectra of the elements (part II)

by H. G. J. Moseley

Extracts from a paper published in the *Philosophical Magazine* in 1914 (*Phil. Mag.* (1914), **27**, 703–713)
The first part of this paper dealt with a method of photographing X-ray spectra, and included the spectra of a dozen elements. More than thirty other elements have now been investigated, and simple laws have been found which govern the results, and make it possible to predict with confidence the position of the principal lines in the spectrum of any element from aluminium to gold. The present contribution is a general preliminary survey, which claims neither to be complete nor very accurate.

[The results obtained with different elements for the frequencies of a particular series of lines in the X-ray spectra are shown in figure 4.2d. This is a much simplified version of the figure referred to in the paragraph below. In figure 4.2d the symbol Z is used for atomic number, which is customary at the present time; Moseley uses the symbol N.]

In the figure the spectra of the elements are arranged on horizontal lines spaced at equal distances. The order chosen for the elements is the order of the atomic weights, except in the cases of A, Co, and Te, where this clashes with the order of the chemical properties. Vacant lines have been left for an element between Mo and Ru, an element between Nd and Sm, and an element between W and Os, none of which are yet known, while Tm, which Welsbach has separated into two constituents, is given two lines. This is equivalent to assigning to successive elements a series of successive characteristic integers. On this principle the integer N for Al, the thirteenth element, has been taken to be 13, and the values of N then assumed by the other elements are given on the left-hand side of the figure. This proceeding is justified by the fact that it introduces perfect regularity into the X-ray spectra. Examination of the figure shows that the values of $v^{\frac{1}{2}}$ for all the lines examined now fall on straight lines.

Now if either the elements were not characterized by these integers, or any mistake has been made in the order chosen or in the number of places left for unknown elements, these regularities would at once disappear. We can therefore conclude from the evidence of the X-ray spectra alone, without using any theory of atomic structure, that these integers are really characteristic of the elements.

Now Rutherford has proved that the most important constituent of an atom is its central positively charged nucleus, and van den Broek has put forward the view that the charge carried by this nucleus is in all cases an integral multiple of the charge on the hydrogen nucleus. There is every reason to suppose that the integer which controls the X-ray spectrum is the same as the number of electrical units in the nucleus, and these experiments therefore give the strongest possible support to the hypothesis of van den Broek. Soddy has pointed out that the chemical properties of the radio-elements are strong evidence that this hypothesis is true for the elements from thallium to uranium, so that its general validity would now seem to be established.

Summary

1 Every element from aluminium to gold is characterized by an integer N which determines its X-ray spectrum. Every detail in the spectrum of an element can therefore be predicted from the spectra of its neighbours.

2 This integer N, the atomic number of the element, is identified with the number of positive units of electricity contained in the atomic nucleus.

3 The atomic numbers for all elements from Al to Au have been tabulated on the assumption that N for Al is 13.

4 The order of the atomic numbers is the same as that of the atomic weights, except where the latter disagrees with the order of the chemical properties.

5 Known elements correspond with all the numbers between 13 and 79 except three. There are here three possible elements still undiscovered.

6 The frequency of any line in the X-ray spectrum is approximately proportional to $A(N - b)^2$, where A and b are constants.

Bakerian lecture: nuclear constitution of atoms
by Sir E. Rutherford
Extracts from a lecture published in the *Proceedings of the Royal Society* in 1920 (*Proc. Roy. Soc.* (1920), **97**, 374–400)

The conception of the nuclear constitution of atoms arose initially from attempts to account for the scattering of α-particles through large angles in traversing thin sheets of matter. Taking into account the large mass and velocity of the α-particles, these large deflexions were very remarkable, and indicated that very intense electric or magnetic fields exist within the atom. To account for these results, it was found necessary to assume that the atom consists of a charged massive nucleus of dimensions very small compared with the ordinarily accepted magnitude of the diameter of the atom. This positively charged nucleus contains most of the mass of the atom, and is surrounded at a distance by a distribution of negative electrons equal in number to the resultant positive charge on the nucleus. Under these conditions, a very intense electric field exists close to the nucleus, and the large deflexion of the α-particle in an encounter with a single atom happens when the particle passes close to the nucleus. Assuming that the electric forces between the α-particle and the nucleus varied according to an inverse square law in the region close to the nucleus, the writer worked out the relations connecting the number of α-particles scattered through any angle with the charge on the nucleus and the energy of the α-particle. Under the central field of force, the α-particle describes a hyperbolic orbit round the nucleus, and the magnitude of the deflexion depends on the closeness of approach to the nucleus. From the data of scattering of α-particles then available, it was deduced that the resultant charge on the nucleus was about $\frac{1}{2}Ae$, where A is the atomic weight and e the fundamental unit of charge. Geiger and Marsden made an elaborate series of experiments to test the correctness of the theory, and confirmed the main conclusions.

It was suggested by van den Broek that the scattering of α-particles by the atoms was not inconsistent with the possibility that the charge on the nucleus was equal to the atomic number of the atom, i.e. to the number of the atom when arranged in order of increasing atomic weight. The importance of the atomic number in fixing the properties of an atom was shown by the remarkable work of Moseley on the X-ray spectra of the elements. He shows that the frequency of vibration of corresponding lines in the X-ray spectra of the elements depended on the square of a number which varied by unity in successive elements. This relation received an interpretation by supposing that the nuclear charge varied by unity in passing from atom to atom, and was given numerically by the atomic number. I can only emphasize in passing the great importance of Moseley's work, not only in fixing the number of possible elements, and the position of undetermined elements, but in showing that the properties of an atom were defined by a number which varied by unity in successive atoms. This gives a new method of regarding the periodic classification of the elements, for the atomic number, or its equivalent the nuclear charge, is of more fundamental importance than its atomic weight.

Atomic emission spectroscopy

Every element in the Periodic Table has a unique electronic structure. Excited atoms of a certain element therefore will give rise to a line spectrum which is different from that of any other element and, consequently, line spectra can be used to identify elements. Furthermore, the intensities of lines in an emission spectrum are proportional to the concentrations of the atoms present in a sample so that a quantitative analysis is possible.

In emission spectroscopy, the spectra can be recorded on a photographic plate and the intensities of the lines measured with a densitometer. In some cases, however, the intensity of radiation at a selected wavelength is measured by allowing it to fall on to a photomultiplier tube. This is a device which converts radiant energy into electrical currents which are proportional to the intensity of the radiation.

Atomic emission spectroscopy is capable of quantitative accuracy comparable to chemical methods and, in addition, it can deal with concentrations down to the part per million level. An important industrial application of atomic emission spectroscopy is to be found in modern steelmaking plants.

The demand for steel in a technologically advanced society like our own is enormous and extremely efficient production methods must be developed to meet it. This in turn means that high capital investment in the plant is necessary and that such plant cannot be allowed to remain idle for longer than necessary during the process of steelmaking.

Oxygen convertors and high power electric furnaces enable steel to be produced so rapidly that time lost waiting for analyses of the melts may amount to a significant proportion of the total heat time. Tons of molten steel must wait for the analysts' report before the operators can proceed to pour it into moulds or, if the analysis is unsatisfactory, modify its composition.

A skilled analytical chemist would require about 45 minutes to identify the major constituents in a sample of the melt and the determination of trace elements would take longer still. During this time the melt is waiting, in danger of atmospheric contamination and with an electric furnace in operation at a running cost of more than £60 per minute. Add to this the fact that skilled plant operators are also standing idle, and you will appreciate the serious economic problem involved. A good solution to this problem is provided by an instrument called an atomic emission spectrometer. The Polyvac, manufactured by Hilger and Watts, is a highly specialized version of such an instrument.

The Polyvac

A small sample of melt is poured into a mould and cooled from red to black heat in running water; at black heat, the temperature is high enough to dry the sample by the time it arrives in the laboratory. Once there, a slice about $\frac{1}{2}$ inch thick is cut from the sample and polished to produce a clean surface. This slice is then placed in a spark source which is flushed with dry argon. The reason for the argon flushing will be explained later. Radiation from the sparked sample passes through a condensing lens and slit into a vacuum chamber where, after collimation, it is dispersed by two fluorite prisms and brought to focus to form a spectrum along a curved focal surface. Figure 4.7b shows the radiation path in the vacuum system of the Polyvac spectrometer. Radiation of selected analytical wavelengths, i.e. those known to be characteristic of elements of interest to the steelmaker, passes through exit slits in the curved focal surface and is deflected by a system of mirrors and lenses onto a row of photomultipliers. Figure 4.7c shows the photomultipliers, exit slits, etc. Each photomultiplier receives radiation characteristic of one element and converts it into an electric current proportional to the intensity of the radiation which is proportional to the amount of element present in the sample. Polyvacs are available with up to 25 photomultipliers so that 25 different elements may be quantitatively determined in a sample simultaneously.

The outputs from the photomultipliers pass into an electronic unit where they are measured and information regarding their magnitude passed on to some form of 'read-out' device. Before we consider read-out devices, look at the photograph of a Polyvac installation shown in figure 4.7d and consider the reason for the vacuum system in the spectrometer.

The elements of interest for steel analysis have spectral lines in a region of the ultra-violet, having wavelength below 200 nm, which is known as the vacuum ultra-violet. This is because such radiation is strongly absorbed by air, particularly by the oxygen present. If the spectrometer were full of air, no radiation of interest would reach the photomultipliers; hence the system is kept under vacuum. The spark source, however, will not operate satisfactorily in a vacuum so it is flushed with argon, a gas which is quite transparent in the vacuum ultra-

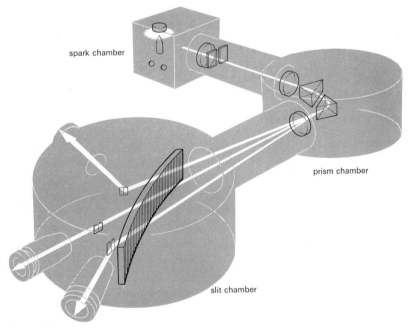

Figure 4.7b
Radiation path in the Polyvac.
Hilger & Watts

Figure 4.7c
Photomultipliers and exit slits in the Polyvac.
Hilger & Watts

violet. In table 4.7b, some of the important elements in steel-making are listed, together with the emission lines used to identify them in the Polyvac. (All values are given in nanometres; 1 nm = 10^{-9} m.)

P	178.3	Al	186.3	Mn	192.1
S	180.7	As	189.1	C	193.1
Si	181.7	Sn	190.0	Mo	202.0
B	182.6	Ti	190.8	Cr	206.5

Table 4.7b

A typical read-out device would be a computer, which converts the information passed to it by the electronic unit into percentage compositions, and an automatic typewriter which prints out the information supplied by the computer.

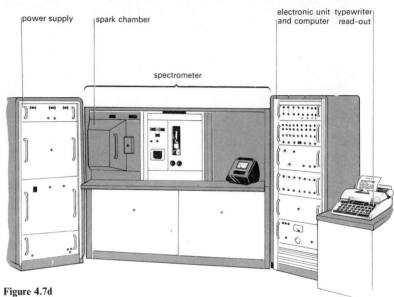

Figure 4.7d
Polyvac installation.
Hilger & Watts

Figure 4.7e shows a typical set of results obtained in this fashion. Let us now consider the time factor in all this, bearing in mind the 45 minutes or so required for a chemical analysis for major components only. The Polyvac supplies a typewritten report on up to 25 elements in one to one-and-a-half minutes from receipt of the sample, which may take perhaps five minutes to reach it from the melt. Figure 4.7f illustrates this in terms of this sequence of operations.

sample no.	Fe	C	Si	Mn	S	P	Cr	Ni	Mo	Ti	V	Al	Cu
18/8 stainless	070.5	.0554	0.430	1.413.	.0177	.0208	18.17	09.01	0.150				
FL 1429	070.5	.0562	0.413	1.432	.0189	.0221	18.22	09.07	0.147				
free cutting steel	096.9	0.374	0.336	1.152	0.379	.0532	0.129	0.115	0.020	.0122	.0233	.0810	0.360
RF 3714	097.0	0.362	0.348	1.174	0.368	.0518	0.123	0.127	0.021	.0135	.0220	.0825	0.348

Figure 4.7e
Computer read-out of Polyvac analytical data.

One further economy in time can be achieved. The data must be conveyed from the laboratory to the plant operator who is standing by with the melt. If a messenger is sent, minutes may elapse before he gets it; if the information is telephoned, errors may occur. The situation is met by coupling the read-out to a teleprinter which operates an automatic typewriter situated at the exact spot where the information is required.

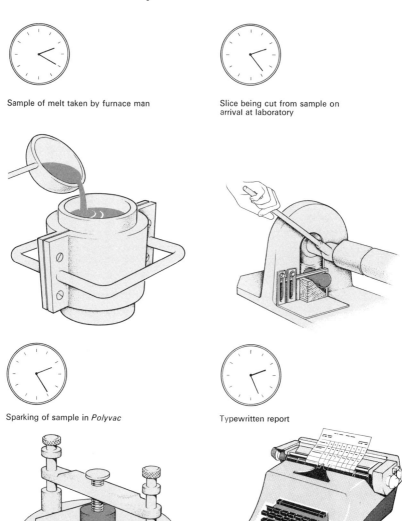

Sample of melt taken by furnace man

Slice being cut from sample on arrival at laboratory

Sparking of sample in *Polyvac*

Typewritten report

Figure 4.7f
Approximate timings for a Polyvac analysis.
Hilger & Watts

The cost of a Polyvac installation is high, being about £30 000 for a single instrument and telex system. In fact, a large steelworks will have several in-
stallations in operation simultaneously but the high cost is fully justified. The
Polyvac can be operated by a single person who need not be a qualified chemist
or skilled technician, the cost of chemical reagents and glassware is eliminated,
and a chemist is relieved of repetitive procedures for more gainful employment.
On average, the cost of a complete analysis is around 6 p per sample and a single
Polyvac can regularly cope with 130 samples per shift.

Steelmakers are still not satisfied with the time of analysis achieved in Polyvac
systems. How could we get a result more quickly? The slowest step in the process
is the conveyance of the sample from the melt to the laboratory. Can a system be
devised in which direct sampling of the molten steel, at temperatures greater than
1000 °C, is carried out? Can seconds be saved in the operation of a Polyvac
itself once it has received the sample? Work on these questions is being actively
undertaken: for example, the electronic unit of the Polyvac measures photocell-
currents and hence line intensities indirectly. The current is used to charge up
a capacitor and charging takes time. Time could be saved if a system could be
devised in which the line intensities could be measured directly. Such a system
lies within the grasp of modern technology.

All large scale industrial operations are beset with scientific problems which, if
solved, could substantially increase their productivity. Perhaps you will solve
some of them yourselves.

The Polyvac is manufactured by Hilger and Watts, to whom we are indebted for
permission to reproduce figures and other data contained in this section.

You will find an account of the operation of a modern steel plant in *The Chemist
in Action.*

Problems
1   On the graph (figure 4.8) the first ionization energy of some elements is
plotted against the atomic number of the elements.
   i   State two of the elements likely to be alkali metals.
   ii   State one of the elements likely to be an inert gas.
   iii   Which one of the elements would you expect to be in the same group of
the Periodic Table as the element C?

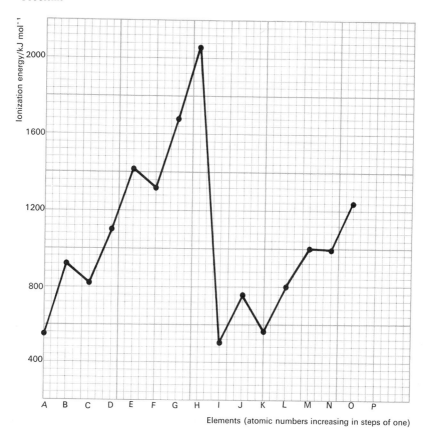

Figure 4.8

iv State briefly how the first ionization energy of the elements varies with the rising atomic number.

v In which of the following ranges should the first ionization energy of the next element, P, lie?

A 1400–1600 kJ mol^{-1}
B 1600–1800 kJ mol^{-1}
C 1800–2000 kJ mol^{-1}
D 2000–2200 kJ mol^{-1}
E 2200–2400 kJ mol^{-1}

2   Here are the first six ionization energies (in kJ mol^{-1}) of two elements X and Y.

$$X(g) \xrightarrow{1402} X^+(g) \xrightarrow{2855} X^{2+}(g) \xrightarrow{4522} X^{3+}(g)$$
$$\xrightarrow{7452} X^{4+}(g) \xrightarrow{9462} X^{5+}(g) \xrightarrow{53\ 170} X^{6+}(g)$$

$$Y(g) \xrightarrow{1315} Y^+(g) \xrightarrow{3387} Y^{2+}(g) \xrightarrow{5317} Y^{3+}(g)$$
$$\xrightarrow{7452} Y^{4+}(g) \xrightarrow{11\ 010} Y^{5+}(g) \xrightarrow{13\ 320} Y^{6+}(g)$$

i   The element Y has an atomic number greater by one than that of X. What would you expect the approximate seventh ionization energy of Y to be? State briefly the reasons for your answer.

ii   The element Z has an atomic number less by one than that of X. The first four ionization energies (in kJ mol^{-1}) of Z are:

$$Z(g) \xrightarrow{1089} Z^+(g) \xrightarrow{2350} Z^{2+}(g) \xrightarrow{4650} Z^{3+}(g) \xrightarrow{6200} Z^{4+}(g)$$

What would you expect the approximate fifth ionization energy of Z to be? State briefly the reasons for your answer.

3   The electron energy levels of a certain element can be represented by $1s^2\ 2s^2\ 2p^6\ 3s^2\ 3p^1$. Sketch a graph showing the general form which you would expect for the first five ionization energies of the element.

4   A certain element X, atomic number 32, forms a chloride, XCl_4. The chloride is a non-conducting liquid at room temperature. Denote the electron energy levels of the element.

5   The electron energy levels of a certain element can be represented by $1s^2 2s^2 2p^6\ 3s^2 3p^6 3d^{10}\ 4s^2 4p^6\ 5s^2$.
i   What is the atomic number of the element?
ii   In which group of the Periodic Table should the element be?
iii   The element forms an ionic bond when it reacts with oxygen. What will be the charge on the ion of the element?

6 The following table shows the first three ionization energies (in kJ mol^{-1}) of elements in the same group of the Periodic Table.

Element	1st ionization energy	2nd ionization energy	3rd ionization energy
A	383	2437	not known
B	409	2667	3881
C	425	3065	4438
D	502	4568	6929
E	527	7314	11 840

i Which of these elements should have the largest atomic number? Give reasons for your answer.

ii In which group of the Periodic Table should the elements be placed? Give reasons for your answer.

Questions 7 to 10 refer to the following table of ionization energies (kJ mol^{-1}) of five elements (the letters are not the symbols for the elements).

Elements	1st ionization energy	2nd ionization energy	3rd ionization energy	4th ionization energy
A	520	7301	11 817	—
B	578	1817	2746	10 813
C	1087	2354	4621	6425
D	496	4566	6917	9547
E	590	1146	4944	6469

7 Which of the elements, when it reacts, is most likely to form a 3+ ion?

8 Which one of the following pairs of elements are likely to be in the same group of the Periodic Table?

B and E
A and D
D and E
B and C
C and E

9 Which of the elements would require the most energy to convert one mole of atoms into ions carrying one positive charge?

10 Which of the elements would require the most energy to convert one of atoms into ions carrying two positive charges?

11   Which of the following would require the *most* energy to convert them completely from the gaseous state into gaseous ions each carrying one positive charge?

- A   1 mole of lithium atoms
- B   1 mole of sodium atoms
- C   1 mole of potassium atoms
- D   1 mole of rubidium atoms
- E   1 mole of caesium atoms

12   Which of the following would require the *least* energy to convert them completely from the gaseous state into gaseous ions each carrying one positive charge?

- A   1 mole of lithium atoms
- B   1 mole of beryllium atoms
- C   1 mole of boron atoms
- D   1 mole of carbon atoms
- E   1 mole of nitrogen atoms

13   Natural silicon consists of a mixture of three isotopes and its atomic number is 14.

Isotope	Isotopic mass	Percentage abundance by numbers of atoms
A	28.0	92.2
B	29.0	4.7
C	30.0	3.1

i   In each of the isotopes how many (a) neutrons, and (b) protons are there in each atom?

ii   Calculate the atomic weight of natural silicon. Show how you arrive at your answer.

iii   State, in the form of numbers and symbols, the energy levels of the electrons in the isotope B.

Topic 5

The halogens and oxidation numbers

If a survey is made of groups III, IV, V, VI, and VII of the Periodic Table, a number of characteristic properties can be seen. This topic begins by reviewing some of these properties.

The elements of group VII, the halogens, are then selected for a more detailed study, and their properties are investigated by means of a number of experiments. During this study, especial attention will be paid to the trends in properties to be seen on going down a group of the Periodic Table.

	Group					
Period	III	IV	V	VI	VII	0
2	B	C	N	O	F	Ne
3	Al	Si	P	S	Cl	Ar
4	Ga	Ge	As	Se	Br	Kr
5	In	Sn	Sb	Te	I	Xe
6	Tl	Pb	Bi	Po	At	Rn

Table 5.1a
The post-transition elements

5.1 The post-transition elements

Before carrying out a study of the halogens, we shall survey briefly the post-transition elements as a whole. In this way the properties of the halogens can be compared with those of their neighbouring elements in the Periodic Table. From such a survey a number of general conclusions can be reached, and trends can be detected. Four of the most important of these general conclusions are now discussed.

1 Non-metallic character

One of the simplest ways of classifying elements is to divide them into two groups, namely, metals and non-metals. We all know roughly what is meant by these terms, but it is a good idea to try to be specific about such ideas. These two categories can be distinguished by physical and chemical properties. Write down in your notebook as many physical points of difference as you can. A convenient way to do this is in the form of a table, as shown: one example is given as a start. Any exceptions to the generalizations that you know can be included as well.

Distinctions between metals and non-metals

Property	Metals	Non-metals
1 Conduction of electricity	Good conductors	Do not conduct, except for carbon (graphite)
2		

Table 5.1b

Now consider chemical points of difference. Are there any distinctive points of difference between the chemical reactions of metallic and non-metallic elements, or between their compounds, such as the chlorides and the oxides, that you studied in Topic 2?

Once differences of this sort have been noted, the various post-transitional elements can be examined to see whether they should be classified as metals or non-metals. Copy out table 5.1a into your notebook and classify the elements as metals by underlining the symbol twice, and non-metals by leaving them un-marked. If they are difficult to classify, underline the symbol once.

The post-transition block of elements contains all the non-metals: it will be noted that non-metallic character *decreases* as one moves to the *left* along a period, and *decreases* as one moves *down* a group. Alternatively, one may say that *metallic* character increases as one moves to the left along a period, and increases as one moves down a group.

2   Structure

Molecular weight determinations show that many of the post-transition elements exist as distinct molecules, each having only a small number of atoms. This is reflected in the low melting and boiling points, and latent heats, of the elements concerned. High values of these constants for the rest of these elements point to more complex structures in their cases.

Make another copy of table 5.1a, and using the *Book of Data* write against each symbol the number of atoms in its molecule. Thus, for nitrogen, for example, write N_2, and for phosphorus P_4. For those elements which exist as giant structures, put *n* in place of a number, thus: C_n.

Can you see any relationship between your classification of the post-transition elements as metals or non-metals, and what you have now noted as the structural pattern for these elements?

3   Size of the atoms

As would be expected from the increasing number of electrons, the size of the atoms increases as a group of the Periodic Table is descended. One way of comparing the sizes of the atoms is by means of their *covalent radii*. Values of this quantity for some elements of groups VI and VII are given in table 5.1c.

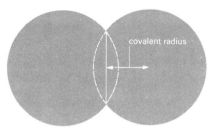

Figure 5.1a
The covalent radius of an atom.

The covalent radius of an atom is half the distance between the nuclei in a diatomic molecule of the element in question (figure 5.1a).

Element	Covalent radius/nm	Element	Covalent radius/nm
Oxygen	0.066	Fluorine	0.064
Sulphur	0.104	Chlorine	0.099
Selenium	0.114	Bromine	0.111
		Iodine	0.128

Table 5.1c
Covalent radii of some p-block elements

4   Electronic structure of the atoms

In their shells of electrons of highest energy, the atoms of the post-transition elements all have one or more electrons in the p sub-level, as well as two electrons in the s sub-level. As one goes from left to right the number of these p electrons increases as is now indicated in table 5.1d for the third short period.

Group	III	IV	V	VI	VII	O
Electrons in highest energy level	Al $3s^23p^1$	Si $3s^23p^2$	P $3s^23p^3$	S $3s^23p^4$	Cl $3s^23p^5$	Ar $3s^23p^6$

Table 5.1d
Electronic structures in the third period of the Periodic Table

For this reason, and because they form a compact block in the Periodic Table, the post-transition elements are sometimes called the p-*block elements.*

The elements of groups I and II have only s electrons in their shell of highest energy, and so are called the s-*block elements*; in a similar manner the transition elements are referred to as d-block elements, and the lanthanides and actinides as f-block elements (figure 5.1b).

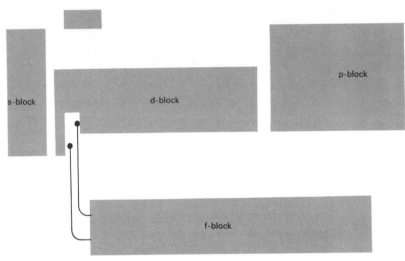

Figure 5.1b
Electronic structure and the Periodic Table.

The first ionization energies of the p-block elements are generally higher than those of the other elements, as can easily be seen from the graph of first ionization energy against atomic number given in Topic 4. The elements at the top right-hand corner of the block have the highest first ionization energies, and partly for this reason they have little or no tendency to form positively charged ions by loss of electrons. They do, however, form negatively charged ions by gaining electrons. In ion formation sufficient electrons are gained for the resulting ions to have the electronic structure of the nearest inert gas. Oxygen and sulphur, for example, acquire two electrons each per atom, and fluorine and chlorine one electron each per atom.

$$O + 2e^- \rightarrow O^{2-}$$
$$(1s^2 2s^2 2p^4) \qquad (1s^2 2s^2 2p^6)$$

$$F + e^- \rightarrow F^-$$
$$(1s^2 2s^2 2p^5) \qquad (1s^2 2s^2 2p^6)$$

5.2 **An experimental investigation of halogen chemistry**

We shall now investigate some of the properties of the halogens. This is in order to gain some factual knowledge of these elements, and to look for trends in behaviour, both within the group of halogens themselves, and between the halogens and the rest of the p-block. Fluorine will be excluded from the experimental investigations because of the difficulties and dangers involved in handling this element.

Experiment 5.2a
An investigation of some reactions of the halogens

Use the halogen elements chlorine, bromine, and iodine in solution (in water for the first two of these and in a solution of potassium iodide in water for iodine, as the solubility of iodine in water is small). Fluorine is too hazardous for use under ordinary laboratory conditions.

Handle the solutions with care. Avoid inhaling the vapours from them, and do not allow them to come into contact with your skin or clothing.

Do the following tests with each solution:

1 Add about 1 cm^3 of tetrachloromethane (carbon tetrachloride) to about 2 cm^3 of each halogen solution. Shake gently and observe the colour of the tetrachloromethane layer. The halogens do not react with tetrachloromethane.

In which of the two solvents, water and tetrachloromethane, are the halogens more soluble? Do they have the same colour in both solvents?

2 To about 1 cm^3 of each halogen solution add sodium hydroxide solution (approximately 2 M) a few drops at a time until no further change takes place. Note what happens. Has reaction occurred?

Record your observations in your notebook in the form of a table, using the following headings. As you have not been able to use fluorine, results for this element are given as an example.

Name	Colour and physical state	Colour of aqueous solution	Effect of adding CCl_4 (colour of CCl_4 layer)	Effect of adding alkali
Fluorine	pale yellow gas	(decomposes water)	–	–
Chlorine etc.				

Halogens react with alkalis, as you have seen. Chlorine reacts with sodium hydroxide, for example, in two ways.

In cold dilute alkali, chloride and hypochlorite ions are formed:

$$Cl_2(aq) + 2OH^-(aq) \longrightarrow Cl^-(aq) + ClO^-(aq) + H_2O(l)$$

On heating, however, hypochlorite ions react together:

$$3ClO^-(aq) \longrightarrow 2Cl^-(aq) + ClO_3^-(aq)$$

So hot concentrated alkali yields chloride and chlorate ions:

$$3Cl_2(aq) + 6OH^-(aq) \longrightarrow 5Cl^-(aq) + ClO_3^-(aq) + 3H_2O(l)$$

Other halogens react with alkalis similarly.

Try to work out, and write in your notebook, the equations and the names of the products formed in the cases of the other two halogens that you have examined.

3   This experiment investigates the relative reactivity of the halogen elements towards the halide anions. Set up four test-tubes containing equal volumes of potassium chloride, potassium bromide, and potassium iodide solution, and water as a control. Add two or three drops of chlorine solution to each. Using colour changes as a guide, have reactions taken place, and what are the products? Would the addition of tetrachloromethane help you in reaching a decision?

Now repeat the experiment using in turn bromine solution and iodine solution. Is there a definite trend in reactivity observable in this experiment?

Draw up a table in which to record your results, similar to the one now given.

Solution added	Action on			
	water	potassium chloride solution	potassium bromide solution	potassium iodide solution
Chlorine solution etc.				

Halogens combine readily with most other elements. Chlorine, for example, combines directly with every other element except carbon, nitrogen, oxygen, and the inert gases; chlorides can be made indirectly for all these except helium, neon, and argon.

You have seen in Topic 2 a number of instances of these reactions, and you will probably be shown some similar reactions of bromine and iodine.

Experiment 5.2b
An investigation of some reactions of the halides
You are now going to examine some of the properties of the binary compounds of the halogens. These compounds are known as *halides*.

For the first part of this experiment use *solutions* of potassium (or sodium) chloride, bromide, iodide, and, if available, fluoride, which are 0.1M with respect to the halide ions. Fluorides are poisonous, so take care in using them. Where you can, attempt to estimate roughly the proportions of the solutions needed for complete reaction.

1 To separate 1 cm^3 portions of the halide solution add 0.1M silver nitrate solution.
2 To the precipitates obtained in (1) add ammonia solution.
3 Obtain a second set of silver halide precipitates and leave them exposed to the light for an hour.

In parts 4, 5, and 6 of this experiment use *solid* potassium (or sodium) chloride, bromide, and iodide. *Do not use fluoride.*

4 Investigate the action of concentrated sulphuric acid on the salts. Put about 0.1 g of the solid salt into a test-tube (about enough to fill the rounded end of the tube if it is 100 × 16 mm) and add about 10 drops of concentrated

sulphuric acid. Warm the reaction mixture *gently* if necessary. Identify as many products as you can, noting the similarities and differences between the reactions. Record and explain your observations as fully as you can.

5   Repeat (4) using phosphoric acid in place of sulphuric acid. Note any difference.

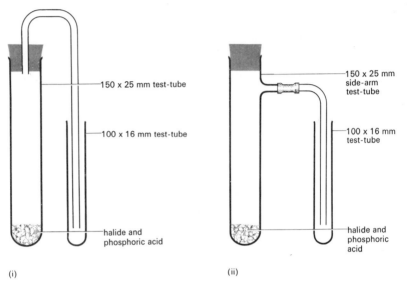

(i)   (ii)

Figure 5.2
Alternative apparatus for making hydrogen halides.

6   Use the reaction in (5) to prepare and collect samples of hydrogen chloride, hydrogen bromide, and hydrogen iodide. The apparatus shown in figure 5.2 is convenient for this purpose. A good yield of gas is obtained if solid 100 per cent phosphoric acid is used. Mix about 2 g halide with an equal quantity of solid phosphoric acid in the (side-arm) test-tube and then cork it securely. Place a *dry* test-tube round the delivery tube and warm the mixture gently until gas is evolved. Collect at least three tubes of gas, corking them when apparently full (when the gas forms copious white fumes at the test-tube mouth). Use the tubes of gas to investigate:

    A   The solubility of the gas in water. Invert a tube of gas in a beaker of water. If the water rises rapidly the gas is readily soluble. Is there a residue of undissolved gas, and if so, what do you suppose it is?

    B   The reaction of the gas with ammonia gas. Hold a drop of fairly concentrated ammonia solution in the mouth of an open test-tube, using a glass tube or rod. What do you observe, and what do you suppose is formed?

c The stability of the gas to heat. Heat the end of a length of nichrome wire or a glass rod to dull red heat, and plunge it into a tube of gas; if no change occurs in the gas, try again with the wire hotter. What do you observe?

Record the properties of these hydrogen halides in a table in your notebook.

Experiment 5.2c
An investigation of some reactions of the halates
1   Heat small separate samples of the potassium halates, $KClO_3$, $KBrO_3$, and KIO_3.
What products can you identify and what differences do you observe during their decomposition?
2   Dissolve *small* separate samples of the potassium halates in dilute sulphuric acid. To each add a *small* amount of sodium nitrite solid and shake to dissolve. Is there any sign of reaction? What do you think the products are? If there is no sign of reaction there may have been reaction but to colourless products. What colourless products are possible? Do you know of any simple test to enable you to decide if such products are present?
3   Repeat (2) but use sodium sulphite in place of sodium nitrite.

Trends in the behaviour of the halogens
From the relatively few reactions you have seen, you will have noticed the similarity which exists between the various halogens. You should also have noticed some trends in their behaviour, including the following:
1   Their relative reactivity, as illustrated by the displacement series found in experiment 5.2a (3).
2   The relative stability, and ease of oxidation, of the hydrogen halides.

Do you see any relationship between the thermal stability and the heat of formation of the hydrogen halides as given in table 5.2a?

HF	HCl	HBr	HI
− 271	− 92.3	− 36.2	+ 26.5

Table 5.2a
Standard heats of formation of the hydrogen halides/kJ mol^{-1}

Trends can also be seen in the properties of the halides. The following tables give examples of these trends.

Look carefully at the data in table 5.2b and table 5.2c, and record the answers to the following questions in your notebook.

1 What is the general trend of (a) the melting point, and (b) the standard heat of formation, as one goes from the fluoride to the iodide of any element?
2 Are there any exceptions to this generalization?
3 What is the general trend of (a) the melting point, and (b) the standard heat of formation, for each halide as one descends a group of elements?
4 Are there any exceptions to this generalization?

Element		Fluoride	Chloride	Bromide	Iodide
Group I	Lithium	845	614	550	449
	Sodium	995	808	750	662
	Potassium	857	772	735	685
	Rubidium	775	717	680	640
	Caesium	682	645	636	621
Group II	Beryllium	800	410	488	510
	Magnesium	1263	714	711	(d)
	Calcium	1418	772	765	740
	Strontium	1450	875	643	515
	Barium	1290	963	850	740(d)

Table 5.2b
Melting points of the halides of groups I and II/°C

Element		Fluoride	Chloride	Bromide	Iodide
Group I	Lithium	-612	-409	-350	-271
	Sodium	-569	-411	-360	-288
	Potassium	-563	-436	-392	-328
	Rubidium	-549	-431	-389	-328
	Caesium	-531	-433	-395	-337
Group II	Beryllium	-1052	-512	-370	-212
	Magnesium	-1100	-642	-518	-360
	Calcium	-1215	-795	-675	-535
	Strontium	-1215	-828	-716	-569
	Barium	-1200	-860	-755	-603

Table 5.2c
Standard heats of formation of the halides of groups I and II/kJ mol^{-1}

Manufacture and uses of the halogens
The halogens have important industrial and other applications. All of them are manufactured commercially and are used for making compounds having many

different uses. Details of the manufacturing processes are given in many text-books, and some background reading concerning the applications of these elements in modern society is given at the end of this topic.

5.3   Oxidation numbers

From the formulae of the compounds that you have met so far, both in your O-level course and in the earlier parts of this course, you will no doubt have realized that elements have some definite 'combining power'. This can be seen, for example, when comparing the formulae of the chlorides and oxides of the elements of period 2.

$LiCl$	$BeCl_2$	BCl_3	CCl_4	NCl_3	
					Cl_2O*
Li_2O	BeO	B_2O_3	CO_2^*	$N_2O_3^*$	

* In these cases only one of several possible compounds has been selected.

After looking at these formulae one might conclude that oxygen had twice the combining power of chlorine. A survey of a large number of compounds, however, shows that it is quite difficult to fix an unambiguous numerical value to this combining power. Assigning an *oxidation number* is a useful way of setting about this.

We can begin by considering ionic compounds. Experiments which measure the quantity of electricity needed to deposit one mole of atoms of various metals and non-metals tell us how many charges are associated with the ions formed by these elements. In this way we are able to write the formulae of ions such as Pb^{2+}, Ba^{2+}, Cu^{2+}, Na^+, Cl^-, Br^-, O^{2-}, etc.

The observation that compounds containing such ions are electrically neutral enables us to write formulae such as $NaBr$, $PbBr_2$, Na_2O, PbO, etc. These formulae are confirmed by experiments to find the relative number of moles of atoms of each element present in these compounds.

In ionic compounds of two elements such as those mentioned above, the charge on the ion of each element is taken as the oxidation number of that element. In $PbBr_2$, therefore, the oxidation number of lead is $+2$ and that of bromine is -1, and in sodium monoxide, Na_2O, sodium and oxygen have oxidation numbers of $+1$ and -2 respectively.

The use of oxidation number can be extended to molecular compounds in the following way. It is found that in all ionic oxides (excluding peroxides, which contain the O_2^{2-} ion), the oxidation number of oxygen is -2. Suppose we give it that number in the molecular compound CO_2. The oxidation number of carbon in this compound must therefore be $+4$. As a check, we can apply this idea to another compound. Chlorine has the oxidation number -1 in all ionic chlorides. Suppose it is also -1 in the molecular compound tetrachloromethane (carbon tetrachloride), CCl_4. The oxidation number for carbon will therefore again be $+4$.

Extensions of this sort enable one to assign an oxidation number to any element in any compound, once the empirical formula of that compound has been determined experimentally. You can try this yourself, using the formula of the hydrides, chlorides, and oxides of the elements of the first, second, and third periods of the Periodic Table (elements hydrogen to argon). Start by copying the chart given as figure 5.3 in your notebook, marking the axes as shown.

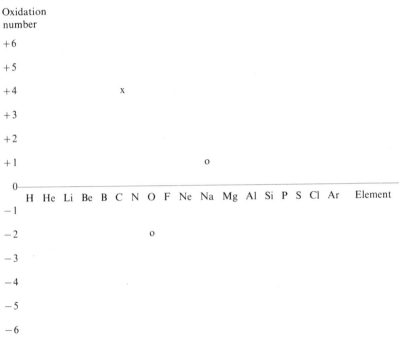

Figure 5.3
Oxidation numbers of the elements H to Ar.

Start with ionic compounds, marking the oxidation numbers of the elements concerned with o. Sodium and oxygen are already filled in as examples. Now extend the idea to molecular compounds, marking the point with x in this case. Carbon is already given as an example.

When you have finished you will find a very interesting pattern has been produced, giving further evidence of the periodicity of properties mentioned in Topic 2.

Is there any relationship between oxidation number and group of the Periodic Table?

These ideas that we have been using can be summarized and extended for further use as a set of rules for assigning oxidation numbers, as now given.

Rules for assigning oxidation numbers

1 The sign given to the oxidation number of an element in a binary compound is decided in the following way. One element is given a positive, and the other a negative oxidation number. In most cases of compounds of metals with non-metals, there is no difficulty in deciding which sign should be given to which element; the metal is given a positive sign and the non-metal a negative one. For many other compounds the signs can be decided using the invariable oxidation numbers given in rule 4. The signs are always relative to other elements. For example, the oxidation number of sulphur in sodium sulphide, Na_2S, is -2; its oxidation number in sulphur dioxide, SO_2, however, is $+4$.

2 The numerical value of the oxidation number of an element in a compound is found using the empirical formula of that compound. The oxidation number of each atom in the formula counts separately, and their algebraic sum is zero. For example, in NaCl the oxidation number of sodium is $+1$ and that of chlorine is -1. $+1-1 = 0$. In PCl_3, if the oxidation number of chlorine is taken as -1, the total chlorine contribution is -3. The oxidation number of phosphorus is $+3$.

It follows that the oxidation number of any uncombined element is zero.

3 The oxidation number of an element existing as a monatomic ion is the charge on that ion. In a polyatomic ion, the algebraic sum of the oxidation numbers of the atoms is the charge on the ion. For example, in compounds containing Al^{3+} ions, the oxidation number of aluminium is $+3$. In the SO_4^{2-} ion, if the oxidation number of oxygen is taken as -2, the total for oxygen is -8 and the oxidation number of sulphur is therefore $+6$.

4 Some elements have invariable oxidation numbers in their compounds, or oxidation numbers that are invariable under certain conditions. They include the following, which have the oxidation numbers given.

Na, K	$+1$
Mg, Ca	$+2$
Al	$+3$
H	$+1$ except in metal hydrides
F	-1
Cl	-1 except in compounds with oxygen and fluorine
O	-2 except in peroxides

You can test your understanding of these rules by trying to answer the questions on oxidation numbers given at the end of this topic.

Oxidation numbers and nomenclature

The Roman numerals used in the naming of compounds of metals are, in fact, the oxidation numbers of these elements. This system of naming is known as the Stock notation, after the chemist who devised it. Its use provides a simple way of distinguishing between similar compounds. For example, in the two oxides of copper, Cu_2O and CuO, the oxidation numbers of copper are 1 and 2 respectively. These compounds are known as copper(I) and copper(II) oxide respectively.

Stock notation is not generally used for non-metals. Compounds such as PCl_3 and PCl_5 are distinguished by the names phosphorus trichloride and phosphorus pentachloride. This latter system is sometimes more convenient in the case of metal compounds and, if so, it is used; for example, for Fe_3O_4, the preferred name is tri-iron tetroxide. (The Stock name for this compound is iron(II) di-iron(III) oxide.)

Oxidation number charts

As mentioned in the rules, oxidation numbers are invariable for some elements, but those of a number of elements, such as the halogens, may have different values in different compounds. The range of oxidation numbers of such elements is very well shown by constructing an oxidation number chart. The formulae of the various compounds can be written on the chart, as shown in the following example.

Oxidation number chart for chlorine

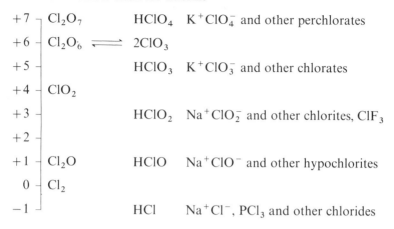

$+7$ ⌐ Cl_2O_7 $HClO_4$ $K^+ClO_4^-$ and other perchlorates

$+6$ ┤ Cl_2O_6 ⇌ $2ClO_3$

$+5$ ┤ $HClO_3$ $K^+ClO_3^-$ and other chlorates

$+4$ ┤ ClO_2

$+3$ ┤ $HClO_2$ $Na^+ClO_2^-$ and other chlorites, ClF_3

$+2$ ┤

$+1$ ┤ Cl_2O $HClO$ Na^+ClO^- and other hypochlorites

0 ┤ Cl_2

-1 ┘ HCl Na^+Cl^-, PCl_3 and other chlorides

Now that you have seen this one example, construct an oxidation number chart for iodine in your notebook. Write in the compounds of iodine you have met in this topic, and add to it as you come across any new compounds of this element.

It is a good plan to draw up such a chart for each element that you study.

Oxidation and reduction

Whether a particular element has been oxidized or reduced in a given reaction can be discovered by finding out if its oxidation number has been increased or decreased in the course of the change. Similarly, if a compound high on the oxidation number chart for a given element is to be made from one lower down, an oxidation reaction will be needed, and vice versa.

Some problems on this will be found at the end of the topic, and instructions for some experiments involving changes of oxidation number are now given.

Experiment 5.3a

Small-scale preparation of potassium perchlorate

When potassium chlorate, $KClO_3$, is heated at its melting point it changes to potassium perchlorate, $KClO_4$, and potassium chloride, KCl. What oxidation number changes occur in this reaction? Write a balanced equation to determine the relative amounts of $KClO_4$ and KCl that will be formed from 2 g of $KClO_3$.

Take a small, clean crucible and weigh into it roughly 2 g of $KClO_3$. Place the crucible on a pipe-clay triangle on a tripod and heat it with a Bunsen burner. Use a very low, colourless flame to start with, so as to avoid cracking the crucible, and then gradually raise the temperature.

Temperature control is important. Too low a temperature will mean the reaction does not take place; too high a temperature will result in the decomposition of the perchlorate and the evolution of oxygen. At the correct temperature (which is about 500 °C), the potassium chlorate will melt, but as the melting point of the perchlorate is higher, the mixture will gradually become pasty. If bubbles of gas appear the temperature must be reduced.

After maintaining the correct temperature for 5–10 minutes, allow the mixture to cool. The cooled solid is likely to be a mixture of KCl, $KClO_3$, and $KClO_4$.

	0°	20°	100°
KCl	27.6	34	56.7
$KClO_3$	3.3	7.4	57
$KClO_4$	0.75	1.8	21.8

Values are given in g per 100 g of solvent.

From this table of solubilities at 100 °C and 0 °C, is fractional crystallization likely to be a successful method of separation? Work out from the quantity of $KClO_4$ hoped for as product, a suitable volume of water to use.

Place the cold crucible and solid contents in a small beaker and pour in the calculated quantity of water, which should be preheated to boiling. The solid should soon separate from the crucible which may be retrieved and the beaker should then be heated to dissolve the remaining solid.

Allow the solution to cool, preferably by standing overnight. Potassium perchlorate is only sparingly soluble in cold water, and will crystallize on cooling in glistening white flakes; the potassium chloride and any unchanged chlorate will remain in solution. Filter off the crystals when the solution is quite cool, and dry them by pressing between filter papers.

Consult a practical book about the properties of $KClO_3$ and $KClO_4$ and test your product for absence of $KClO_3$. (*Warning:* do not test with concentrated sulphuric acid because the reaction is highly dangerous and uncontrollable.)

How might you purify your product further if it proves to be impure?

Experiment 5.3b

The reaction between sodium thiosulphate and iodine

Dissolve about ten large crystals of sodium thiosulphate ($Na_2S_2O_3$, $5H_2O$) in about 10 cm^3 of water in one test-tube, and two or three *small* crystals of iodine in about 10 cm^3 of potassium iodide solution in another test-tube. Divide each solution into two roughly equal parts.

Add half of the sodium thiosulphate solution slowly to half of the iodine solution, shaking the test-tube at intervals during the addition to see that the solutions are well mixed.

What do you see happen?

What do you suppose happened to the iodine?

Now repeat the operation using the second part of each solution, but this time put a few drops of starch solution into the iodine solution before adding the solution of sodium thiosulphate. Record carefully what you see happening at each stage.

The equation for the reaction is:

$$2Na_2S_2O_3 + I_2 \longrightarrow 2NaI + Na_2S_4O_6$$

or, expressed ionically,

$$2S_2O_3^{2-}(aq) + I_2(aq) \longrightarrow 2I^-(aq) + S_4O_6^{2-}(aq)$$

The compound having the formula $Na_2S_4O_6$ is known as sodium tetrathionate.

What is the oxidation number of the iodine (a) before, and (b) after the reaction?

Starch reacts with iodine (but not iodide ions) to form an intensely coloured compound, useful in the detection of iodine present in minute quantities.

The reaction between iodide and iodate ions

The value of an idea such as that of oxidation numbers can be assessed by using it to make predictions, and then testing these predictions experimentally.

In the next experiment you can use the oxidation number idea to predict the number of moles which react together, and then test your predictions by experiment.

The reaction is between solutions of potassium iodide and potassium iodate.

You can discuss how to make the predictions in class; guidance on how to do the experimental work is given here.

Experiment 5.3c
To test predictions made about the reaction between acidified aqueous solutions of potassium iodide and potassium iodate

1   Mix about $1 cm^3$ 0.1M potassium iodate (KIO_3) solution with about $1 cm^3$ M potassium iodide (KI) solution and add about $1 cm^3$ M hydrochloric acid. Note what happens. What is the solid product? Devise and carry out a test to check your answer.

2   Using a measuring cylinder place about $10 cm^3$ of approximately M potassium iodide solution in a $250 cm^3$ conical flask. This will be sufficient to provide a large excess of potassium iodide for the titration to be undertaken. Acidify this solution with about the same volume of approximately M hydrochloric acid.

Using a burette, add exactly $5 cm^3$ of 0.1M potassium iodate solution. Iodine will be formed and held in solution.

Find out how many moles of iodine atoms have been formed by titrating the iodine against accurately 0.1M sodium thiosulphate solution. Place the last-named solution in a burette and run it into the conical flask, $1–2 cm^3$ at a time. When the colour of the iodine has nearly gone, add $1–2 cm^3$ of starch solution and continue the addition, now adding 1–2 drops at a time. Note the volume of sodium thiosulphate solution required to remove the colour of the iodine completely.

Repeat the whole operation as a check on your accuracy.

Calculation
1   Calculate the number of moles of sodium thiosulphate present in the volume of solution that you have run out of the burette.

2   From the equation below, work out how many moles of iodine atoms must therefore have reacted with the sodium thiosulphate.

$$2S_2O_3^{2-}(aq) + I_2(aq) \longrightarrow 2I^-(aq) + S_4O_6^{2-}(aq)$$

3   Now work out how many moles of iodine atoms were present in the $5 cm^3$ of 0.1M potassium iodate solution that you started with. This will tell you what fraction of the iodine calculated in (2) came from the iodate, and hence what fraction from the iodide. You can now tell how many moles of iodide ions are required for each mole of iodate ions.

Does this number confirm your prediction?

Why do you suppose that an excess of potassium iodide was used? If you cannot think of a reason, try doing the reaction with only the reacting quantities of iodide, iodate, and acid. This may help you to find, or confirm, your answer.

Background reading
Sources of the halogens

All the halogens are in use commercially, either as the free element or in compounds. They are obtained from a variety of sources and extracted by a number of different methods depending on convenience and cost. Some information is summarized in table 5.4.

Halogen	Abundance (parts per million by weight)		Source	Relative cost of sodium halide (reagent grade)	Chemical process to obtain the free element
	in rocks	in the sea			
Fluorine	700	1.4	fluorite, CaF_2, e.g. Derbyshire 'Blue John'	5	Electrolysis of a solution of potassium fluoride in anhydrous hydrogen fluoride.
Chlorine	200	19 000	rock salt, NaCl, and sea water	1	Electrolysis of an aqueous saturated solution of sodium chloride.
Bromine	3	67	sea water	2.5	Oxidation of bromide (aq) by chlorine.
Iodine	0.3	0.05	caliche, $NaNO_3$, containing $NaIO_3$	10	Reduction of iodate (aq) by sodium hydrogen sulphite.

Table 5.4

Fluorine compounds occur in a number of rocks but these compounds are so widely dispersed that deposits which can be worked economically are rare. By contrast iodine, with much the lowest overall abundance of the halogens, occurs in extensive deposits in the Chilean desert. The principal mineral is caliche, or sodium nitrate. It contains iodine compounds at a concentration of 1500 parts per million, high enough to make the extraction of the element economically feasible.

Sea water is potentially a good source of chemicals, and chlorine and bromine are amongst the elements whose compounds are obtained from this source.

The halogens are found in a variety of locations other than those of commercial importance and an explanation of some of these freaks of nature poses difficult questions for geochemists.

The occurrence of hydrogen fluoride and hydrogen chloride gases in nature is surprising, as they are so reactive. They are usually associated with volcanic action. In the Valley of Ten Thousand Smokes in Alaska for example over one million tonnes of hydrogen chloride and nearly a quarter of a million tonnes of hydrogen fluoride are emitted every year!

Geochemists are also puzzled by the Chilean desert deposits. On a desolate plateau just inland from the Pacific lies a deposit of soluble salts 30 kilometres wide and stretching for over 500 kilometres (the distance between Leeds and Plymouth) that is quite unique in the world. How did iodine come to be present in such a high concentration? How can the high oxidation number of the elements be explained, nitrogen as nitrate, iodine as iodate, and chromium as chromate? These questions cannot be answered with assurance.

The biologist also finds difficult questions posed by the halogens, for a number of plants and animals are capable of making special hoardings of halogen compounds. The *Laminariaceae* seaweeds take in iodide ions from the sea to the extent of increasing the sea water value (0.05 parts per million iodine) to 800 parts per million in fresh wet weed. Certain species of marine snail, called *Purpura aperta*, found principally in the Eastern Mediterranean take in bromide ions to produce compounds which are fine dyes, Tyrian purple or dibromo-indigo. The tea plant, *Camellia sinensis*, takes in fluoride ions from the soil to the extent that dried tea leaves contain about 100 parts per million of fluorine, which results in a concentration of about 1 part per million in a cup of tea.

Manufacture and uses of the halogens

The method of obtaining the halogens as free elements from the sources cited depends on the nature of the source and the reactivity of the free element.

The source of fluorine is fluoride ions so the problem is to remove electrons by an oxidation reaction

$$2F^- \longrightarrow F_2 + 2e^-$$

This reaction cannot be conducted in aqueous solution because of the reactivity of fluorine, nor is it possible to find a sufficiently powerful oxidizing agent. The reaction is therefore conducted as the electrolysis of potassium fluoride in solution in anhydrous hydrogen fluoride, using a carbon anode with a mild steel cathode at a potential difference of between 8 and 11 volts.

Fluorine is used to make uranium(VI) fluoride gas:

$$UF_4(s) + F_2(g) \longrightarrow UF_6(g)$$

The controlled diffusion of this gas makes possible the separation of the isotopes of uranium for use in atomic power stations. Some fluorine is used as an oxidant in rocket motors (see *The Chemist in Action*, Chapter 2 'The aerospace industry').

Hydrogen fluoride is more important commercially and is used to make a range of fluorocarbon compounds which find application as refrigerants, aerosol propellents, and fire-extinguisher fluids.

Fully fluorinated hydrocarbons are inert substances, very resistant to chemical attack. A good example is the polymer polytetrafluoroethylene (PTFE), $(CF_2CF_2)_n$, a white solid with a waxy feel. Its inertness results in applications being found for it in chemical plants and laboratories for joint rings, gland packings, and sleeves. It is also an excellent electrical insulator, and has interesting surface properties which give it a very low coefficient of friction. It is made into bearings which need no lubrication and its anti-stick properties have been exploited in the printing industry, where it is used to coat rollers in presses, and in making non-stick saucepans for cooking. The use of other fluorinated hydrocarbons as anaesthetics is described in Chapter 8 of *The Chemist in Action*.

Chlorine is less reactive than fluorine and can therefore be obtained by the electrolysis of aqueous solutions. Saturated sodium chloride solution is used in a cell which has carbon anodes with a mercury cathode and operates at about 5 volts.

Chlorine is required in the manufacture of a number of familiar domestic materials, including dry-cleaning solvents, PVC plastic (polyvinyl chloride), and bleaches. Chlorinated hydrocarbons such as DDT and BHC are widely used as insecticides.

Bromine is obtained by adding chlorine to sea water after acidification. The bromide ions are thereby oxidized to bromine. Bromine is used to make 1,2-dibromoethane which is added to petrol at the same time as tetraethyl lead (see Topic 2, page 36). It is also used for the manufacture of dye-stuffs, and of photographic film and printing paper.

The mineral source of iodine is as iodate ions, in which iodine has a positive oxidation number. A reduction process is therefore required to obtain this element. After concentration by aqueous extraction, treatment with sodium hydrogen sulphite produces free iodine:

$$2IO_3^-(aq) + 5HSO_3^-(aq) \longrightarrow I_2(s) + 5SO_4^{2-}(aq) + 3H^+(aq) + H_2O(l)$$

The antiseptic virtues of iodine should be familiar through its use as 'tincture of iodine'. Iodine is an essential component of animal diet and iodide is therefore added to animal foodstuffs. Small amounts of the element are used in the organic chemical industry and in photography.

The halogens in the control of fire

The control of fire is an important task for the scientist and in this, the role of the chemist is to provide agents which will either extinguish fire once it has started or, perhaps more importantly, prevent normally inflammable materials from catching fire at all. Compounds of the halogens play an important part in both fire extinguishing and flameproofing.

The three essential components of a fire are *heat*, which provides the necessary activation energy to start the chemical reaction involved and may also volatilize the fuel, the *fuel* itself, and *oxygen*.

To extinguish a fire, one or more of these three essential components must be removed or separated from the others. Two examples are: blowing out a burning oil well by means of a powerful explosion; and cooling and simultaneously smothering a fire by means of water, CO_2, foams, or inert powders. Certain halogens and their compounds, however, are able to extinguish flames or prevent combustion when present in such small amounts that they cannot possibly be acting as coolants or smotherers.

This suggests that flame inhibition by halogens and their compounds involves interaction of the inhibitor with some of the chemical reactions taking place in the flame itself and has given rise to what is known as the chemical theory of flame prevention.

At the present time, little is known about the complex chemical reactions involved in flame suppression although a great deal of research is being carried out into these problems. However, it is known that in the flames of burning organic substances free radicals such as OH, CH_2O, HCO, and CH_3 are present. These may react with halogen atoms or with hydrogen halides evolved from the inhibitors in such a way as to stop the normal reactions in the flame. The successful action of an inhibitor is thus dependent on the ease with which halogen atoms and hydrogen halides are liberated from it at the temperature of the flame. The thermal stability of the inhibitor will be related to the strengths of the bonds in the molecules concerned.

In selecting a suitable compound for fire extinguishing, factors such as toxicity, corrosiveness, and storability have to be taken into account. In table 5.5 you will find some currently used compounds listed.

Agent	Formula	Relative effectiveness	Lethal concentration (ppm) un-decomposed vapour	decomposed vapour
Tetrachloromethane[1]	CCl_4	12.5	28 000	300
Bromochloromethane[2](CB)	CH_2BrCl	4.5	65 000	4000
Bromochlorodifluoromethane[3]	$CBrClF_2$	2.9	324 000	7650
Bromotrifluoromethane[3]	$CBrF_3$	5.6	800 000	14 000

Table 5.5

1 In general use, high toxicity of the decomposed vapour.
2 Developed by Germans in World War II, for use in German armed forces. Now used on crash trucks and as supplement to foam.
3 Low toxicity types for use in aircraft extinguishing systems.

Fire proofing

Let us now turn to the problem of preventing fire by producing materials which are flame resistant. Nowadays, a large number of synthetic materials such as polymers and man-made fibres are in widespread use and many are inflammable. Expanded polystyrene foam is used for packaging and for thermal and sound insulation in buildings; polyester resins reinforced with glass fibres are used for vehicle bodies, road and rail tankers, small boats including ship's lifeboats, and for building panels; polyethylene and polystyrene are used in packaging and for the fabrication of household articles; PVC and other polymers are widely used in the electrical industry for cable sheaths, printed circuit boards, and so on. Natural materials, such as wood, rubber, fibreboards, paper, and cellulose-based fabrics are all inflammable, yet are employed in situations where fire risks are often high. The chemist can reduce hazards associated with the use of such materials by devising treatments to reduce their inflammability.

Again we find that compounds of the halogens are predominant in this field although, here, they are often used in conjunction with agents, known as synergists, such as antimony oxide, whose presence greatly enhances the activity of the halogen compound.

An effective flameproofing agent must retard ignition and prevent the spread of the flame. Such an agent is called a 'retardant'. The mechanisms of flameproofing which have been suggested are based on the decomposition of the fire retardant under burning conditions in any of the following ways.

1 Coating theory; the fire retardant fuses to form an impervious barrier which prevents the access of oxygen. Examples are the glassy flux from borax-boric acid mixtures, and the fused melt from antimony oxide.

2 Gas theory; on being heated, the flameproofing material produces large volumes of incombustible gases such as ammonia or nitrogen which dilute the oxygen supply.

3 Thermal theory; the fire retardant changes endothermically, for example by fusion or sublimation, or undergoes an endothermic reaction with the inflammable substrate. The absorption of the heat reduces the temperature of the flame and it goes out. For example, calcium carbonate is used in fire-resistant paints where it decomposes endothermically to liberate carbon dioxide.

Flame proofing of a polymer for example may be carried out by additive methods, in which halogen compounds and synergists are added to the polymer itself, or to the monomer system before polymerization takes place. Alternatively, an intrinsically flame-resistant polymer can be made in which fire retardant molecules are incorporated in the actual polymer chain. Natural materials such as wood and fibres can be treated with surface coatings and paints which incorporate fire retardants.

Any additive must meet stringent requirements and these often raise considerable problems for chemists working in the field. Thus, the flame retardant must remain stable during fabrication processes and under the conditions of use of the fabricated material, that is, it must not decompose or give rise to staining or toxic vapours. Finally, it must be active in small enough concentrations to make its use economical.

Chlorine compounds are used as additives; for example ICI produce a range of chlorinated hydrocarbons known as 'Cerechlors' which have average carbon chain lengths between C_{12} and C_{24} and a chlorine content varying between 52 and 70 per cent. These compounds in addition to their flame retardant properties can also act as plasticizers. The 'Cerechlors' may be incorporated in paints and other surface coatings. Chlorinated rubber, such as ICI's 'Alloprene' can also be used as a surface coating and incorporated in adhesives.

Copolymers which are flame retardant have been made by copolymerizing vinylidene chloride with methyl methacrylate and both vinylidene chloride and vinyl chloride have been incorporated in man-made fibres.

Up to now, bromine compounds have not been widely used as fire retardants. This is because such compounds are generally more reactive chemically than chlorine derivatives. Fire retardants containing bromine may decompose in fabricated articles under normal conditions of use, giving rise to toxic vapours or spoiling the appearance of a finished article. Others may be stable enough at room temperature but will decompose in fabrication processes. The addition of bromine compounds to monomers prior to polymerization can also lead to problems since their presence may interfere with the normal polymerization reaction.

However, chemists are actively seeking bromine compounds which will be suitable for fire retardant applications. You may like to find out something about this for yourself.

The halogens in human metabolism

As has been seen, the halogens are powerful oxidizing agents. In all living cells there are a large number of complex organic molecules which are extremely sensitive to even mild oxidizing systems, and for this reason the halogens in general are toxic to life, except at very low concentrations. However, their reduction products, that is the halides, occur in many living systems at varying concentrations.

The element fluorine is too powerful an oxidizing agent to occur free in nature and it is extremely toxic to all living systems. On the other hand, the fluorides are widely distributed in the plant and animal kingdom. They are present in some of the food that we eat, and the fluoride ion is easily absorbed and slowly excreted. It is concentrated by some obscure mechanism in the supporting structures such as the bones. In man, there is a high concentration in teeth, especially in the enamel. At the present time there appears to be no known specific role for the fluoride ion in metabolism.

A large number of dental studies have shown that fluoride ions in low concentrations in drinking water can effectively arrest the development of tooth decay in children. As a result of these studies, fluoride ions are now added to the drinking water supply in certain areas in this country. However, the concentration of the fluoride ion must be critically controlled because at slightly higher concentrations the ion causes mottling of the dental enamel, and at much higher concentrations the ion is toxic to life. But at the correct concentration in drinking water it is beneficial to the healthy development of teeth.

Chlorine, like fluorine, is very toxic to living systems and was used as a war gas in the First World War. This halogen is sufficiently soluble in water to be useful as an antibacterial solution and for this reason it is often added to the drinking water at concentrations of 0.1 to 0.5 parts per million.

The chloride ion is the principal anion found in the fluid which bathes our body cells (the extracellular fluid); blood plasma forms a significant proportion of this fluid. In this extracellular fluid the chloride ion plays an important role in the maintenance of the osmotic equilibrium between the intracellular fluid and the extracellular fluid. The concentration of chloride ions in the blood plasma is about 365 mg per 100 cm^3 which closely resembles the concentration of chloride in sea water. From this fact attempts have been made to draw conclusions that life originated in the sea.

Apart from the abundance of the chloride ion in blood plasma, it is also present in sweat and saliva. Consequently during bouts of hard physical activity, or if we have to live in a hot climate, situations in which we would sweat more than usual, it is essential that we increase the intake of salt to compensate for the increased losses of sodium and chloride ions. Muscular cramp is one of the first symptoms of this salt deficiency.

Bromine is also too powerful an oxidizing agent to be encountered in living systems but the bromide anion occurs in small amounts. This ion is readily absorbed from the diet and, unlike the fluoride or the chloride ion, the bromide ion exhibits a highly specific effect on the central nervous system. Bromides depress the higher centres of the brain so that at the correct dosage the effect is one of sedation, but at higher dosage the effect is drowsiness and sleep. Bromides, unlike the fluorides, are not concentrated in any one tissue in the body and are eliminated in much the same way as the chlorides, namely by urinary excretion.

The element iodine is essential for man and all other mammals, and we derive much of our daily requirement from small amounts of iodide ions which are present in common salt as a trace contaminant. In mammals the iodide is concentrated by a small endocrine gland which is located in the throat and called the thyroid. The iodide trapping mechanism in the thyroid is not fully understood but after the anion is concentrated it is then subjected to an oxidation-reduction reaction and converted from iodide to iodine. The iodine is then involved in a series of reactions which eventually yield the hormone thyroxine. This hormone is then secreted by the thyroid gland and circulates via the blood; it is picked up by almost all cells and tissues. Thyroxine influences the rate of metabolism of body tissues, in particular the rate of oxygen uptake. In certain communities where the drinking water is low in iodide and the diet does not contain any other sources of iodide, there is the possibility of iodine deficiency disease developing. This can be prevented by supplying such communities with common salt to which iodide has been added.

Problems

1 Complete the following ionic half-equations by inserting the appropriate number of electrons on one side or the other.

i $\quad Li \longrightarrow Li^+$

ii $\quad Ba \longrightarrow Ba^{2+}$

iii $\quad \frac{1}{2}Cl_2 \longrightarrow Cl^-$

iv $\quad S \longrightarrow S^{2-}$

v $\quad Fe^{2+} \longrightarrow Fe^{3+}$

vi $\quad Cu^{2+} \longrightarrow Cu^+$

vii $\quad 2O^{2-} \longrightarrow O_2^{2-}$

2 What will be the oxidation number of each of the following elements as a result of the changes represented by the half equations?

i $\quad Fe^{3+} + e^- \longrightarrow$

ii $\quad Fe - 2e^- \longrightarrow$

iii $\quad Cu^+ - e^- \longrightarrow$

iv $\quad Zn - 2e^- \longrightarrow$

v $\quad O_2 + 4e^- \longrightarrow$

vi $\quad Cl_2 + 2e^- \longrightarrow$

vii $\quad Hg^{2+} + e^- \longrightarrow$

viii $\quad Al^{3+} + 3e^- \longrightarrow$

ix $\quad Br^- - e^- \longrightarrow$

x $\quad O^{2-} - 2e^- \longrightarrow$

3 What is the oxidation number of the named element in each of the following compounds?

i	Strontium in $SrCl_2$	*xi*	Chromium in $Cr_2O_7^{2-}$
ii	Rubidium in Rb_2O	*xii*	Manganese in $MnSO_4$
iii	Silver in $AgBr$	*xiii*	Manganese in Mn_2O_3
iv	Arsenic in As_2O_3	*xiv*	Manganese in MnO_2
v	Carbon in CCl_4	*xv*	Manganese in K_2MnO_4
vi	Carbon in CO	*xvi*	Manganese in $KMnO_4$
vii	Silicon in SiO_2	*xvii*	Vanadium in $VSO_4, 7H_2O$
viii	Chromium in $CrCl_3$	*xviii*	Vanadium in $NaVO_3$
ix	Chromium in CrO_3	*xix*	Vanadium in $VOCl_2$
x	Chromium in CrO_4^{2-}	*xx*	Uranium in UO_2Br_2

4 Name the compounds the formulae of which are given below, showing the oxidation number of the metal.

i	CrF_3	*xvii*	Mn_2O_3
ii	CrI_2	*xviii*	MnO_3
iii	$CoBr_2$	*xix*	Mn_2O_7
iv	Co_2O_3	*xx*	$Pt(CN)_2$ (the formula for sodium
v	$CoI_2, 6H_2O$		cyanide is NaCN)
vi	$CuSO_4, 5H_2O$	*xxi*	TlF_3
vii	Cu_2O	*xxii*	$TlClO_3$
viii	$Fe(OH)_3$	*xxiii*	TlI_4
ix	FeS	*xxiv*	TiP (the formula for calcium
x	$Ga(NO_3)_3$		phosphide is Ca_3P_2)
xi	$GaCl_2$	*xxv*	UF_6
xii	GeS	*xxvi*	CuCNS (the formula for sodium
xiii	$GeBr_4$		thiocyanate is NaCNS)
xiv	$PbCO_3$	*xxvii*	$Ce_2(SeO_4)_3$ (the formula for
xv	$PbCl_4$		potassium selenate is K_2SeO_4)
xvi	$MnCO_3$		

5 Consider the first element in each of the following reactions and state whether its oxidation number goes:

A – up,   B – down,   C – remains the same

i $Ag^+(aq) + Cl^-(aq) \longrightarrow AgCl(s)$

ii $Zn(s) + 2H^+(aq) \longrightarrow Zn^{2+}(aq) + H_2(g)$

iii $2Sr(s) + O_2(g) \longrightarrow 2SrO(s)$

iv $2Na(s) + Cl_2(g) \longrightarrow 2NaCl(s)$

v $Cl_2(g) + 2Na(g) \longrightarrow 2NaCl(s)$

vi $Co^{2+}(aq) + \frac{1}{2}Cl_2 \longrightarrow Co^{3+} + Cl^-(aq)$

vii $H_2 + Cl_2 + aq \longrightarrow 2HCl(aq)$

viii $I^-(aq) + \frac{1}{2}Br_2(aq) \longrightarrow \frac{1}{2}I_2(aq) + Br^-(aq)$

ix Cl^- (at anode) $\longrightarrow \frac{1}{2}Cl_2 + e^-$

x $\frac{1}{2}O_2(g) + H_2(g) \longrightarrow H_2O(g)$

xi $BaCl_2(s) + aq \longrightarrow Ba^{2+}(aq) + 2Cl^-(aq)$

xii $Cu^{2+}(aq) + Cu(s) \longrightarrow 2Cu^+(aq)$

xiii $Zn(s) + Pb^{2+}(aq) \longrightarrow Zn^{2+}(aq) + Pb(s)$

xiv $Cu^{2+}(aq) + 2e^-$ (at cathode) $\longrightarrow Cu(s)$

xv $Hg(l) \longrightarrow Hg(g)$

6 In each of the following changes state the initial and final oxidation number of the named element.

   i   Manganese in:

$$MnO_4^-(aq) + 8H^+(aq) + 5e^- \longrightarrow Mn^{2+}(aq) + 4H_2O(l)$$

   ii   Copper in:

$$CuO(s) + H_2(g) \longrightarrow Cu(s) + H_2O(l)$$

   iii   Sulphur in:

$$SO_2(g) + \tfrac{1}{2}O_2(g) \longrightarrow SO_3(g)$$

   iv   Sulphur in:

$$Na_2SO_3(aq) + \tfrac{1}{2}O_2(g) \longrightarrow Na_2SO_4(aq)$$

   v   Nitrogen in:

$$NO(g) + \tfrac{1}{2}O_2(g) \longrightarrow NO_2(g)$$

   vi   Nitrogen in:

$$N_2O_4(l) \longrightarrow 2NO_2(g)$$

   vii   Gold in:

$$AuCl_3(s) \longrightarrow Au(s) + 1\tfrac{1}{2}Cl_2(g)$$

   viii   Iron in:

$$[Fe(CN)_6]^{4-}(aq) + \tfrac{1}{2}Cl_2(aq) \longrightarrow [Fe(CN)_6]^{3-}(aq) + Cl^-(aq)$$
(the cyanide ion is CN^-)

   ix   Bromine in:

$$BrO_3^-(aq) + 6H^+(aq) + 6e^- \longrightarrow Br^-(aq) + 3H_2O$$

   x   *a*, manganese; *b*, sulphur in:

$$Mn^{2+}(aq) + S^{2-}(aq) \longrightarrow MnS(s)$$

7 In each of the following state whether each named element is:

A – oxidized,   B – reduced,   C – neither oxidized nor reduced.

   i   Iodide ions in:

$$2I^-(aq) + Cl_2(aq) \longrightarrow I_2(aq) + 2Cl^-(aq)$$

   ii   Iron(III) ions in:

$$2Fe^{3+}(aq) + Sn^{2+}(aq) \longrightarrow 2Fe^{2+}(aq) + Sn^{4+}(aq)$$

iii   Sodium ions in:

$$NaOH(aq) + HCl(aq) \longrightarrow NaCl(aq) + H_2O(l)$$

iv   Sodium in:

$$Na(s) + H_2O(l) \longrightarrow NaOH(aq) + \tfrac{1}{2}H_2(g)$$

v   Aluminium in:

$$Al(s) + OH^-(aq) + H_2O(l) \longrightarrow AlO_2^-(aq) + 1\tfrac{1}{2}H_2$$

vi   Hydrogen in:

$$Ca(s) + H_2(g) \longrightarrow CaH_2(s)$$

vii   Hydrogen in:

$$H_2(g) + Br_2(g) \longrightarrow 2HBr(g)$$

8   Use the oxidation number method to balance the following equations:

i $Cl_2(aq) + OH^-(aq) \longrightarrow Cl^-(aq) + ClO^-(aq) + H_2O(l)$

ii $Zn(s) + Fe^{3+}(aq) \longrightarrow Zn^{2+}(aq) + Fe^{2+}(aq)$

iii $Al(s) + H^+(aq) \longrightarrow Al^{3+}(aq) + H_2(g)$

iv $Fe(s) + Fe^{3+}(aq) \longrightarrow Fe^{2+}(aq)$

v $Sn(s) + HNO_3(l) \longrightarrow SnO_2(s) + NO_2(g) + H_2O(l)$

vi $Cu^{2+}(aq) + I^-(aq) \longrightarrow CuI(s) + I_2(aq)$

vii $SO_2(aq) + Br_2(aq) + H_2O(l) \longrightarrow H^+(aq) + SO_4^{2-}(aq) + Br^-(aq)$

viii $As_2O_3(s) + I_2(aq) + H_2O(l) \longrightarrow As_2O_5(aq) + H^+(aq) + I^-(aq)$

ix $MnO_4^-(aq) + H^+(aq) + Fe^{2+}(aq) \longrightarrow Mn^{2+}(aq) + Fe^{3+}(aq)$
$$+ H_2O(l)$$

x $[Fe(CN)_6]^{4-}(aq) + Cl_2(aq) \longrightarrow [Fe(CN)_6]^{3-}(aq) + Cl^-(aq)$

9   A solution containing 24.8 g of sodium thiosulphate ($Na_2S_2O_3, 5H_2O$) in one cubic decimetre was prepared. 23.6 cm^3 of this solution reacted exactly with 25.0 cm^3 of an aqueous solution of iodine.

i   What is the molarity of the sodium thiosulphate ($S_2O_3^{2-}(aq)$) solution?

ii   What is the molarity of the iodine ($I_2(aq)$) solution?

iii   What is the concentration of the iodine solution in grammes per cubic decimetre of iodine?

10 20 cubic decimetres of air, contaminated with chlorine, were bubbled through an excess of aqueous potassium iodide. The iodine so formed re-acted exactly with 45.0 cm^3 of 0.100M thiosulphate ($S_2O_3^{2-}$(aq)). Calculate the volume of chlorine in the sample of air. (All measurements at room temperature and pressure. Assume that 1 mole of gas has a volume of 24 cubic decimetres under these conditions.)

11 25.0 cm^3 of 0.0200M potassium dichromate ($Cr_2O_7^{2-}$(aq)) solution were added to an excess of acidified potassium iodide solution. The resulting solution of iodine reacted exactly with 30.0 cm^3 0.100M sodium thiosulphate solution. Use this information to deduce the equation for the reaction between dichromate and iodide ions.

Topic 6

The s-block elements; and the acid-base concept

This topic considers the properties of the s-block elements and during this investigation attention will be directed to two main features:

1 The similarities and differences which exist between the elements of group I and group II of the Periodic Table.
2 The trends in properties to be found in group II.

The metallic elements of group I and group II form compounds in which the oxidation number of the element does not vary. The oxidation number of the alkali metals (group I) in all their compounds is $+1$, and that of the alkaline earth metals (group II) is $+2$. In this investigation we look mainly at the group II metals (magnesium, calcium, strontium, and barium only; not beryllium as its compounds are too toxic to handle), but some group I compounds are included for purposes of comparison.

6.1 An investigation of alkali and alkaline earth metal chemistry

You may not have time to deal with all the elements and compounds suggested for this investigation. You will be told which of them you are to investigate; other members of the class will deal with those that you do not study. Observations and conclusions can be exchanged later.

Experiment 6.1a
An investigation of the flame colours of the s-block metals

Many elements give characteristic colours when their compounds are placed in a Bunsen burner flame. Chlorides are the most satisfactory compounds to use for studying these colours because they are usually volatile at temperatures attainable in the Bunsen burner flame.

Clean a platinum or nichrome wire by heating it in a non-luminous Bunsen flame, dipping it into a little concentrated hydrochloric acid (in a crucible or small watchglass), and heating it again. Continue this until the wire imparts no colour to the flame.

Pour the impure acid away and take a fresh portion. Dip the clean wire into the acid and then into a small portion of powdered compound on a watchglass. Use chlorides where possible, otherwise nitrates or carbonates. Hold the wire so that the powdered solid is in the edge of the flame and note any coloured flame which results. The colour disappears fairly quickly but can be renewed by dipping the wire into acid again and reheating. Observe the flame through a diffraction grating or direct vision spectroscope and identify as many coloured lines as you can. You may need to take a fresh sample of solid for this and to do the experiment in a darkened corner of the laboratory. The wire must be cleaned before examining a new compound, and a fresh portion of acid will be needed for this.

Record in your notebook the coloured lines that you see for each element that you examine.

The significance of the flame colours

You will remember that in Topic 4 it was pointed out that spectra can be used to gain information about the distribution of electrons in energy levels.

The spectra that you have seen in the experiment you have just done are of help in determining the energy levels of the electrons in these elements. The light emitted as a 'flame colour' is caused by electrons dropping back to lower energy levels, after a transfer to higher levels brought about by the heat of the Bunsen burner flame. Each line in the spectrum is caused by electrons moving between specific energy levels. This is because particular quantities of energy give rise to light of a particular wavelength. The sodium yellow line is due to a $3p \rightarrow 3s$ transition; the lithium red to $2p \rightarrow 2s$; and the calcium orange line to $5s \rightarrow 4p$.

All possible transitions cannot be seen by eye because the visible region is only a small part of the total spectrum, and certain transitions give rise to 'light' in other regions such as the ultraviolet. (Coloured photographs of the spectra of some alkali and alkaline earth metals are given on the inside of the front cover of this book.)

Figure 6.1a is a chart which represents the energy levels of electrons in atoms. You will remember from Topic 4 that the energy levels are filled in order of increasing energy; that there is a maximum number of electrons that can be accommodated in any one energy level; and that electrons are grouped in pairs having opposite 'spin'.

In figure 6.1a the energy levels are set out as in Topic 4, but 'boxes' are drawn to represent energy levels, instead of straight lines. Each 'box' can hold two electrons of opposite spin, and the correct number of boxes are shown to accommodate all the electrons that can be held in each energy level.

Make some copies of figure 6.1a in your notebook and fill them in so as to show the energy levels of the electrons in, say, sodium and magnesium. Use arrows to represent electrons, as was done in Topic 4.

Remember that electrons do not pair until an energy level is half full. The boxes on a given level should therefore *each* be given one arrow to represent one electron before a second arrow is placed in any of them on that level.

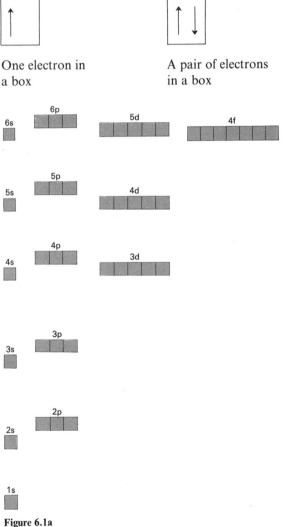

One electron in   A pair of electrons
a box   in a box

Figure 6.1a
The electronic configuration of a gaseous . . . atom.

Experiment 6.2
   The preparation of aluminium potassium sulphate
Aluminium potassium sulphate is an example of a type of double salt known as
an alum. Alums have the formula

$$M^+M^{3+}(SO_4^{2-})_2, 12H_2O$$

where M^+ is a metal ion bearing a single charge, such as Na^+, K^+, (but not Li^+)
or the ammonium ion NH_4^+; and M^{3+} is a metal ion carrying a charge of 3
units, such as Al^{3+} or Cr^{3+}. The alums crystallize in octahedra.

This preparation is begun by adding aluminium metal to potassium hydroxide
solution. The equation for the reaction which takes place is

$$2Al(s) + 2OH^-(aq) + 2H_2O(l) \longrightarrow 2AlO_2^-(aq) + 3H_2(g)$$

The product is potassium aluminate, $KAlO_2$.

When this reaction is complete, dilute sulphuric acid is added to form aluminium
potassium sulphate:

$$AlO_2^-(aq) + 4H^+(aq) \longrightarrow Al^{3+}(aq) + 2H_2O(l)$$

Procedure
Measure out $10 \, cm^3$ of 2M potassium hydroxide solution and place it in a
150×25 mm test-tube. Tear up some household aluminium foil into small pieces
and place a few pieces in the alkali. Work out the approximate weight of
aluminium that you will need from the equation, and then take a slight excess.
Go on adding the pieces of aluminium as the reaction proceeds.

The reaction will be complete in about thirty minutes. Remove any large pieces
of aluminium which may remain (a few small pieces will not matter) and carefully add some 2M sulphuric acid. Again, the approximate quantity can be
calculated, and a slight excess added.

A precipitate will form, and this should be stirred with a glass rod. Transfer the
mixture to a $100 \, cm^3$ beaker and heat it until it is boiling, when most of the
precipitate will dissolve. Test this mixture with indicator paper; a slight excess
of acid is desirable.

Filter the hot mixture into a crystallizing dish, cover the dish, and leave it to
cool, preferably overnight.

If enough of the residual saturated solution remains when cool, transfer it to a 150×25 mm test-tube, remove the crystals, and dry them on filter paper or blotting paper. Select the most perfect crystal, tie a small loop of cotton round it, and suspend it in the saturated solution from a piece of wire or a wooden splint placed across the top of the test-tube. If this is left in a cool place for a few days, the crystal will grow.

Quite large crystals may be grown in this way, but to ensure success a good deal of care should be taken (see, for example, Nuffield Chemistry Background Book *Growing Crystals*, and *Crystals and Crystal Growing*, by Alan Holden and Phylis Singer, Heinemann, London, 1961).

Questions on the preparation

1   Write down the formula of aluminium potassium sulphate.

2   Is the potassium hydroxide solution behaving as an acid? Discuss this statement in the light of the various definitions of an acid you have met in class.

3   Discuss the role of the sulphuric acid in the same way.

4   Why do you suppose an excess of sulphuric acid is desirable before the final crystallization?

5   Examine your crystals closely. Can you pick out any octahedra? Are the crystals merely sections of octahedra, and if so, what section?

6   Look up in a textbook both double salts and alums. You may like to try to prepare some more alums when you have found out more about them.

7   Find out what crystal overgrowths are, and try to prepare some.

8   Why do you suppose there are no alums containing lithium?

Background reading
The industrial uses of sodium metal

Sodium is present in a range of useful compounds such as $NaOH$, Na_2CO_3, $NaCN$, and Na_2O_2, but we shall be concerned here with uses of the metal itself. One of the principal chemical uses of sodium is in the preparation of a lead-sodium alloy employed in the manufacture of antiknock additives for gasoline, such as tetraethyl lead (page 35).

$$4Pb/Na + 4C_2H_5Cl \longrightarrow (C_2H_5)_4Pb + 3Pb + 4NaCl$$

The metal is also employed as a powerful reducing agent in the preparation of other organic compounds.

There are two interesting examples of the use of sodium metal in modern technology which at first sight might seem remarkable in view of the reactive nature of the element; these are the use of liquid sodium as a coolant, and the use of the solid metal as an electrical conductor.

Liquid sodium

Fast nuclear reactors, such as the one at Dounreay in Scotland, consist essentially of a core, in which the nuclear reaction takes place, surrounded by a system whereby a cooling liquid is circulated past the core.

The operating temperature of the reactor, about 600 °C, is too high to use normal molecular liquid coolants so a search had to be made for alternative liquids. Molten salts might have provided a solution to the problem but many have rather high melting points and the melts themselves are often corrosive. Liquid sodium however has very suitable properties for a coolant in nuclear reactors. The fast reactor now being built at Dounreay will contain some 700 tonnes of pure liquid sodium continuously circulating in its cooling system.

In table 6.3, some physical properties of liquid sodium, and those of two other possible cooling liquids, water and mercury, are compared.

Physical property	Sodium	Water	Mercury
Melting point/°C	97.8	0	-39
Boiling point/°C	890	100	357
Thermal conductivity/J s^{-1} cm^{-1} K^{-1}	0.786	0.0067	0.079
Specific heat capacity/J g^{-1} K^{-1}	1.34	4.18	0.13
Density/g cm^{-3}	0.89	0.97	13.5
Viscosity/centipoise	0.39	0.36	1.4

Figures other than melting and boiling points are for sodium at 250°C, water at 80°C, and mercury at 50 °C.

Table 6.3
Physical properties of liquid sodium, water, and mercury

As can be seen from the table, the low melting point and high boiling point of sodium make it suitable for use as a coolant. The high thermal conductivity is also essential for a liquid whose purpose is to transfer heat energy. Finally, the viscosity of liquid sodium is very close to that of water; this means that it is a mobile fluid and is easily circulated around the reactor core. Since sodium is cheap to produce, costing approximately 15 p a kilogramme, it would appear that the problems of cooling nuclear reactors are completely solved by the use of this material. Unfortunately, liquid sodium is a solvent, and this property must be taken into account!

Properties of liquid sodium

In the reactor, liquid sodium comes into contact with metals used in forming the core and cooling system; with carbon from the same sources; and with gases such as hydrogen, oxygen, and nitrogen which may gain access to the system owing to leaks of air or moisture. The liquid metal may dissolve any of these. Once dissolved, reaction may take place in the liquid sodium just as reactions occur in the molecular solvents with which you are familiar.

Corrosion by liquid sodium is largely attributable to dissolved impurities and, in view of the importance of liquid sodium as a coolant in the nuclear age, it is important to understand the nature of species dissolved in the liquid and the types of reaction which can occur in it.

At its melting point, the electrical conductivity of sodium is high; the value is 10.4×10^4 ohm^{-1} cm^{-1}. This implies that electrons are freely available as charge carriers in the system and such electrons are said to be present in a *conduction band*. It is known that each atom of sodium, in the solid or liquid state, provides one electron as a charge carrier. This electron is the one in the 3s energy level. Thus we may consider the separate 3s energy levels of the individual atoms as combining in the solid or liquid metal to form a conduction band. You will appreciate that the low ionization energy of sodium is an important factor here in making such electrons available for conduction. It is interesting to note that the mechanisms of electrical and thermal conductivity of metals are both dependent upon the electrons in the conduction bands. Hence the high electrical conductivity of sodium is allied to its good thermal conduction properties. Copper is another metal in common use both as an electrical conductor and as a heat transfer material, e.g. in the construction of radiators in cars and central heating systems.

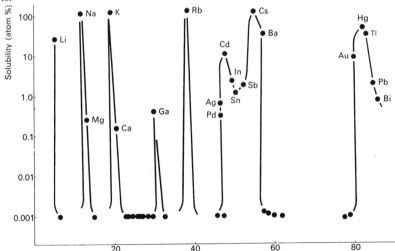

Figure 6.3
Solubilities of metals in liquid sodium at 200 °C.

Solutions of metals in liquid sodium

The solubilities of metals in liquid sodium vary widely and the known values are given in figure 6.3. An interesting feature is the correlation that exists between solubility and the position of the solute metal in the Periodic Table. The top row of the transition metals (scandium to zinc, atomic numbers 21–30) are virtually insoluble but the solubility increases suddenly, at gallium, that is, as soon as the atom of the solute metal includes an electron in the outer p-energy level. The transition metals of higher atomic weight in the second and third rows of the Periodic Table, namely Zn, Nb, Mo, and W are also insoluble. Such metals, together with those of the first transition series, form the constructional materials for nuclear reactors.

Solutions of non-metals in liquid sodium

Many non-metals such as hydrogen, oxygen, and the halogens dissolve in liquid sodium. It has been suggested that, in the case of hydrogen, the following reaction occurs with electrons being drawn from the higher energy levels of the conduction band;

$$H_2(g) + 2e^- \text{ (from conduction band of sodium)} \longrightarrow 2H^-$$

Temperature/$^\circ$C	% weight of H_2 dissolved
250	0.00042
300	0.0022
330	0.01
450	0.20

Table 6.4
Solubility of hydrogen in liquid sodium

The solubility of hydrogen in liquid sodium decreases if oxygen is already present in solution. Oxygen dissolves and withdraws electrons from the conduction band, probably as follows,

$$O_2(g) + 4e^- \text{ (from conduction band)} \longrightarrow 2O^{2-}$$

Since the conduction band contains a limited number of electrons in its higher energy levels, the electron withdrawal by O_2 leaves fewer conduction band electrons over for any H_2 which comes along later. There is, in effect, a competition for electrons between dissolved non-metal elements, those with higher electron affinities being better at withdrawing electrons than species like H_2.

Although the solubilities of these elements are small at 300°, they increase rapidly with temperature (see table 6.4); it is these solubilities, at the higher temperatures at which a fast reactor operates, which are of industrial importance. At 500 °C, the solubilities of hydrogen and oxygen in pure sodium are 13.4 and 0.11 atoms per cent respectively.

Solid sodium as an electrical conductor

Polyethylene insulated sodium conductors may offer a new approach to electrical conductors with considerable economic advantages and an improved performance.

Sodium is light in weight, cheap and plentiful, and compares well as an electrical conductor to the two metals most widely used for that purpose, namely copper and aluminium. A practical sodium conductor is produced by extruding a tube of polyethylene and simultaneously filling it with molten sodium. The polyethylene provides insulation, forms a barrier against air and water, and imparts the necessary mechanical strength to the cable. From the purely mechanical point of view, cables of this kind have distinct advantages over those of copper or aluminium. They are flexible down to temperatures as low as $-40\,°C$; they can recover their original dimensions after considerable stretch, unlike copper which does not recover but breaks after about 20 per cent elongation; and their light weight is an advantage for overhead power lines.

Copper and aluminium conductors are 6.8 and 2.3 times as expensive as an equivalent sodium cable.

Extensive trials currently being carried out in the United States indicate that sodium conductors, for both overhead and underground supplies, are proving reliable in operation, and that workmen used to installing high voltage cables can soon acquire the techniques of handling sodium cables.

Lithium, potassium, and calcium have also been considered as cable materials but for reasons which include cost, and either high or low melting points, they have been abandoned in favour of sodium.

Calcium and magnesium in agriculture

An adequate supply of calcium compounds in the soil is essential for healthy plant growth. Calcium itself is an essential constituent of plants, and it is also important for another reason. Calcium compounds are the principal factor in controlling the pH of the soil, and this affects the ability of plants to absorb nutrients through the roots.

Figure 6.4 shows the way in which absorption of some nutrients varies with pH. Where the shaded strip is broad the absorption is good, and where it narrows it is poor.

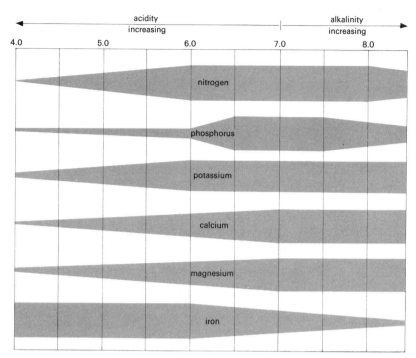

Figure 6.4
The influence of soil pH on plant nutrient supply.

From figure 6.4 it can be seen that the ability of plants to absorb nitrogen, phosphorus, potassium, and calcium decreases rapidly if the pH becomes more acid than about 6.0. On the other hand, if the pH becomes more alkaline than 7.0, the ability to absorb iron (and also manganese) decreases rapidly. For most plants the best pH is about 6.5.

The pH of natural soils may vary from 3.5 to 8.5 and the most important method of bringing it to the optimum value of about 6.5 is by ensuring the presence of an appropriate quantity of calcium compounds.

Figure 6.5 shows in the foreground two barley plots. The one on the right contains soil with low calcium content; it has become acid in reaction and the barley crop has failed there. The plot on the left has been well limed, has a good calcium content and a pH of about 6.5, and the barley crop is good there.

Figure 6.5
Barley on a limed plot (left), and an unlimed plot (right).
Photo, Rothamsted Experimental Station

Calcium is removed from soils by the leaching effect of rainwater. The use of fertilizers which are acidic in reaction, such as ammonium sulphate, also contributes to the dissolving and disappearance of the element. In addition, calcium is removed when the crops are harvested; the harvesting of potatoes (at a yield of 7 tonnes per acre) results in the removal of 2.3 kg of calcium oxide per acre, and the harvesting of lucerne hay (at a yield of $3\frac{1}{2}$ tonnes an acre) results in the removal of 98 kg of calcium oxide an acre. Thus, even if a soil initially has an appropriate calcium content it is usually necessary to replenish it.

Calcium is usually applied as limestone, $CaCO_3$, calcium hydroxide, $Ca(OH)_2$, or calcium oxide, CaO. The last of these must not be applied direct to growing crops; it must be left, usually in lump form, in the open to weather for some time. Why do you think calcium oxide is not applied direct, and what happens during the weathering process?

Magnesium is an essential plant nutrient. A magnesium ion occupies the centre of a chlorophyll molecule, and photosynthesis depends upon chlorophyll. Magnesium ions are also necessary for the functioning of several plant enzymes, the catalysts for biochemical reactions. Certain crops often suffer from severe magnesium deficiency; examples are some varieties of apples, glasshouse tomatoes, cauliflowers, oats, potatoes, and sugar beet.

The correct amount of calcium and magnesium to be applied should be determined by a proper soil analysis. Too little or too much is harmful (figure 6.4).

Since grazing animals obtain their food from grass it is necessary for the grass to provide all the nutrients required. If the levels of calcium and phosphate in the grass are not correct young growing stock fail to attain full size, and the milk yield of dairy cows will be restricted.

Cattle and sheep feeding on lush spring pastures are liable to suffer from 'staggers' if the grass fails to supply sufficient magnesium. In its acute form the condition can be swiftly fatal. One precaution is to 'feed' the animals a weighted bullet made of magnesium alloy. The bullet, which is given orally by means of a special 'gun', lodges in the beast's rumen. There it breaks down slowly, continuously releasing an effective dose of magnesium throughout the first critical weeks at pasture when the risk of 'staggers' is greatest. The bullets, which are three inches long and one inch in diameter for cows, and rather smaller for sheep, last five to six weeks.

Calcium and magnesium in human metabolism

The total amount of calcium in a typical adult human body is about 1 kilogramme. 99 per cent of this calcium is in the bones of the skeleton, mainly in the form of calcium phosphate.

The intracellular fluid also contains calcium, and this calcium is continually exchanging with that in the bone. Under normal conditions there is a dynamic equilibrium. Under abnormal conditions excess calcium may be deposited on the bones; or conversely too much calcium may leave the bones, which will become lighter and more fragile. The latter situation occurs in laying hens if they are given too little calcium-containing food or grit; in order to continue producing calcium carbonate for the eggshell, calcium dissolves from the bones and is transferred in the blood to the site where it is required for making eggshell. Laying hens must be provided with plenty of food containing calcium.

Calcium is absorbed in the upper part of the intestine, and for this absorption vitamin D is essential. An inadequate supply of vitamin D results in a calcium deficiency disease of bones known as rickets. In addition to making possible the absorption of calcium, vitamin D also controls the equilibrium between bone calcium and blood calcium.

A typical dietary intake of calcium is about 1 g per day and the amount of this which is actually absorbed by the intestines is about 200 mg. The amount excreted in the urine is also about 200 mg per day, so that for a non-growing adult body the amount excreted in the urine equals the amount absorbed by the gut under normal conditions.

The greatest requirements for calcium are in pregnancy and in youth while the body is growing. The mature foetus contains about 30 g of calcium, which has been transferred to it by the mother through the placenta. A baby of 1 year old contains about 100 g of calcium, a gain of 70 g. The milk and other food must provide more than this since absorption of calcium is always incomplete. It is essential that expectant and nursing mothers should have an adequate and balanced diet containing plenty of vitamin D and calcium. Vitamin D is also manufactured in the skin under the action of sunlight, so that plenty of sunlight received in the open air will result in an adequate supply of the vitamin.

The main sources of calcium in the diet are milk and cheese, and in the United Kingdom bread to which calcium compounds have been added. The main natural source of vitamin D is by the action of direct sunlight on the skin. In some countries, including the United Kingdom and the United States, certain foods are fortified with the vitamin, for instance margarine, dried milk, and infant cereals. Mothers are also encouraged to give babies a recommended amount of cod liver oil each day, since this is rich in vitamin D. Absence of vitamin D leads to rickets, but excessive quantities can also be harmful.

A range of defects of calcium metabolism occur in humans. The kidneys may fail to reabsorb calcium properly, and too much calcium will then be excreted in the urine, leading in time to calcium deficiency. The kidneys may fail to excrete H^+ ions, leading to acidity in the body and alkalinity in the urine; alkalinity in the urine may then lead to deposition of kidney stones consisting of insoluble calcium salts.

Magnesium is an essential component of the human body. It is necessary for the functioning of certain hormones, for instance insulin which controls sugar metabolism. It is also necessary in other processes, an abnormally low level leading to fits; in these instances the disease can usually be alleviated by feeding the patient with a diet containing a few grammes of a suitable magnesium compound each day.

Every kidney patient today has his blood analysed for its concentration of the ions, Na^+, K^+, Mg^{2+}, Ca^{2+}, HCO_3^-, PO_4^{3-}, and for its urea concentration. Large hospitals now have automated instruments for doing the analysis quickly.

Thus a knowledge of the chemistry and biochemistry of calcium and magnesium is necessary for physicians to alleviate or cure a number of diseases, and growing knowledge will lead to increased control over them.

Problems

* Indicates that the *Book of Data* is needed.

*1   What is the ratio of the price of calcium to the price of magnesium (1968 prices)

   i   per gramme,
   ii per mole of atoms,
   iii per cm^3?
   iv 'Calcium compounds are nearly twice as abundant as magnesium compounds, but calcium is more expensive than magnesium.' Make a comprehensive list of possible reasons for this anomaly.

*2   Suppose you wanted to make as much pure calcium oxalate as possible, given a solution of 1.9 g of sodium oxalate in 50 g of water (20 °C) and a supply of 0.1M calcium chloride solution.

   i   Calculate the minimum volume of calcium chloride solution which you would require. Give an equation for the reaction.
   ii Describe the procedure.
   iii What is the maximum weight of calcium oxalate you could expect to obtain?
   iv. Give all the reasons why you are unlikely to obtain in practice the weight given in answer to (*iii*).

3   The element M reacts with the element Q to form a compound MQ_2. M is in group II of the Periodic Table. The compound is a solid at room temperature and has an ionic structure.

   i   In which group of the Periodic Table would you expect to find Q?
   ii Give the symbols, showing the electrical charges, of the two different ions of which MQ_2 is composed.
   iii Write a half ionic equation to show the change which occurs to M when it combines with Q.
In this change is M acting as

   A An oxidizing agent
   B A reducing agent
   C An acid
   D A base
   E None of these
   iv Write a half ionic equation to show the change which occurs to Q when it reacts with M.

In this change is Q acting as
- A   An oxidizing agent
- B   A reducing agent
- C   An acid
- D   A base
- E   None of these

4   The element X reacts with cold water to form a hydroxide and hydrogen:

$$X(s) + H_2O(l) \longrightarrow XOH(aq) + \tfrac{1}{2}H_2(g)$$

i   In which group of the Periodic Table would you expect to find X?

ii   The hydroxide XOH is ionic; give the symbols, showing the electrical charges, of the two ions of which it is composed.

iii   Write a half ionic equation to show the change which occurs to X when it reacts with water.
How would you classify the change which takes place in X?
- A   reduction
- B   oxidation
- C   acid/base
- D   radioactive decay
- E   not a chemical change

iv   Write a half ionic equation to show the change which occurs to the water when it reacts with X.
How would you classify this change?
- A   reduction
- B   oxidation
- C   acid/base
- D   radioactive decay
- E   not a chemical change

5   The element Y forms a compound with hydrogen of formula HY which dissolves in water to form a solution of pH less than 7.

i   In which group of the Periodic Table would you expect to find Y?

ii   Write the symbols, showing the electrical charges, for the ions which you would expect to find in an aqueous solution of HY.

iii   Write a half ionic equation to show the change which occurs to HY when it dissolves in water.
Which of the following applies to HY when it dissolves?
- A   It has been reduced
- B   It has been oxidized
- C   It has acted as an acid
- D   It has acted as a base
- E   It has decomposed

iv Write a half ionic equation to show the change which occurs to water when HY dissolves in it.
Which of the following applies to water when HY dissolves in it?
- A It has been reduced
- B It has been oxidized
- C It has acted as an acid
- D It has acted as a base
- E It has decomposed

Questions 6–15
In this set of questions classify each numbered reaction as one of the lettered types.
- A oxidation
- B reduction
- C acid/base
- D ionic precipitation

6 $Cu^{2+}(aq) + e^- \longrightarrow Cu^+(aq)$

7 $NH_3(g) + HBr(g) \longrightarrow NH_4Br(s)$

8 $Pb^{2+}(aq) + 2I^-(aq) \longrightarrow PbI_2(s)$

9 $Cl^-(aq) \longrightarrow \frac{1}{2}Cl_2(g) + e^-$

10 $2MnO_4^{2-}(aq) \longrightarrow 2MnO_4^-(aq) + 2e^-$

11 $Ca^{2+}(aq) + CO_3^{2-}(aq) \longrightarrow CaCO_3(s)$

12 $Na^+(aq) + Cl^-(aq) \longrightarrow NaCl(s)$

13 $H_2Se(g) + 2NaOH(aq) \longrightarrow Na_2Se(aq) + 2H_2O(l)$

14 $2H^+(aq) + 2e^- \longrightarrow H_2(g)$

15 $2H_2O(l) \longrightarrow H_3O^+(aq) + OH^-(aq)$

*16 Describe concisely how you would distinguish between each member of the following pairs of substances. Choose the quickest practicable methods.

You are advised to construct your answer in the light of the *general* properties of the elements and compounds in each group of the Periodic Table, rather than by the properties of particular elements and compounds. You should describe briefly any practical tests which may be necessary, but you should not need to write more than four or five lines on each part of the question.

    i A group I element and a group II element.
   ii A chloride of a group I element and a chloride of a group V element.
  iii A chloride of a group I element and an iodide of a group I element.
  iv A carbonate of a group I element and a carbonate of a group II element.
   v A hydroxide of a group I element and a hydroxide of a group III element.

*17 Give an account of the general trends in (*i*) the solubility and (*ii*) the stability of the hydroxides, carbonates, and sulphates of sodium, magnesium, calcium, strontium, and barium. To what extent do the patterns in solubility follow the position of the elements in the Periodic Table?

*18 Make a general comparison of the melting points and the latent heats of fusion (per mole of atoms) of the following elements: sodium, magnesium, calcium, strontium, barium, phosphorus (white), sulphur, chlorine, bromine, and iodine.

Topic 7
Energy changes and bonding

In nearly all chemical reactions, whether in the laboratory when for example solutions of an acid and an alkali are mixed, or in a power station when coal or oil is burning, there is an energy change. The study of these energy changes is as much a part of chemistry as the study of the changes of materials in a reaction or of the structure of substances.

There are several aspects of energy changes which could be investigated.

In this topic we shall conduct some experiments to seek answers to two questions.

1 It is sometimes possible to convert a substance A into another substance B by several different routes. Does the route by which a chemical change takes place make any difference to the overall energy change?
2 When a chemical reaction occurs, there is a change in the nature and perhaps the number of bonds between atoms. Can we attribute to specific bonds specific contributions to overall energy changes?

The answers to these questions will lead us to important conclusions concerning energy changes. Their investigation will start in this topic, and be continued in later topics.

7.1 Energy changes – definitions
If we are to investigate energy changes, we must state under what conditions the changes are measured in order that the results can be compared.

The energy change most frequently used is called the *enthalpy change*, and this can be considered to be the heat that would be exchanged with the surroundings if the reaction occurred in such a way that the temperature and pressure of the system before and after the reaction were the same.

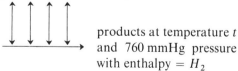

energy interchange with the surroundings

reactants at temperature t and 760 mmHg pressure with enthalpy $= H_1$ $\longrightarrow$ products at temperature t and 760 mmHg pressure with enthalpy $= H_2$

Enthalpy change = heat of reaction, $\Delta H = H_2 - H_1$

= energy interchange with the surroundings

Normally we insulate the system from its surroundings, and allow the heat of the reaction to change the temperature of the system. We then calculate how much heat would have to be put into or taken from the system to bring it back to its initial temperature. This amount of heat is the enthalpy change.

The enthalpy change is the energy exchange with the surroundings *at constant pressure*. This means that if the volume of the system changes as in the reaction

$$Zn(s) + 2H^+(aq) \longrightarrow Zn^{2+}(aq) + H_2(g)$$

the system will have to expand *against the atmosphere* and work will have to be done by the system. Thus the enthalpy change includes this energy as well as the energy changes due to changes in bonding. If the reaction is conducted *at constant volume*, this work does not have to be done, and the energy change under these conditions is called the *internal energy change*.

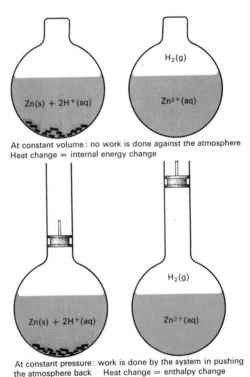

At constant volume: no work is done against the atmosphere
Heat change = internal energy change

At constant pressure: work is done by the system in pushing the atmosphere back   Heat change = enthalpy change

Figure 7.1a
Reactions at constant volume and pressure.

As it is more convenient to carry out reactions in the laboratory under conditions of constant pressure (in open beakers and test-tubes), we normally refer to enthalpy changes rather than internal energy changes. In reactions in which there is no volume change, enthalpy change = internal energy change; and when there is a volume change, the difference is less than 5 per cent for reactions in which the numerical value of ΔH is greater than 40 kJ.

If the reaction is *exothermic*, that is, if heat is given out from the system to the surroundings during the reaction, then the enthalpy of the reactants, H_1, must be greater than that of the products, H_2, so the enthalpy change

$$\Delta H = H_2 - H_1$$

is negative.

Conversely if the reaction is *endothermic*, that is, if heat is taken in to the system from the surroundings during the reaction, then H_2 must be greater than H_1 and the enthalpy change

$$\Delta H = H_2 - H_1$$

is positive.

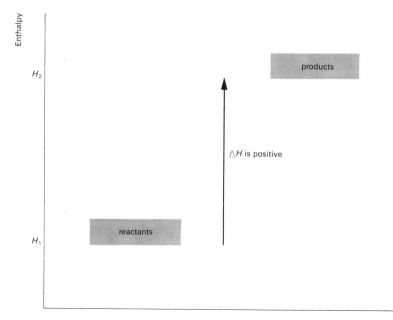

Figure 7.1b   Reaction path
Endothermic reaction.

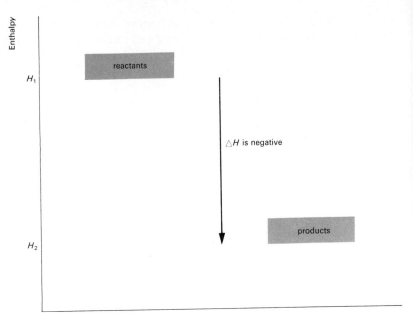

Figure 7.1c
Exothermic reaction.

For the enthalpy change

$$C(\text{graphite}) + O_2(g) \longrightarrow CO_2(g); \qquad \Delta H = -393.5 \text{ kJ mol}^{-1}$$

the value of ΔH indicated is for the amounts *shown in the equation*, that is, for one mole of carbon atoms, one mole of oxygen molecules, O_2, and for one mole of carbon dioxide molecules, CO_2. Normally a *standard* enthalpy change is quoted which refers to that at 760 *mm mercury pressure* and at some particular temperature, and it is then given the symbol $\Delta H^{\ominus}$. The superscript $^{\ominus}$ indicates *standard* at some temperature *which must be stated*; the normal temperature used is 298 K and this is shown as a subscript to the symbol, thus: $\Delta H^{\ominus}_{298}$. This symbol also means that the substances must be in the physical state normal at 298 K and one atmosphere, that is solid carbon*, gaseous oxygen, and gaseous carbon dioxide. If there are any solutions involved, then the standard condition is unit activity, which approximates to one mole per cubic decimetre.

> * In the case of elements or compounds which can exist in different forms such as carbon, the most stable form is chosen as the standard. In this case the most stable form is graphite.

It is not possible to find the standard enthalpy change of formation of, say, carbon dioxide, at 298 K *directly*, because carbon and oxygen do not react at this temperature. However, it is possible to determine a value at a temperature at which they do react, and calculate a value at 298 K from this result.

> $\Delta H^{\ominus}_{298}$ is the *standard enthalpy change* for a reaction. It refers to the amounts shown in the equation, at a pressure of 760 mm mercury, at a temperature of 298 K, with the substances in the physical states normal under these conditions.
> Solutions must have unit activity approximating to 1 mol dm^{-3}.

For certain reactions we have a particular shorthand which saves us from having to write the full equation. The reaction quoted above is one such that the enthalpy change for it refers to the formation of carbon dioxide. The symbol $\Delta H^{\ominus}_{f,298}[CO_2(g)] = -393.5 \text{ kJ mol}^{-1}$ refers to 1 *mole of carbon dioxide* being formed in the standard state. $\Delta H^{\ominus}_{f,298}[CO_2(g)]$ is called the *standard heat of formation of carbon dioxide.*

Notice the use of the word *heat* in this definition. Since the enthalpy change is apparent as heat, the term 'heat of formation' is frequently used in place of 'enthalpy change of formation'.

> $\Delta H^{\ominus}_{f,298}$, the *standard heat of formation* of a substance refers to the formation of one mole of the substance under the standard conditions.

One result of this definition of the standard heat of formation of a substance is that the heat of formation of an element in its standard state is zero:

For $Hg(l) \longrightarrow Hg(l);$ $\Delta H^{\ominus}_{f,298}[Hg(l)] = 0$

   $Hg(l) \longrightarrow Hg(g);$ $\Delta H^{\ominus}_{f,298}[Hg(g)] = +60.84 \text{ kJ mol}^{-1}$

Another energy change that is particularly defined is the *standard heat of atomization of an element*, $\Delta H^{\ominus}_{at,298}$. This refers to the enthalpy change when *one mole of gaseous atoms* is formed from the element in the defined physical state (normally the standard state)

   $\frac{1}{2}H_2(g) \longrightarrow H(g);$ $\Delta H^{\ominus}_{at,298}[H_2(g)] = +218 \text{ kJ mol}^{-1}$

> $\Delta H^{\ominus}_{at,298}$ the *standard heat of atomization* of an element refers to the formation of one mole of gaseous atoms from the element in the defined physical state under standard conditions.

The standard heat of combustion of a substance, $\Delta H^{\ominus}_{combustion,298}$, is defined as the enthalpy change that occurs when one mole of the substance undergoes complete combustion under standard conditions. For a compound containing carbon, for example, complete combustion means the conversion of the whole of the carbon to carbon dioxide, as shown in the following equation.

$$C_{12}H_{22}O_{11}(s) + 12O_2(g) \longrightarrow 12CO_2(g) + 11H_2O(l);$$

(sucrose) $\Delta H^{\ominus}_{combustion,298} = -5647 \text{ kJ mol}^{-1}$

$\Delta H^{\ominus}_{combustion,298}$, the *standard heat of combustion* of a substance refers to the complete combustion of one mole of the substance under standard conditions.

So far we have only been concerned with enthalpy *changes*. Absolute enthalpies cannot be determined, only differences between them. For this reason the value of zero is assigned to the enthalpy of an element *in its standard state* at 298 K.

Examples:

$H^{\ominus}_{298}[O_2(g)] = 0$
$H^{\ominus}_{298}[C(graphite)] = 0$
$H^{\ominus}_{298}[Na(s)] = 0$
$H^{\ominus}_{298}[Cl_2(g)] = 0$

Other enthalpies are calculated from these reference points.

Note on units: the joule is the internationally accepted unit of energy. It has only recently been adopted and many of the other books that you consult may use the calorie. These are easily related:

4.18 joules = 1 calorie*

7.2   Measurement of standard heats of formation

Having discussed the various energy changes which can be measured, and the conditions under which these measurements should be made, it is now possible to carry out some experiments. The object of the experiments in this topic is to find out the heat changes which take place during certain chemical reactions, and then to use the values obtained to answer the questions mentioned at the beginning of the topic.

* A table for the conversion of calories to joules is given in the *Book of Data.*

Experiment 7.2a
The conversion of solid sodium hydroxide to sodium chloride
solution

This experiment is designed to give an answer to the question 'Does the route by which a chemical change is carried out make any difference to the overall energy change?'

The experiment involves the conversion of solid sodium hydroxide to a solution of common salt by two different routes.

Solid sodium hydroxide can be dissolved in water, and this solution then neutralized with hydrochloric acid; or the hydrochloric acid may be diluted and then the solid sodium hydroxide added:

Route 1 $NaOH(s) + aq \longrightarrow NaOH(aq, 4M)$
           $NaOH(aq, 4M) + HCl(aq, 4M) \longrightarrow H_2O(l) + NaCl(aq, 2M)$
Route 2 $HCl(aq, 4M) + aq \longrightarrow HCl(aq, 2M)$
           $HCl(aq, 2M) + NaOH(s) \longrightarrow H_2O(l) + NaCl(aq, 2M)$

In both cases the reactants are the same, $NaOH(s)$ and $HCl(aq, 4M)$; and so are the products, $H_2O(l)$ and $NaCl(aq, 2M)$.

Procedure
Route 1. Place $50 \, cm^3$ of distilled water in one measuring cylinder and $50 \, cm^3$ of 4.0M hydrochloric acid in another.

Weigh a weighing bottle and cap, and then weigh into it 4.00 g of sodium hydroxide flakes. It will be sufficiently accurate if you measure this to ± 0.02 g; this means that the weight which you use should be between 3.98 g and 4.02 g. Keep a stock of flakes in a sealed bottle while weighing and do not allow the flakes to come into contact with your skin.

Place $25 \, cm^3$ of water in an insulated calorimeter, and record its temperature. Record the temperature of the acid. Tip the solid sodium hydroxide into the water, in three or four quick stages, stir with the thermometer, and record the temperature as soon as all the solid has dissolved. Immediately add $25 \, cm^3$ of the acid, stir, and record the temperature.

Pour the hot solution down the sink and wash the calorimeter with water.

Route 2. Once again weigh out 4.00 g of solid sodium hydroxide correct to ± 0.02 g. Place the remaining $25 \, cm^3$ of distilled water into the calorimeter, and record its temperature. Also record the temperature of the acid. Add the remaining 25 cm^3 of acid to the water, stir with the thermometer, and record the temperature. Immediately add the solid sodium hydroxide in three or four quick stages, stir, and record the temperature as soon as all the solid has dissolved.

Treatment of the results

Find the temperature rise for each stage of each reaction. If the water and the acid were not at the same temperature you can make a correction for this as shown in the following example.

> Temperature of water $= 18.0\,°C$
> Temperature of acid $= 20.0\,°C$
> Temperature after mixing $= x\,°C$

If *no reaction had occurred* when the water and acid were mixed, the temperature of the mixture (25 cm^3 of acid and 25 cm^3 of water) would have been 19 °C. Thus the rise in temperature due to the reaction is $(x-19)\,°C$.

Calculate the enthalpy changes that accompanied these four reactions. Remember that an enthalpy change is *per mole* of sodium hydroxide and/or hydrochloric acid and is the heat exchange with the surroundings if the system returns to its original temperature after the reaction. In the above example the temperature change would be from $x°$ back to 19 °C, that is, $(x-19)\,°C$. A specimen calculation for water in a glass calorimeter is given below. The specific heat capacity of the solution can be taken as approximately $4.2\,J\,g^{-1}\,°C^{-1}$.

Specimen calculation

Mass of glass calorimeter $= 530\,g$
Specific heat capacity of glass calorimeter $= 0.84\,J\,g^{-1}\,°C^{-1}$
Volume of water in calorimeter $= 600\,cm^3$
Specific heat capacity of water $= 4.2\,J\,g^{-1}\,°C^{-1}$
(Corrected) temperature change $= 12.0\,°C$
Heat capacity of glass calorimeter $= (530 \times 0.84)\,J\,°C^{-1}$
Heat capacity of the water $= (600 \times 4.2)\,J\,°C^{-1}$

$\therefore$ Total heat capacity $= 445 + 2520\,J\,°C^{-1}$
$= 2965\,J\,°C^{-1}$

$\therefore$ Heat change in the reaction $= 2965 \times 12.0\,J$
$= 35\,580\,J$

Suppose that the heat had been produced by a reaction involving 0.040 mole of a substance. Then the heat change per mole would be

$$35\ 580 \times \frac{1.00}{0.04} = 889\ 500\ \text{J mol}^{-1}$$

$$= 890\ \text{kJ mol}^{-1}\ (\text{3 significant figures})$$

When you are using a thin plastic or expanded polystyrene cup as a calorimeter the mass of this is so small that the heat capacity of the calorimeter may be neglected, in comparison with the mass of water or solution. If you are using a polythene bottle of appreciable mass you should weigh it to ± 1 g, and use a value of 2.5 J g^{-1} °C^{-1} for the specific heat capacity.

Calculate the enthalpy change for the reaction

$$\text{NaOH(s)} + \text{HCl(aq, 4M)} + \text{aq} \longrightarrow \text{H}_2\text{O(l)} + \text{NaCl(aq, 2M)}$$

in each case and compare the two results. You should now be able to answer the question posed at the beginning of the experiment.

Hess's Law

If the law of conservation of energy applies to chemical processes, then if one set of substances is converted to another, by whatever route, the total energy change must be the same.

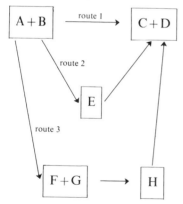

If this were not so, and the energy changes were different in different routes, it would be possible to change $A + B$ into $C + D$ by route 1, say, and then change $C + D$ back into exactly the same quantity of the same substances $A + B$ by, say, route 3, having at the same time an overall gain or loss of energy. According to the law of conservation of energy this is impossible: *the energy change must be the same by whatever route we travel from* $A + B$ *to* $C + D$. In 1840, G. H. Hess discovered this particular application of the law of conservation of energy experimentally, and it is generally referred to as Hess's Law.

Hess's Law: The total energy change accompanying a chemical change is independent of the route by which the chemical change takes place.

The great value of Hess's Law is that it can be used to calculate enthalpy changes that cannot be determined directly by experiment. Some examples of this will make the ideas clear.

The standard heats of formation of water and of carbon dioxide can be measured directly by calorimetry experiments, and these two values are known to a very high degree of accuracy. An account of the way in which such measurements are made is given in the background reading at the end of this section.

$$C(graphite) + O_2(g) \longrightarrow CO_2(g); \quad \Delta H^{\ominus}_{f,298} = -393.5 \text{ kJ mol}^{-1}$$
$$H_2(g) + \tfrac{1}{2}O_2(g) \longrightarrow H_2O(l); \quad \Delta H^{\ominus}_{f,298} = -285.9 \text{ kJ mol}^{-1}$$

$\Delta H^{\ominus}_{f,298}$ values have also been determined directly for many other oxides.

The standard heat of formation of ammonia has been determined by decomposing ammonia gas to nitrogen and hydrogen in a calorimeter, using a heated catalyst.

$$NH_3(g) \longrightarrow \tfrac{1}{2}N_2(g) + 1\tfrac{1}{2}H_2(g); \quad \Delta H^{\ominus}_{298} = +46.39 \text{ kJ mol}^{-1}$$

From this, $\Delta H^{\ominus}_{f,298}$ for ammonia is -46.39 kJ mol^{-1}.

But there are very many compounds for which the standard heats of formation cannot be determined directly. Methane is an example, because graphite and hydrogen do not combine directly. Nor is it possible to decompose methane readily to carbon and hydrogen. The equation $C(graphite) + 2H_2(g) \rightarrow CH_4(g)$ which represents the reaction by which $\Delta H^{\ominus}_{f,298}$ for methane is defined, is a reaction which cannot be performed in a calorimeter.

$\Delta H^{\ominus}_{f,298}$ for diborontrioxide, B_2O_3, is another example of a standard heat of formation which cannot be measured directly. This is because it has so far proved impossible to burn the element boron completely in oxygen; a protective coating of B_2O_3 forms around particles of unchanged boron.

In these, and in many hundreds of other instances, values for the standard heats of formation have been obtained in indirect ways, by means of calculations using Hess's Law.

Indirect methods for obtaining standard heats of formation

An indirect method of wide applicability is to determine the heat of combustion of the compound, and the heats of combustion of the constituent elements; from an *energy cycle* the heat of formation of the compound can then be found.

Methane will serve as an example. First, write down the equation for the standard heat of formation (that is, from the elements in their standard states).

$$C(\text{graphite}) + 2H_2(g) \xrightarrow{\Delta H^{\ominus}_{f,298}} CH_4(g)$$

Now write in the equations for burning both the reactants and the methane in oxygen, as now shown.

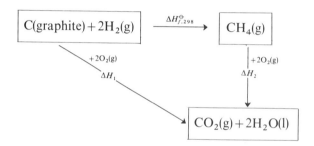

ΔH_1 is made up of the standard heat of combustion of graphite, and twice the standard heat of combustion of hydrogen.

$$C(\text{graphite}) + O_2(g) \longrightarrow CO_2(g); \quad \Delta H^{\ominus}_{\text{combustion},298} = -393.5\,\text{kJ}$$

$$2H_2(g) + O_2(g) \longrightarrow 2H_2O(l); \quad 2 \times \Delta H^{\ominus}_{\text{combustion},298} = 2 \times (-285.9)\,\text{kJ}$$
$$= -571.8\,\text{kJ}$$

Therefore

$$\Delta H_1 = -393.5 + (-571.8)$$
$$= -965.3\,\text{kJ}$$

ΔH_2 is the standard heat of combustion of methane

$$\Delta H_2 = \Delta H^{\ominus}_{\text{combustion},298} = -890.3\,\text{kJ mol}^{-1}$$

We may now substitute these values in the earlier diagram.

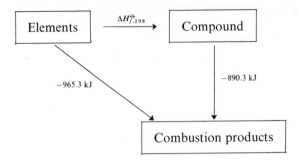

The overall energy change in going from elements to combustion products must be the same whatever the route. Equating the two routes,

$$-965.3 = \Delta H^{\ominus}_{f,298} - 890.3$$

$$\Delta H^{\ominus}_{f,298} = -75.0 \text{ kJ mol}^{-1}$$

The cycle of operations can be represented by an *energy level diagram*.

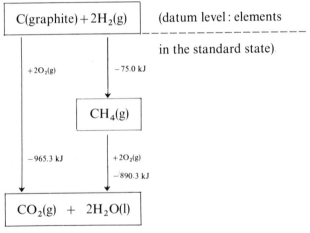

Route 1   Route 2

Hydrogen sulphide will serve as another example. Hydrogen does not combine with solid sulphur; therefore the equation

$$H_2(g) + S(s) \longrightarrow H_2S(g)$$

which represents the conditions for obtaining the standard heat of formation also represents a reaction which does not take place. Even if hydrogen gas is bubbled into molten sulphur, only traces of hydrogen sulphide are formed, and clearly this is not an experiment which can be conducted in a calorimeter. $\Delta H^{\ominus}_{f,298}$ can, however, be obtained from combustion experiments.

First, write down the formation equation, then the burning of the elements, and the burning of hydrogen sulphide. The standard heat of combustion of hydrogen sulphide gas, H_2S, is

$$\Delta H^{\ominus}_{combustion,298}[H_2S(g)] = -561.4 \text{ kJ mol}^{-1}$$

Calculate the standard heat of formation of $H_2S(g)$, given that

$$\Delta H^{\ominus}_{combustion,298}[H_2(g)] = -285.9 \text{ kJ mol}^{-1}$$

$$\Delta H^{\ominus}_{combustion,298}[S(s)] = -296.6 \text{ kJ mol}^{-1}$$

Methanol, CH_3OH, cannot be prepared directly from its elements. Given that the standard heat of combustion of methanol is $\Delta H^{\ominus}_{combustion,298} = -726.3 \text{ kJ mol}^{-1}$, calculate the standard heat of formation of the compound.

$$C(graphite) + 2H_2(g) + \tfrac{1}{2}O_2(g) \longrightarrow CH_3OH(l)$$

The other standard heats of combustion that are needed have been given earlier in this topic.

Indirect methods for obtaining other enthalpy changes
Other enthalpy changes which cannot be determined experimentally, can be calculated using Hess's Law.

Experiment 7.2b
Evaluating an enthalpy change that cannot be measured directly
If anhydrous copper(II) sulphate powder is left in the atmosphere it slowly absorbs water vapour giving the hydrated solid.

$$CuSO_4(s) \xrightarrow{\text{water vapour}} CuSO_4.5H_2O(s)$$

anhydrous   hydrated

Heat is evolved in the reaction. Consider how this enthalpy change might be measured directly, using either water vapour or liquid water. Either way presents difficulties.

Could the value be obtained indirectly?

Two reactions for which the enthalpy change could readily be measured are (1) dissolving anhydrous copper(II) sulphate in water, and (2) dissolving hydrated copper(II) sulphate crystals in water.

The equations for these changes are:

1   $CuSO_4(s) + aq \longrightarrow CuSO_4(aq)$; enthalpy change ΔH_1
2   $CuSO_4, 5H_2O(s) + aq \longrightarrow CuSO_4(aq)$; enthalpy change ΔH_2

The reaction with the unknown enthalpy change is

$$CuSO_4(s) + 5H_2O(l) \longrightarrow CuSO_4, 5H_2O(s); \text{ enthalpy change } \Delta H_3.$$

You can measure ΔH_1 and ΔH_2 directly, and then calculate ΔH_3.

Procedure
1   Determine the enthalpy change when 0.025 mole anhydrous copper(II) sulphate is dissolved in 50 cm^3 water.
2   Determine the enthalpy change when 0.025 mole hydrated copper(II) sulphate is dissolved in 50 cm^3 water. Note that 50 g water less (0.025 × 5) mole water which will come from the hydrated salt, should be measured out. Assume that the specific heat capacity of the solution is equal to the specific heat capacity of water.
3   Calculate the enthalpy changes ΔH_1 and ΔH_2 for one mole of copper(II) sulphate, and from these values calculate ΔH_3.

What is the percentage error introduced by weighing the solids to ±0.02 g?

What is the percentage error involved in the values for the temperature changes, as a result of the accuracy with which the thermometer can be read?

In this experiment is there any point in weighing the solids to a greater accuracy than ±0.02 g?

Uses of standard heats of formation
Standard heats of formation can be used to calculate the enthalpy change which will take place in a reaction. Ethylene gas reacts with bromine to form 1,2-dibromoethane.

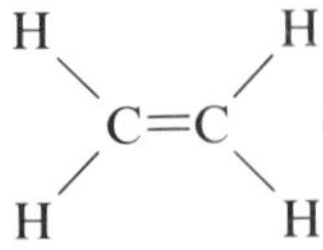

$(g) + Br_2(l) \longrightarrow CH_2BrCH_2Br(l)$

The standard heats of formation are:

$$\Delta H^{\ominus}_{f,298}[C_2H_4(g)] = +52.3 \text{ kJ mol}^{-1}$$

$$\Delta H^{\ominus}_{f,298}[CH_2BrCH_2Br(l)] = -80.7 \text{ kJ mol}^{-1}$$

$$\Delta H^{\ominus}_{f,298}[Br_2(l)] = 0 \text{ (by definition)}$$

ΔH for the reaction can be calculated as follows.

First write down the equation in which you are interested:

$$C_2H_4(g) + Br_2(l) \longrightarrow CH_2BrCH_2Br(l)$$

Then add the formation equations, from the same elements to both sides of the equation:

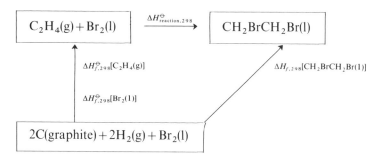

The total enthalpy change must be the same by whatever route the 1,2-dibromo-ethane is formed (whether it is formed 'direct' from its elements, or through the intermediate of ethylene gas).

Therefore

$$\Delta H^{\ominus}_{f,298}[CH_2BrCH_2Br(l)] = \Delta H^{\ominus}_{f,298}[C_2H_4(g)] + \Delta H^{\ominus}_{reaction,298}$$

that is,

$$-80.7 = +52.3 + \Delta H^{\ominus}_{reaction,298}$$

$$\Delta H^{\ominus}_{reaction,298} = -133.0 \text{ kJ}$$

Thus

$$C_2H_4(g) + Br_2(l) \longrightarrow CH_2BrCH_2Br(l); \qquad \Delta H^{\ominus}_{298} = -133.0 \text{ kJ}$$

Stability in chemistry

It is almost meaningless to describe a substance as stable or unstable unless it is made quite clear in what sense the term stable is used and what set of conditions is being referred to.

Meanings of 'stable'

Chemists usually mean one of two things when they call a substance stable. They may mean *energetic* stability or they may mean *kinetic* stability.

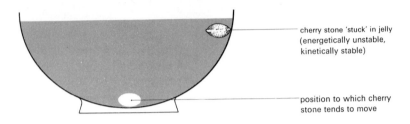

cherry stone 'stuck' in jelly (energetically unstable, kinetically stable)

position to which cherry stone tends to move

Figure 7.2a
A cherry stone in a bowl of jelly.

Figure 7.2a shows a cherry stone on the side of a bowl of jelly. If it were not for the fact that the jelly had set, the cherry stone would naturally move to the bottom of the bowl, its position of lowest energy. It may indeed be moving very slowly in that direction, but the jelly presents a 'barrier' which makes such movement imperceptibly slow. If, however, we find a way of turning the jelly into a liquid the cherry stone will then be relatively unimpeded and will move to its equilibrium position, at the bottom of the bowl. The less viscous the jelly becomes the more rapidly will the stone be allowed to move. One way of rendering the jelly less viscous is of course to warm it.

We may say that the stone embedded in the solid jelly is *energetically* unstable since it tends to descend, but *kinetically* stable, since its movement is actually prevented. The same language can be used of chemical substances as the following examples show.

    1   *Ozone*, $O_3(gas)$
For the change

$$O_3(g) \longrightarrow 1\tfrac{1}{2}O_2(g); \qquad \Delta H^{\ominus}_{298} = -142 \text{ kJ}$$

It can be seen that 142 kJ of heat is evolved per mole of O_3 decomposing to oxygen. Ozone is however a gas which can be kept at low pressures for long periods of time if the temperature is not raised much above room temperature and if no catalyst is present. At high temperatures, or if a catalyst is present, the decomposition of ozone to oxygen is speeded up. Oxygen is more stable than ozone since oxygen has the lowest energy. Ozone can therefore be said to be energetically unstable with respect to decomposition to oxygen but, at low temperatures and pressures and in the absence of catalysts, kinetically stable.

2 Fuel-air mixtures

Many fuels are energetically unstable but kinetically stable with respect to reaction with the oxygen in the air. For example, for the burning

$$C_8H_{18}(l) + 12\tfrac{1}{2}O_2(g) \longrightarrow 8CO_2(g) + 9H_2O(l); \quad \Delta H^{\ominus}_{298} = -5498 \text{ kJ}$$

octane
(petrol)

much energy (5498 kJ per mole of octane burnt) can be released in the combustion but, unless a spark is applied, petrol is safe to handle in air.

Many other substances are, under normal circumstances, kinetically stable, although energetically unstable in air, a fact which all fire insurance companies depend upon for their continuing profitability.

Stable with respect to what reaction?

Important as it is to distinguish clearly between kinetic and energetic stability, this is not in itself enough to make the term stable unambiguous: it is necessary to say with respect to what reaction a substance is stable. The following example will help to make this clear.

3 Hydrogen peroxide, H_2O_2

Hydrogen peroxide has a standard heat of formation of -210 kJ mol^{-1}. This does not, however, enable us simply to say 'hydrogen peroxide is energetically stable.' It *is* stable relative to decomposition to the elements hydrogen and oxygen:

$$H_2O_2(l) \longrightarrow H_2(g) + O_2(g); \quad \Delta H^{\ominus}_{298} = +210 \text{ kJ}$$

It is *not* energetically stable, however, relative to decomposition into water and oxygen:

$$H_2O_2(l) \longrightarrow H_2O(l) + \tfrac{1}{2}O_2(g); \quad \Delta H^{\ominus}_{298} = -96.2 \text{ kJ}$$

This also happens to be a case where the change takes place at room temperature rather slowly so that we may say that hydrogen peroxide, although energetically unstable with respect to decomposition into water and oxygen, has a certain degree of kinetic stability.

This subject of kinetic and energetic stability will be discussed further in later topics (see especially Topics 14–17).

$\Delta H^{\ominus}_{f,298}$ as a measure of energetic stability

One assumption we have been making concerning energetic stabilities is that strongly exothermic substances will be energetically stable with respect to most reactions and strongly endothermic substances energetically unstable. This is a useful rule of thumb for pure substances *at room temperatures.* At high temperatures the situation becomes more complicated and, as we shall indicate in a later section, there are in fact a number of exceptions to this rule even at room temperature.

Standard heats of formation and the Periodic Table

How do standard heats of formation change across, and down, the Periodic Table? Below are given the values, to the nearest unit, for the oxides across the period sodium to chlorine. In each case the oxide having the element in its highest oxidation state is quoted.

	Oxide						
	Na_2O	MgO	Al_2O_3	SiO_2	P_4O_{10}	SO_3	Cl_2O_7
$\Delta H^{\ominus}_{f,298}$ kJ mol^{-1}	−416	−602	−1670	−911	−2984	−395	+250 (approximately)
kJ per mole of oxygen atoms	−416	−602	−557	−456	−298	−132	+36 (approximately)

Table 7.2a

In order to compare the enthalpy changes it is necessary to refer each to a constant quantity of oxygen (since the number of moles of oxygen atoms in one mole of oxide varies across the Table). A convenient reference quantity is one mole of oxygen atoms. In the second row of figures a definite trend may be seen.

Plotted on a graph the trend is clear (see figure 7.2b).

Considering the relationship between heat of formation and stability, what do you consider the likely relative stabilities of magnesium oxide, aluminium oxide, and dichlorine heptoxide might be? Find out the uses of magnesium oxide and aluminium oxide, and see whether these are in accordance with your prediction about likely stability. Look in a reference book for information concerning the stability of the oxides of chlorine.

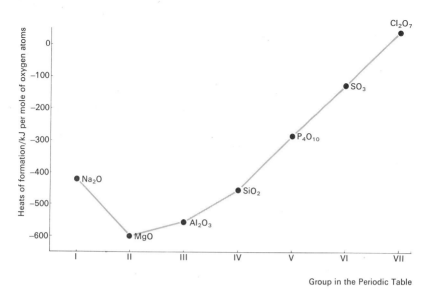

Figure 7.2b
Heats of formation of oxides in the period Na to Cl. Values in kilojoules per mole of oxygen atoms.

Standard heats of formation of oxides in selected groups of the Periodic Table are given below.

Group I		Group II		Group IV	
Li_2O	−596	BeO	−611	CO_2	−394
Na_2O	−416	MgO	−602	SiO_2	−911
K_2O	−362	CaO	−636	GeO_2	−551
Rb_2O	−330	SrO	−590	SnO_2	−581
Cs_2O	−318	BaO	−558	PbO_2	−277

Table 7.2b
$\Delta H^{\ominus}_{f,298}/\text{kJ mol}^{-1}$ (to nearest unit)

It can be seen from these values that there is a steady trend as one descends group I, but there is no similar trend in the other two groups mentioned. It is significant that carbon dioxide exists as CO_2 molecules, and is a gas at room temperature, whereas the rest of the group IV oxides are solids. SiO_2 and GeO_2 are giant molecules and SnO_2 and PbO_2 are largely ionic. The inconsistency in $\Delta H^{\ominus}_{f,298}$ may therefore be attributed to structural factors.

The standard heats of formation of the hydrides of group VII of the Periodic Table are given below.

Group VII

HF	-271
HCl	-92
HBr	-36
HI	$+26$

Table 7.2c

$\Delta H^{\ominus}_{f,298}/\text{kJ mol}^{-1}$ (to nearest unit)

How do you think the likely stability of the halogen hydrides will compare with one another?

Look back at the results you obtained for an experiment on this, and see if your predictions are verified. The experiment was in Topic 5, experiment 5.2b.

7.3   Bond energy terms

Can heats of combustion give information about bond contributions?

One way of attempting to answer this question would be to find a series of substances which are closely related to each other, and which differ from each other by some fixed unit of structure; then by studying such substances it might be possible to see whether that fixed unit of structure makes any consistent contribution to the overall energy situation.

An example would be the series of alcohols:

$CH_3CH_2CH_2OH$	Propan-1-ol
$CH_3CH_2CH_2CH_2OH$	Butan-1-ol
$CH_3CH_2CH_2CH_2CH_2OH$	Pentan-1-ol
$CH_3CH_2CH_2CH_2CH_2CH_2OH$	Hexan-1-ol
$CH_3(CH_2)_6OH$	Heptan-1-ol
$CH_3(CH_2)_7OH$	Octan-1-ol

Each compound differs from the rest by one —CH_2— unit. The situation will be made very clear if you examine structural models of these compounds.

In this series, the question can be posed 'does the —CH_2— group make any consistent contribution to the energy situation?' One method of finding out would be to burn the alcohols and measure the enthalpy change per mole of each.

Experiment 7.3

Can heats of combustion give information about bond contributions?

Find the heat of combustion of one member of the series of alcohols. Proceed as follows, using the combustion calorimeter with which you have been provided.

A suitable apparatus is illustrated in figure 7.3a and instructions for this apparatus are now given; there are several other forms of instrument that can be used instead.

to water pump

water

air

Figure 7.3a

First the apparatus must be calibrated. It would be possible to carry out the experiment by calculating the heat capacity of the calorimeter and the water in it, and then calculating the amount of heat it absorbs from the temperature rise when a certain mass of alcohol is burnt. Corrections have then to be made for cooling, so it is more accurate first to calibrate the apparatus by burning an alcohol, the standard heat of combustion of which is known accurately, and noting the temperature rise. From this experiment the heat capacity of the apparatus can be determined. This will include a cooling correction for *that particular temperature rise under those particular conditions*. The experiment can then be repeated with the alcohol of unknown standard heat of combustion, in such a way that the temperature rise is as nearly as possible the same as that in the first experiment. The cooling correction in each experiment is then about the same and the unknown standard heat of combustion can be calculated, using the heat capacity just determined. Propan-1-ol is a suitable alcohol for calibration:

$$C_3H_7OH(l) + 4\tfrac{1}{2}O_2(g) \longrightarrow 3CO_2(g) + 4H_2O(l);$$

$$\Delta H^{\ominus}_{combustion,298} = -2010 \text{ kJ mol}^{-1}$$

Place water in the calorimeter up to the level shown in figure 7.3a and mount it on its stand. Attach a water pump to the top of the copper spiral outlet, and adjust the pump so as to draw a moderately rapid stream of air through the spiral.

Almost fill the small spirit lamp with the alcohol to be used for calibration; light it, and adjust the length of the wick so as to give a flame about $1\tfrac{1}{2}$–2 cm high. Place the lamp under the calorimeter, and watch its behaviour. If the flame remains reasonably steady but slowly goes out this probably indicates that insufficient air is being supplied; adjust the water pump so that more air is drawn through. If the flame is very unsteady and goes out, it may indicate that too great a rush of air is being drawn in; adjust the water pump accordingly.

When you have adjusted the height of the wick and the flow of air so that the burner remains alight with a good flame, extinguish the burner and place the cap over the wick. Weigh it on a balance reading to 0.001 g.

Stir the water in the calorimeter, and read and record the temperature using a thermometer reading to 0.1 °C.

Remove the cap, light the spirit lamp, and without delay place it under the calorimeter. Stir the calorimeter periodically. When a rise in temperature of between 10 °C and 11 °C has been obtained blow out the lamp, remove it from the stand and replace the cap on the lamp. Stir the water thoroughly and read the temperature.

Note the maximum temperature.

Reweigh the spirit lamp as soon as possible after extinguishing it.

Refill the calorimeter with water and repeat the experiment with an alcohol, the standard heat of combustion of which is unknown. Try to ensure that the temperatures are as close to those in the first experiment as possible. The spirit lamp will have to be alight for a different time to produce the same temperature rise.

Treatment of results

1   Calculate the amount of heat evolved in the first experiment from the weight of propan-1-ol which was burnt and its standard heat of combustion.

2   Next, calculate the heat capacity of the calorimeter and its contents. This is the amount of heat required to raise its temperature by one degree C. It can be calculated from the amount of heat evolved in (1) since:

> Heat evolved by lamp = heat absorbed by calorimeter = heat capacity
> of calorimeter × rise in temperature of calori-
> meter.

The heat capacity should be expressed in $kJ \, °C^{-1}$.

3   Now calculate the heat evolved in the second experiment using the relationship

> Heat evolved by lamp = heat capacity × temperature rise

4   Finally calculate the amount of heat evolved per mole of alcohol from the weight of alcohol which was burnt, and the amount of heat found to be evolved by burning that weight of alcohol.

A consideration of sources of error

1   *Heat losses*

Do you think the method has taken heat losses from the calorimeter into account satisfactorily?

2   *Combustion products*

The term 'standard heat of combustion' implies complete combustion, in this instance to carbon dioxide and water. Can we be sure that this has occurred? What might have been formed instead, to some extent?

Do you think that the various sources of error will tend to make your value higher or lower than the true one?

Compare your results with those of others in the class, and see whether they provide any answer to the questions with which this experiment began.

In your work on the standard heats of combustion, you will have found that the difference in value between successive alcohols in a homologous series is about the same. It therefore appears that the

$$
-\overset{\displaystyle H}{\underset{\displaystyle H}{\overset{|}{\underset{|}{C}}}}-\quad \text{group}
$$

contributes that number of kJ in combustion energy; a reasonable assumption then is that each C—H bond and each C—C bond has a definite amount of energy associated with it, as have the C—O and O—H bonds present in the alcohols.

Taking the simplest hydrocarbon, CH_4, it seems reasonable to assume that the energy associated with the C—H bonds must be reflected in the total amount of energy required to break the molecule into its constituent atoms.

$$
H-\overset{\displaystyle H}{\underset{\displaystyle H}{\overset{|}{\underset{|}{C}}}}-H(g) \quad \longrightarrow \quad C(g)+4H(g); \qquad \Delta H^{\ominus}_{298} = +1662 \text{ kJ mol}^{-1}
$$

If the bonds are equal in strength, then the bond energy of one C—H bond should be $\frac{1662}{4} = 415 \text{ kJ mol}^{-1}$. (Obtaining the value of $\Delta H^{\ominus}_{298}$ for this reaction involves a straightforward calculation using an energy cycle.)

A value for the energy of a C—C bond can be obtained from the values of the enthalpy changes for two reactions involving the atomization of two alkanes containing such bonds:

$$
H-\overset{\displaystyle H}{\underset{\displaystyle H}{\overset{|}{\underset{|}{C}}}}-\overset{\displaystyle H}{\underset{\displaystyle H}{\overset{|}{\underset{|}{C}}}}-\overset{\displaystyle H}{\underset{\displaystyle H}{\overset{|}{\underset{|}{C}}}}-\overset{\displaystyle H}{\underset{\displaystyle H}{\overset{|}{\underset{|}{C}}}}-H \quad \longrightarrow \quad 4C(g)+10H(g); \qquad \Delta H^{\ominus}_{298} = +5165 \text{ kJ mol}^{-1}
$$

butane

$$
H-\overset{\displaystyle H}{\underset{\displaystyle H}{\overset{|}{\underset{|}{C}}}}-\overset{\displaystyle H}{\underset{\displaystyle H}{\overset{|}{\underset{|}{C}}}}-\overset{\displaystyle H}{\underset{\displaystyle H}{\overset{|}{\underset{|}{C}}}}-\overset{\displaystyle H}{\underset{\displaystyle H}{\overset{|}{\underset{|}{C}}}}-\overset{\displaystyle H}{\underset{\displaystyle H}{\overset{|}{\underset{|}{C}}}}-H \quad \longrightarrow \quad 5C(g)+12H(g); \qquad \Delta H^{\ominus}_{298} = +6337 \text{ kJ mol}^{-1}
$$

pentane

In order to atomize one molecule of butane it is necessary to break 3 C—C bonds and 10 C—H bonds; for pentane 4 C—C bonds and 12 C—H bonds must be broken.

Denoting a C—H bond energy by $E(C—H)$ and a C—C bond energy by $E(C—C)$ we have:

for butane, $3E(C—C) + 10E(C—H) = +5165$
for pentane, $4E(C—C) + 12E(C—H) = +6337$

Solving these simultaneous equations, we obtain:

$E(C—H) = +412 \text{ kJ mol}^{-1}$
$E(C—C) = +347 \text{ kJ mol}^{-1}$

The bond energy $E(C—Cl)$ has been determined using several compounds. Below are shown the compounds and the values obtained for them:

Compound		$E(C—Cl)/\text{kJ mol}^{-1}$
Cl—C—Cl (with Cl above and Cl below)	tetrachloromethane	+327
H—C—Cl (with H above and H below)	chloromethane	+335
H—C—C—Cl (with H H above and H H below)	chloroethane	+342

From these examples it can be seen that the bond energy value depends upon the compound from which it was determined; that is, the *environment* of the bond affects the value. The X—Y bond energy will vary somewhat, depending upon the nature of the other atoms or groups of atoms which are attached to X and Y.

But if an *average* bond energy is taken this can often be very useful. Tables have therefore been prepared giving average bond energies. Average bond energies per mole of bonds are also known as *bond energy terms*, and are denoted by the symbol $\bar{E}$. Some examples are given on the next page.

(Average bond energies)

Bond	$\bar{E}$		Bond	$\bar{E}$
C—H	413		N—H	391
C—C	346		P—H	322
C=C	610		As—H	247
C—Cl	339		O—H	463
C—Br	284		S—H	344
C—I	218			

Table 7.3b
Bond energy terms, $\bar{E}/\text{kJ mol}^{-1}$

A fuller table is given in the *Book of Data*.

The sum of the bond energy terms for a compound is approximately equal to the enthalpy change involved in the atomization of that compound *from the gaseous state*.

Having read this explanation of bond energy terms, carry out the following exercises.

1   From the bond energy terms given above obtain an approximate value for the energy needed to atomize one mole of the alcohol propan-1-ol, $CH_3CH_2CH_2OH$ (in the gaseous state).

$$\boxed{3C(g) + 8H(g) + O(g)}$$

$$\uparrow \Delta H$$

$$\boxed{CH_3CH_2CH_2OH(g)}$$

2   Make out a table showing the bond energy terms for the hydrides across the Periodic Table, C—H, N—H, O—H, and F—H, and then insert the vertical series F—H, Cl—H, Br—H, and I—H. What are the trends in the ease of breaking the bonds, and what information can you deduce from them?

Background reading
Accurate experimental thermochemistry
Now that you have carried out some thermochemical determinations yourself, you may be interested to read how such measurements can be carried out very accurately.

Since energy changes yield important information concerning bonding and structure a sustained effort has been devoted to improving the accuracy of their measurement; these efforts began when Berthelot first studied energy changes in combustion in 1869, and they are continuing at the present time.

Some idea of the accuracy which is obtainable with present-day apparatus and procedure is given by the following example.

Measurements with a flame calorimeter similar in principle to the heat of combustion apparatus, using a hydrogen flame burning in oxygen, give for the reaction

$$H_2(g) + \tfrac{1}{2}O_2(g) \longrightarrow H_2O(l)$$

a value of $\Delta H^{\ominus}_{f,298} = 286.022 \pm 0.040 \text{ kJ mol}^{-1}$. Calculate the percentage error in this value. Then compare it with the percentage error in the relatively simple calorimetry experiments which you have conducted.

Thermochemical determinations of this type are made up of two main groups of measurements: those relating to the calorimetry itself (the measurement of the heat changes) and those relating to the chemistry of the reactions taking place. Each introduces possible errors.

The calorimetry

In order to provide constant conditions under which heat losses may take place, the calorimeter is surrounded by a container filled with liquid maintained at a constant temperature by means of a thermostat.

Corrections for heat losses can be eliminated if the thermochemical determination is made in conjunction with electrical calibration. The chemical reaction is allowed to take place, and an accurate record of temperature change and time is kept before, during, and after the reaction. The temperature is plotted against the time. Then the calorimeter is emptied and refilled but this time the chemical reactants are replaced by an electric heating coil. The current is adjusted so as to produce a heating curve identical with that obtained for the chemical reaction.

In figure 7.3b, B–C is the reaction time, and C–D is the cooling time. The known current and time enable the electrical energy supplied to be calculated; since the conditions are the same as for the chemical experiment the electrical value is equal to the energy change in the chemical reaction. The total electrical energy supplied includes both the heat loss and the heat absorption due to temperature rise; therefore when the chemical energy change is obtained from it, as described, this automatically includes the heat loss correction.

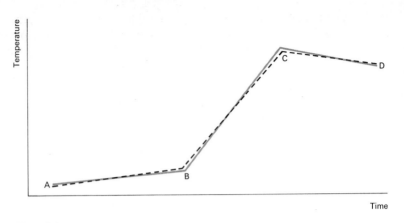

Figure 7.3b
Heating and cooling curves: for calibration and actual determination. (The two curves are almost coincident, and have been separated here for visibility.)

Figure 7.3c shows a cross-section of a typical bomb calorimeter used for accurate thermochemical determinations. It should be noted that the reaction to be studied is carried out in a sealed container known as a 'bomb'. It therefore takes place at constant volume, and not at constant pressure. The energy change measured is thus the change in internal energy, and not ΔH. ΔH may be calculated from the results obtained.

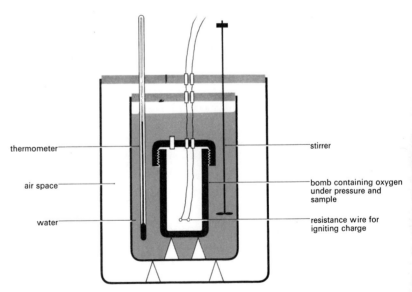

Figure 7.3c
A bomb calorimeter.

The chemistry of the reaction

There are two points of particular importance to be considered when finding the heat of a reaction. One of these is the purity of the reactants, and the other is the extent to which the reaction has taken place. It is clear that if the test material, or one of the other reactants, is not pure, a source of error is introduced; in the instances of combustion reactions this is particularly true if one of the impurities is incombustible. An example is the presence of moisture in an alcohol whose heat of combustion is to be determined.

It is also clear that if the reaction does not proceed to completion an error will be introduced. An example is the incomplete combustion of an alcohol, with the presence of some carbon monoxide in the final gases.

Accurate chemical analysis is therefore an essential factor in accurate thermo-chemistry. There must be an analysis of the starting-materials, for purity; and there must be an analysis of the products to determine the extent of the reaction.

The science of calorimetry, the design of calorimeters and techniques of using them, has advanced to the stage that further improvements in thermochemical measurements are limited by the extent to which pure reagents can be obtained, and by the accuracy with which the extent of the reaction can be determined. Further advances must be sought first in these areas.

A recent extension of calorimetry

Until very recently all thermochemical determinations in oxidation reactions were made using oxygen or an oxygen compound as the oxidizing agent. This was convenient since oxygen in the form of gas or in solid oxidizing agents is readily available in a high degree of purity.

The most powerful oxidizing agent, however, is fluorine, and use of this element would open up new lines of investigation. The chief disadvantages have been the extremely corrosive nature of the substance, leading to extensive attack of the calorimeters, and the difficulty of preparing really pure fluorine. Each of these problems is now being to some extent overcome; calorimeters made of nickel and of the plastic polytetrafluoroethylene are proving resistant to fluorine, and methods for preparing small quantities of 99.9 per cent pure fluorine have been developed. Accuracies in thermochemical determination of ± 0.3 per cent were being achieved in the early 1960s; from your calculation earlier in this section you will see that this is not yet quite so accurate as the older oxygen combustion method.

Many problems still remain to be overcome; but a new field of thermochemistry is opening up, not only in oxidation reactions but also in the energetics of fluorine compounds.

7.4   The Born–Haber cycle: lattice energies

Just as it is often useful to know the heat of formation of a molecular compound from atoms in the gaseous state, so also it is often useful to know the heat of formation of an ionic crystal from ions in the gaseous state. This latter quantity is known as the 'lattice energy' of the compound.

The *lattice energy* of an ionic crystal is the standard heat of formation of the crystal lattice from its constituent ions in the gas phase.

$$Na^+(g) + Cl^-(g) \longrightarrow Na^+Cl^-(s); \qquad \Delta H_{298}^{\ominus} = \text{lattice energy}$$

The direct determination of lattice energies is not possible, but values can be obtained indirectly by means of an energy cycle, known as a Born–Haber cycle.

The cycle is analogous with that for obtaining the heat of formation of molecules from atoms, and may be seen as a triangular two-route process.

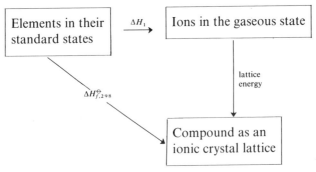

In the case of sodium chloride this is:

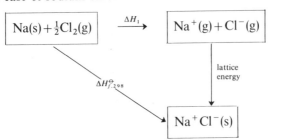

The standard heat of formation of sodium chloride can be measured directly, by the reaction of sodium with chlorine in a calorimeter. If the energy required to convert sodium metal into gaseous ions, and chlorine molecules into gaseous ions can be obtained, ΔH_1 will be known, and then it is possible to obtain a value for the lattice energy.

ΔH_1 has to be obtained in stages. Taking the sodium first,

$$Na(s) \xrightarrow[\substack{\text{standard heat} \\ \text{of atomization}}]{} Na(g) \xrightarrow[\substack{\text{ionization} \\ \text{energy}}]{-e^-} Na^+(g)$$

The two energy values required are the *standard heat of atomization of* sodium, relating to the conversion of solid sodium into gaseous sodium consisting of separate atoms:

$$Na(s) \longrightarrow Na(g); \qquad \Delta H^{\ominus}_{at,298} = +108.3 \text{ kJ mol}^{-1}$$

and the *ionization energy*, relating to the conversion of gaseous atoms into gaseous ions:

$$Na(g) \longrightarrow Na^+(g) + e^-; \qquad \Delta H^{\ominus}_{298} = +500 \text{ kJ mol}^{-1}$$

Taking the chlorine we have

$$\tfrac{1}{2}Cl_2(g) \xrightarrow[\substack{\text{standard heat} \\ \text{of atomization}}]{} Cl(g) \xrightarrow[\substack{\text{electron} \\ \text{affinity}}]{+e^-} Cl^-(g)$$

The two energy values required are the *standard heat of atomization* of chlorine, relating it to the conversion of gaseous chlorine molecules into gaseous chlorine atoms,

$$\tfrac{1}{2}Cl_2(g) \longrightarrow Cl(g); \qquad \Delta H^{\ominus}_{298} = +121.1 \text{ kJ mol}^{-1},$$

and the *electron affinity* which is the energy change occurring when a chlorine atom accepts an electron and becomes a chloride ion,

$$Cl(g) + e^- \longrightarrow Cl^-(g); \qquad \Delta H^{\ominus}_{298} = -364 \text{ kJ mol}^{-1}$$

Each of these can be determined experimentally, although the determination of electron affinity is difficult.

The only other value to place in the cycle is $\Delta H^{\ominus}_{f,298}[Na^+Cl^-(s)]$:

$$Na(s) + \tfrac{1}{2}Cl_2(s) \longrightarrow Na^+Cl^-(s); \qquad \Delta H^{\ominus}_{f,298} = -411 \text{ kJ mol}^{-1}$$

and then the lattice energy can be determined.

From figure 7.4a it can be seen that the lattice energy is

$$-[(121.1 + 500 + 108.3 + 411) - 364] \text{ kJ mol}^{-1} = 776.4 \text{ kJ mol}^{-1}$$

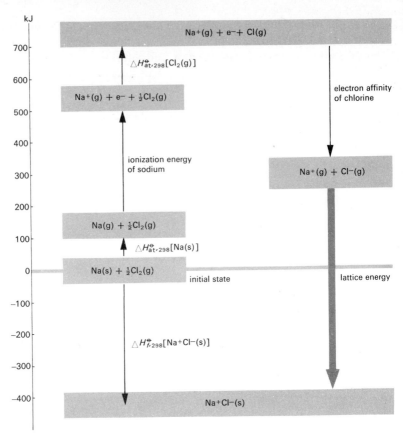

Figure 7.4a
Energy level diagram for the formation of sodium chloride.

Theoretical values for lattice energies

Since a lattice energy is the energy change involved in bringing well-separated electrostatic charges together to form a lattice it should be possible to make an estimate of the magnitude of this energy change, using the principles of electrostatics. The calculations are done on the basis of ions being charged spheres in contact. It is assumed that the ions are spherical, separate entities; each with its charge distributed uniformly around it. Some of the values that have been calculated are given in table 7.4 in the column headed 'Theoretical value'.

Compound	Theoretical value	Experimental value (Born-Haber Cycle)
NaCl	− 766.1	− 776.4
NaBr	− 730.5	− 735.9
NaI	− 685.7	− 688.3
KCl	− 692.0	− 697.8
KBr	− 666.5	− 672.3
KI	− 630.9	− 631.8
AgCl	− 768.6	− 916.3
AgBr	− 758.5	− 907.9
AgI	− 735.9	− 865.4
ZnS	− 3427	− 3615

Table 7.4
Lattice energies/kJ mol^{-1}

Examine the values for the alkali metal halides, and compare the theoretical values with the Born–Haber experimental values. For one or two, calculate the approximate percentage discrepancy between theoretical and experimental value.

The model on which the theoretical values are based is a very simple one; it regards an ionic crystal as made up of discrete spherical ions, each with its electrostatic charge distributed evenly around it. The excellent agreement between the theoretical and experimental values is strong evidence that the simple model of an ionic crystal is a good one, in the instances of the alkali metal halides.

Now examine the corresponding values for the silver halides. Calculate the approximate percentage difference between the theoretical and the experimental values. Do you think that the ionic model accurately represents the bonding situation in the silver halides? If not, some other model is required.

Bonding will be discussed again later, in Topic 8, 'Structure and bonding'.

Lattice energy and stoichiometry
Can energy considerations give any indication of an expected formula for a compound?

Would energy calculations be able to show, for instance, which of the three formulae, MgCl, $MgCl_2$, and $MgCl_3$ would be the most likely for magnesium chloride?

It would be expected that the compound which has the most negative standard heat of formation would be the most stable. If it is assumed that Mg^+Cl^- would have a sodium chloride lattice structure, and $Mg^{3+}(Cl^-)_3$ a structure similar to $AlCl_3$, then a reasonable estimate of the lattice energies for the hypothetical crystals MgCl and $MgCl_3$ may be made. Born–Haber cycles can then be constructed, and values obtained for the standard heats of formation of these hypothetical crystals.

The quantities necessary for drawing the cycles are:

		$\Delta H^{\ominus}_{298}/kJ$
AMg =	heat of atomization of magnesium $Mg(s) \rightarrow Mg(g)$	+146
IE_1 =	1st ionization energy of magnesium, $Mg(g) \rightarrow Mg^+(g)+e^-$	+736
IE_2 =	2nd ionization energy of magnesium, $Mg^+(g) \rightarrow Mg^{2+}(g)+e^-$	+1448
IE_3 =	3rd ionization energy of magnesium, $Mg^{2+}(g) \rightarrow Mg^{3+}(g)+e^-$	+7740
ACl =	heat of atomization of chlorine, $\frac{1}{2}Cl_2(g) \rightarrow Cl(g)$	+121
2ACl =	$2 \times$ heat of atomization of chlorine, $Cl_2(g) \rightarrow 2Cl(g)$	+242
3ACl =	$3 \times$ heat of atomization of chlorine, $1\frac{1}{2}Cl_2(g) \rightarrow 3Cl(g)$	+363
EA =	electron affinity of chlorine, $Cl(g)+e^- \rightarrow Cl^-(g)$	−364
2EA =	$2 \times$ electron affinity of chlorine, $2Cl(g)+2e^- \rightarrow 2Cl^-(g)$	−728
3EA =	$3 \times$ electron affinity of chlorine, $3Cl(g)+3e^- \rightarrow 3Cl^-(g)$	−1092
LE_1 =	estimated lattice energy for MgCl, approximately	−753
LE_2 =	lattice energy for $MgCl_2$	−2502
LE_3 =	estimated lattice energy for $MgCl_3$, approximately	−5440

The stages involved in the first two processes are:

MgCl

$$Mg(s)+\tfrac{1}{2}Cl_2(g) \xrightarrow{AMg} Mg(g)+\tfrac{1}{2}Cl_2(g) \xrightarrow{IE_1} Mg^+(g)+\underset{\downarrow ACl}{e^-}+\tfrac{1}{2}Cl_2(g)$$

$$Mg^+Cl^-(s) \xleftarrow{LE_1} Mg^+(g)+Cl^-(g) \xleftarrow{EA} Mg^+(g)+e^-+Cl(g)$$

MgCl₂

$$Mg(s)+Cl_2(g) \xrightarrow{AMg} Mg(g)+Cl_2(g) \xrightarrow{IE_1} Mg^+(g)+e^-+Cl_2(g)$$

$$Mg^{2+}(g)+2Cl^-(g) \xleftarrow{2EA} Mg^{2+}(g)+2e^-+2Cl(g) \xleftarrow{2ACl} Mg^{2+}(g)+2e^-+\underset{\uparrow IE_2}{Cl_2(g)}$$

$$\downarrow LE_2$$

$$Mg^{2+}(Cl^-)_2(s)$$

$MgCl_3$

Construct the sequence for $MgCl_3$ yourself.

The cycles are represented to scale in figure 7.4b.

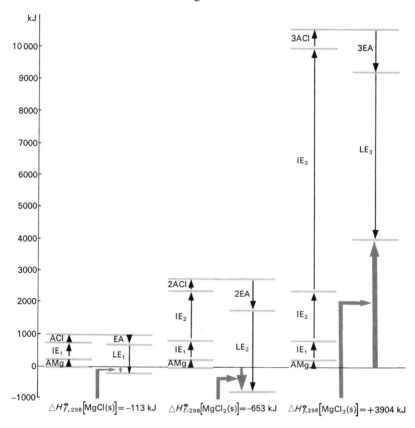

Figure 7.4b
Born–Haber cycles for chlorides of magnesium.

If you add up the component quantities for the various stages, you will find that the standard heats of formation for the compounds, as obtained from the cycles, are:

$$\Delta H^{\ominus}_{f,298}[MgCl(s)] = -113 \text{ kJ mol}^{-1}$$

$$\Delta H^{\ominus}_{f,298}[MgCl_2(s)] = -653 \text{ kJ mol}^{-1}$$

$$\Delta H^{\ominus}_{f,298}[MgCl_3(s)] = +3904 \text{ kJ mol}^{-1}$$

From these it can be seen that the formation of MgCl is just exothermic, and the compound is energetically stable with respect to its elements. The formation of $MgCl_2$ is even more exothermic, and so the compound would also be energetically stable with respect to its elements. The formation of $MgCl_3$ however is highly endothermic, and so it would be extremely unstable with respect to its elements.

MgCl may be stable with respect to its elements but inspection of the data shows that it is not stable with respect to $MgCl_2$.

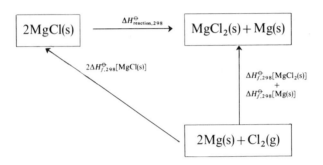

The enthalpy change for the reaction:

$$2Mg(s) + Cl_2(g) \longrightarrow MgCl_2(s) + Mg(s)$$

must be the same whether the reaction occurs 'directly' or through the intermediate of 2MgCl(s); therefore

$$\Delta H^{\ominus}_{f,298}[MgCl_2(s)] + \Delta H^{\ominus}_{f,298}[Mg(s)] = \Delta H^{\ominus}_{reaction,298} + 2\Delta H^{\ominus}_{f,298}[MgCl(s)]$$

$$-653 + 0 = \Delta H^{\ominus}_{reaction\ 298} + [2 \times (-113)]$$

$$\Delta H^{\ominus}_{reaction,298} = -427\ kJ$$

that is, for the reaction

$$2MgCl(s) \longrightarrow MgCl_2(s) + Mg(s); \Delta H_{298} = -427\ kJ$$

So MgCl is energetically unstable with respect to $MgCl_2$ and Mg, and thus the stable compound under standard conditions is $MgCl_2$.

If the cycles are examined in the scale diagram, it will be seen that the largest single contributions in each cycle are made by the ionization energies and the lattice energy, and that these two are always opposite in sign. Broadly speaking, therefore, the magnitude of the standard heat of formation depends upon the result of competition between ionization energies and lattice energy. If it requires more energy to ionize the metal than is returned as lattice energy, then the compound will not be formed.

Calculations of this type can be done for many other classes of compound. For instance it may be shown that the formation of NaO would not be energetically favoured, while the formation of Na_2O would be energetically favoured. From your study of the alkali metal oxides you will recall that Na_2O exists while NaO does not.

Thus energetic considerations help us to understand the stoichiometry of compounds.

Problems

* indicates that the *Book of Data* may be used.

*1   The purpose of this set of questions is to determine the standard enthalpy change at 298 K when 7 g of calcium oxide react with water:

$$CaO(s) + H_2O(l) \longrightarrow Ca(OH)_2(s)$$

   i   What is $\Delta H^{\ominus}_{f,298}$ of the product in kJ mol^{-1}?
   ii   What is the sum of $\Delta H^{\ominus}_{f,298}$ (kJ mol^{-1}) of the reactants?
   iii   Which of the two, reactants or product, has the higher energy content?
   iv   By how much does the enthalpy of the product differ from that of the reactants?
   v   What is $\Delta H^{\ominus}_{298}$ (kJ mol^{-1}) for the reaction given above?
   vi   What is the weight of 1 mole of calcium oxide?
   vii   What fraction of a mole is 7 g of calcium oxide?
   viii   What is the heat transfer when 7 g of calcium oxide react with water (state whether the reaction is exothermic or endothermic)?

*2   The purpose of this set of questions is to determine the standard enthalpy change at 298 K when 1.2 cubic decimetres of ammonia (at 1 atmosphere pressure) react with hydrogen bromide:

$$NH_3(g) + HBr(g) \longrightarrow NH_4Br(s)$$

   i   What is $\Delta H^{\ominus}_{f,298}$ (kJ mol^{-1}) of the product?
   ii   What is the sum of $\Delta H^{\ominus}_{f,298}$ (kJ mol^{-1}) of the reactants?
   iii   Which of the two, reactants or product, has the higher energy content?
   iv   By how much does the enthalpy of the product differ from that of the reactants?
   v   What is $\Delta H^{\ominus}_{298}$ (kJ) for the reaction given above?

vi   What is the approximate volume (2 figures) of 1 mole of ammonia at 25 °C and 1 atmosphere?

vii   What fraction of a mole is the 1.2 cubic decimetres of ammonia?

viii   What is the standard enthalpy change at 298 K when the 1.2 cubic decimetres of ammonia react with hydrogen bromide (state whether the reaction is exothermic or endothermic)?

*3   Calculate $\Delta H_{298}^{\ominus}$ for the following reactions.

i $CH_3OH(l) + O_2(g) \longrightarrow CO_2(g) + 2H_2O(g)$

ii $SO_2(g) + 2H_2S(g) \longrightarrow 3S(s) + 2H_2O(l)$

iii $Fe(s) + 2Ag^+(aq) \longrightarrow Fe^{2+}(aq) + 2Ag(s)$

iv $Ba^{2+}(aq) + SO_4^{2-}(aq) \longrightarrow BaSO_4(s)$

v $N_2O(g) + Cu(s) \longrightarrow CuO(s) + N_2(g)$

vi $NH_4Cl(s) \longrightarrow NH_3(g) + HCl(g)$

vii $NaCl(s) \longrightarrow Na^+(g) + Cl^-(g)$

viii $NH_4Cl(s) \longrightarrow NH_4^+(aq) + Cl^-(aq)$

ix $Mg(s) + \frac{1}{2}O_2(g) \longrightarrow MgO(s)$

x $Mg^{2+}(g) + O^{2-}(g) \longrightarrow MgO(s)$

*4   Calculate the standard enthalpy change at 298 K when the following changes take place. State whether the heat is given to or taken from the surroundings.

i   8.0 g of iron are added to an excess of a solution of a copper(II) salt in water (Fe^{2+}(aq) ions are formed).

ii   24 cubic decimetres (25 °C, 1 atmosphere) of an equimolar mixture of hydrogen and carbon monoxide are burned in oxygen. Assume that the water produced is liquid.

iii   1 mole of hydrogen molecules (H_2) is changed into 2 moles of hydrogen atoms (H).

iv   1 mole of gaseous sodium (Na) is converted to 1 mole of gaseous sodium ions, Na^+(g).

*5   Calculate $\Delta H_{298}^{\ominus}$ for the following reactions.

i $NH_3(g) \longrightarrow N(g) + 3H(g)$

ii $PH_3(g) \longrightarrow P(g) + 3H(g)$

iii $AsH_3(g) \longrightarrow As(g) + 3H(g)$

iv $SbH_3(g) \longrightarrow Sb(g) + 3H(g)$

What generalization do your answers indicate about the energies of the bonds in the hydrides of group V elements?

*6 Calculate $\Delta H_{298}^{\ominus}$ for the following reactions.

   i $HF(g) \longrightarrow H(g) + F(g)$

  ii $HCl(g) \longrightarrow H(g) + Cl(g)$

 iii $HBr(g) \longrightarrow H(g) + Br(g)$

  iv $HI(g) \longrightarrow H(g) + I(g)$

What generalization do your answers indicate about the energies of the bonds in the hydrides of group VII elements?

*7 Calculate $\Delta H_{298}^{\ominus}$ for each of the following reactions.

   i $2Al(s) + 3ZnO(s) \longrightarrow Al_2O_3(s) + 3Zn(s)$

  ii $2Al(s) + 3CaO(s) \longrightarrow Al_2O_3(s) + 3Ca(s)$

 iii $Zn(s) + CaO(s) \longrightarrow ZnO(s) + Ca(s)$

  iv $3Zn(s) + Al_2O_3(s) \longrightarrow 2Al(s) + 3ZnO(s)$

   v $3Ca(s) + Al_2O_3(s) \longrightarrow 2Al(s) + 3CaO(s)$

  vi $Ca(s) + ZnO(s) \longrightarrow Zn(s) + CaO(s)$

8 Suppose you were told the enthalpy changes of the following two reactions:

$$2Fe(s) + 1\tfrac{1}{2}O_2(g) \longrightarrow Fe_2O_3(s)$$

$$Ca(s) + \tfrac{1}{2}O_2(g) \longrightarrow CaO(s)$$

What further information (if any) would you require in order to calculate the enthalpy changes of each of the following reactions?

   i $3Ca(s) + Fe_2O_3(s) \longrightarrow 3CaO(s) + 2Fe(s)$

  ii $Ca(s) + CuO(s) \longrightarrow CaO(s) + Cu(s)$

 iii $2Fe(s) + 3CuO(s) \longrightarrow Fe_2O_3(s) + 3Cu(s)$

9 Suppose you were given the following information:

Enthalpy change of reaction $2H_2(g) + O_2(g) \longrightarrow 2H_2O(l)$

Enthalpy change of reaction $Cu(s) + \tfrac{1}{2}O_2(g) \longrightarrow CuO(s)$.

What further information (if any) would you require to calculate the enthalpy change of reaction:

$$CuO(s) + H_2(g) \longrightarrow Cu(s) + H_2O(g)?$$

*10 The heat of formation of sodium chloride over the temperature range 98–808 °C is about -414 kJ mol^{-1}, whereas the heat of formation over the temperature range 808–892 °C is about -385 kJ mol^{-1}. Explain why the heat of formation of sodium chloride changes abruptly at 808 °C by about 29 kJ mol^{-1}.

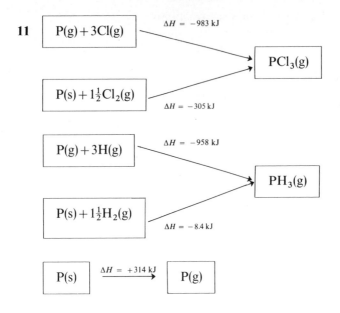

11

$P(g) + 3Cl(g)$ $\Delta H = -983$ kJ

$P(s) + 1\frac{1}{2}Cl_2(g)$ $\Delta H = -305$ kJ

$PCl_3(g)$

$P(g) + 3H(g)$ $\Delta H = -958$ kJ

$P(s) + 1\frac{1}{2}H_2(g)$ $\Delta H = -8.4$ kJ

$PH_3(g)$

$P(s)$ $\xrightarrow{\Delta H = +314 \text{ kJ}}$ $P(g)$

Use the information given above to calculate the (mean) bond energies of the following bonds:

  i   P—Cl in PCl_3
 ii   P—H in PH_3
iii   Cl—Cl in Cl_2
 iv   H—H in H_2

*12   Calculate the lattice energy for calcium oxide.

13   Ethanol and dimethyl ether have the same molecular formula, C_2H_6O. The standard heat of combustion at 298 K of ethanol (gas) is -1402 kJ mol^{-1} and that of dimethyl ether (gas) is -1456 kJ mol^{-1}.
The purpose of the following set of questions is to arrive at an explanation of the difference between the two heats of combustion.

  i   Write equations (showing $\Delta H^{\ominus}_{298}$) for the burning of (a) ethanol(g), and (b) dimethyl ether(g).
 ii   From (i) deduce $\Delta H^{\ominus}_{298}$ in kJ mol^{-1} for the hypothetical change:

   dimethyl ether(g) $\longrightarrow$ ethanol(g)

iii   Suppose ΔH_1 is the heat transfer for the change:

   C_2H_6O(ethanol gas) $\longrightarrow$ $2C(g) + 6H(g) + O(g)$

and ΔH_{11} is the heat transfer for the change:

$$C_2H_6O(\text{dimethyl ether gas}) \longrightarrow 2C(g) + 6H(g) + O(g)$$

Calculate $\Delta H_1 - \Delta H_{11}$.

iv Which of the following is the best explanation of the difference between ΔH_1 and ΔH_{11} in question (*iii*)?

A   Dimethyl ether is more volatile than ethanol.

B   The products of the two changes are not the same.

C   The ratio of weights of carbon, hydrogen, and oxygen in the two reactants is not the same.

D   The bond energies between carbon, hydrogen, and oxygen in the two reactants are not the same.

E   1 mole of ethanol weighs more than 1 mole of dimethyl ether.

14   Which of the heats of the reactions represented by the following equations would be the bond energy H—Cl?

A $HCl(g) \longrightarrow H(g) + Cl(g)$

B $2HCl(g) \longrightarrow H_2(g) + Cl_2(g)$

C $HCl(g) \longrightarrow \frac{1}{2}H_2(g) + \frac{1}{2}Cl_2(g)$

D $HCl(g) \longrightarrow H^+(aq) + Cl^-(aq)$

E $HCl(g) \longrightarrow H^+(g) + Cl^-(g)$

Topic 8
Structure and bonding

The chemical and physical properties of materials are strongly influenced by their structure at an atomic level. Therefore to understand the properties of materials it is necessary to understand their structures. This applies as much to naturally occurring substances such as rocks and minerals, and the constituents of living organisms such as cells, muscles and bone, as it does to man-made substances such as alloys for high-speed turbine blades and polymers for drip-dry textile fibres. Indeed, before a new synthetic material can be designed it is necessary to have some knowledge of how molecular structure affects properties; then by synthesizing an appropriate structure it is possible to produce a substance with predetermined properties—a 'tailor-made' material.

In studying the structures of materials one is at once led to consider also what holds the structures together, that is the bonding and the interatomic or inter-ionic forces. Such bonding and forces are a principal factor in controlling the chemical and physical properties of materials.

8.1 Some physical methods for determining structure

A very wide range of methods is available for obtaining information about the structures of substances; almost any physical phenomenon which is affected by a material can be made to yield evidence of the structure of that material. The principal categories of phenomena are those in which:

1 Electromagnetic radiation is emitted or absorbed by matter (giving rise to emission or absorption spectra);

2 Electromagnetic radiation or a stream of particles such as electrons interacts with matter to give diffraction patterns (X-ray diffraction, electron diffraction);

3 Matter produces effects upon an electric or a magnetic field (measurement of dipole moments, nuclear magnetic resonance).

Each of these phenomena gives different information about a substance, and when the evidence from several of them is added together it is often possible to obtain a detailed knowledge of its structure.

In this course one of these physical methods, X-ray diffraction, will be dealt with in some detail, and two others, electron diffraction and infra-red absorption spectrometry, will be mentioned briefly.

X-ray diffraction

When a beam of X-rays falls on a collection of particles of atomic size the rays
are affected in such a way that they travel outwards in all directions from every
particle, in a manner similar to the ripples produced when a handful of pebbles
is scattered onto the surface of a quiet pond. In the pond, some of the ripple
crests will meet and reinforce each other to give a larger ripple, while some
crests will meet troughs and neutralize each other; in the same way, some of the
X-ray crests will coincide and reinforce, while some will be neutralized by
troughs (see figure 8.1a).

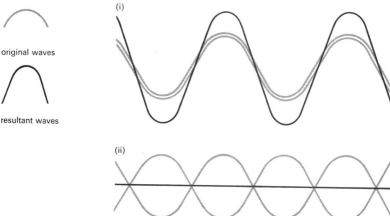

original waves

resultant waves

(i)

(ii)

Figure 8.1a
Interference between waves. (*i*) Waves in phase; reinforcement (stronger signal). (*ii*) Waves out of
phase; destruction (no signal).

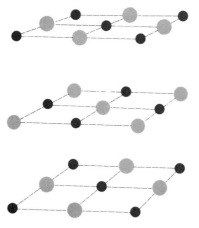

Figure 8.1b
Crystal planes.

In the instance of a beam of X-rays falling onto a crystal composed of regular layers of atoms or ions it is possible to calculate the conditions under which reinforcement will occur. Figure 8.1b shows some of the crystal planes in a crystal such as sodium chloride, and figure 8.1c represents a beam of X-rays being directed onto the crystal.

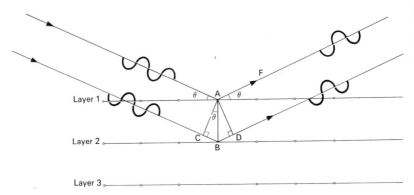

Figure 8.1c
Bragg diffraction, or 'reflection'.

When X-rays are diffracted by the layers, they behave as though they are being reflected. Two waves in phase are shown approaching the layers, and it can be seen that after 'reflection' the ray emerging along BD will have travelled farther than the ray emerging along AF. If the waves are again to be in phase, this path difference must equal a whole number of wavelengths, $n\lambda$. The path difference is $CB + BD$.

Therefore, for reinforcement $CB + BD = n\lambda$

Now, $CB = AB \sin \theta$ and $BD = AB \sin \theta$

$$CB + BD = AB \sin \theta + AB \sin \theta = n\lambda$$

$$2AB \sin \theta = n\lambda$$

$$2d \sin \theta = n\lambda$$

using d to replace AB and represent the distance apart of the crystal planes.

The relationship $2d \sin \theta = n\lambda$ is known as the Bragg diffraction equation.

From the equation it can be seen that for a given wavelength of X-rays, λ, and a given crystal face, with a value of d, maximum reinforcement will occur only at certain values of θ. If these positions are found experimentally and θ is measured, for X-radiation of known wavelength λ, then the value of the crystal plane separation, d, can be calculated.

Figure 8.1d shows the results of an experiment conducted by Sir Lawrence Bragg using a crystal of sodium chloride. The peaks indicate a marked increase in signal strength in the X-ray detector, and they therefore represent reinforcement of the waves; the figures on the horizontal axis of the graph represent 2θ in degrees.

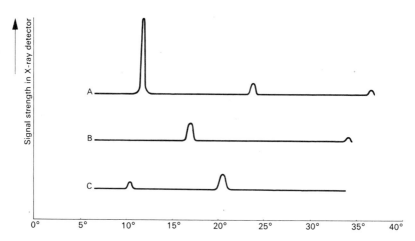

Figure 8.1d
A simplified Bragg spectrometer trace for sodium chloride.

With sodium chloride, three different traces are obtainable; each represents a different set of crystal planes. Obtain a ball-and-spoke model of a sodium chloride crystal and examine it for planes; it should be possible to find three different sets.

Why are there several peaks for each set of planes? How is this represented in the Bragg equation?

The wavelength of the X-rays, λ, which produced the peaks in figure 8.1d was 0.0585 nm.

Use the given wavelength and the traces in figure 8.1d to calculate the distance apart of the planes in one of the sets present in sodium chloride.

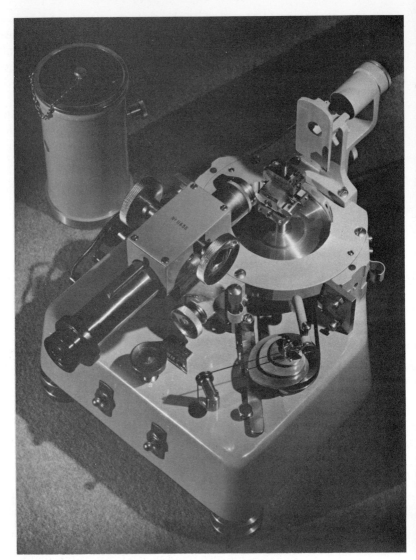

Figure 8.1e
An X-ray goniometer.
Photo, Unicam Instruments Ltd

The single-crystal X-ray goniometer

Figure 8.1e shows a modern device for the X-ray study of crystals; it is called an X-ray goniometer. A single crystal of the substance under study is mounted on the projection on the goniometer head which occupies the central circular well of the instrument. (The goniometer head is shown enlarged in figure 8.1f.) The X-ray beam is directed at the crystal through the support on the righthand side of the instrument. The microscope on the lefthand side of the instrument is used for observing the crystal when mounting it on the goniometer head; the crystal must be placed directly in the X-ray beam, and it must have its faces in the correct orientation. The cylindrical drum in the background contains the X-ray film, which may partially or wholly surround the crystal, when placed over the central well.

Figure 8.1f
The goniometer head, and X-ray beam collimator.
Photo, Unicam Instruments Ltd

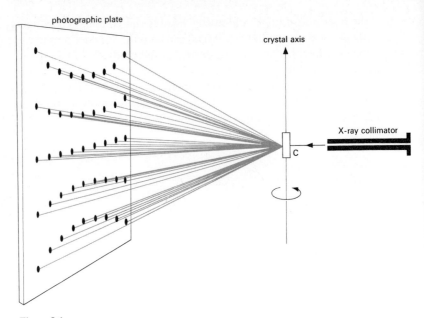

Figure 8.1g
The production of an X-ray diffraction pattern by a rotating crystal.

Figure 8.1f gives an enlarged view of the goniometer head. This consists of a central spindle which can be rotated, and two arcs with vernier attachments. The crystal is placed on top of the projection, and it may then be set in any required orientation with respect to the impinging X-ray beam. Only a small crystal is necessary (1 millimetre long by 0.5 millimetre wide is adequate). The X-ray beam enters the apparatus through the narrow collimating channel which is seen on the righthand side of the photograph.

The paths of X-rays during an experiment are shown schematically in figure 8.1g. The crystal, C, is slowly rotated; and as successive crystal planes come to the correct angle for reinforcement an X-ray beam strikes the photographic plate and a spot appears.

If a flat film is used, then the spots appear in curved patterns; but if a film bent into the form of a cylinder with the crystal at its centre is used, then the spots appear in straight line patterns when the film is opened.

Figure 8.1h shows a cardboard model of an ideal crystal of urea. What do you notice about the end faces?

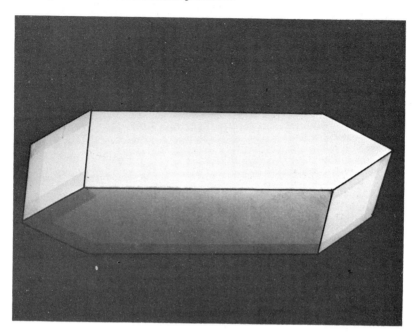

Figure 8.1h
A cardboard model of an ideal crystal of urea.

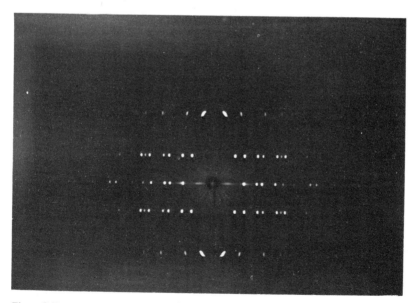

Figure 8.1j
An X-ray diffraction photograph of a single crystal of urea.

Figure 8.1j shows an X-ray diffraction rotation photograph of a crystal of urea, taken with an instrument similar to that in figure 8.1e. Was the photograph taken using a flat film or a film bent into the form of a cylinder?

The problem is to work backwards from the spots in the photograph to the arrangement of atoms or ions which produced them. The distance apart of the spots readily gives the separation of the repeating units, and thus it gives the dimensions of the 'unit cell' (see section 8.2) of the crystal. The exact positions of the atoms or ions are much more difficult to determine. Use is made not only of the distance apart of the spots but also of their relative intensities.

The procedure is to set up a 'trial structure', that is, all the available chemical and physical information is used and a 'likely' structure or 'trial structure' is proposed. Calculations are done to determine the X-ray diffraction pattern which the trial structure would produce, and this is then compared with the actual one obtained experimentally. The differences between the two will suggest how the trial structure should be modified in order to be closer to the actual structure; a further calculation is then done to obtain the diffraction pattern of the modified structure. This is compared with the actual pattern and the modification process is repeated again and again until agreement is obtained.

When X-rays fall on a crystal it is the electrons around the atoms or ions in the crystal which produce the scattering of the rays. The scattering of X-rays by an ion or atom is directly proportional to the number of electrons which it possesses. Since the arrangement of atoms or ions in a crystal is repetitive, or periodic, the electron density pattern is also periodic. The electron density may be represented either by an *optical synthesis pattern* in which regions of high electron density are shown as white and those of low electron density as dark, or they may be represented by an *electron density contour map*.

Examples of *optical synthesis patterns* for urea are shown in figures 8.1k and 8.1l. In figure 8.1k the molecules of urea are seen from above and can be seen to be arranged in an orderly pattern. The molecular structure of urea is

$$O=C \begin{matrix} \diagup N \diagup^{\displaystyle H} \diagdown_{\displaystyle H} \\ \diagdown N \diagup^{\displaystyle H} \diagdown_{\displaystyle H} \end{matrix}$$

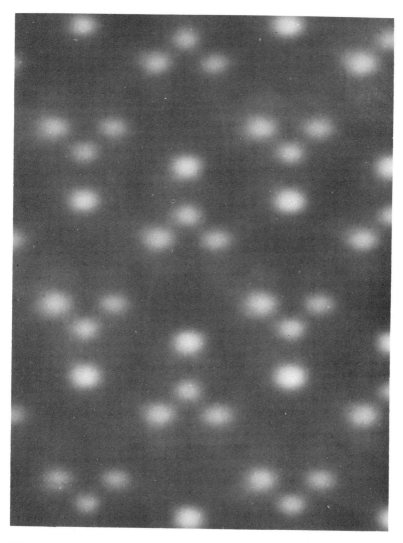

Figure 8.1k
An optical synthesis structure of a crystal of urea.

In figure 8.1k how many atoms in each molecule are visible? Which atoms are they? Why are the other atoms not visible?

In order to answer the last question you will have to recall what it is in an atom that actually scatters X-rays. You will then have to decide whether you think the missing atoms would be likely to scatter X-rays well or poorly.

Look carefully at the arrangement of the blobs, and sketch on a piece of paper the structural formulae, properly orientated, for several adjacent molecules. What do you notice?

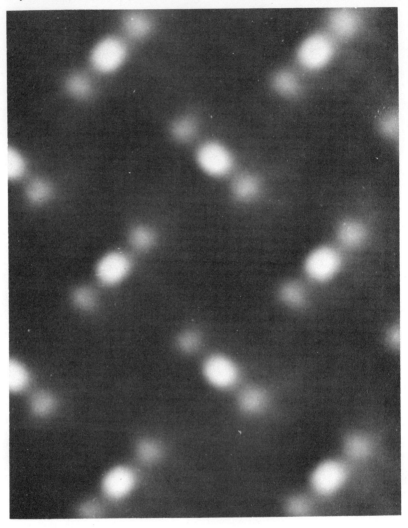

Figure 8.1l
An optical synthesis structure of a crystal of urea, at right-angles to the direction in figure 8.1k.
Photos 8.1h to 8.1l, Professor Dame Kathleen Lonsdale, University College, London

Figure 8.1l shows the molecules viewed from the side. What does the photograph tell us about the shape of the molecule? Why is the central spot for each molecule so much brighter than any of the other spots in either photograph?

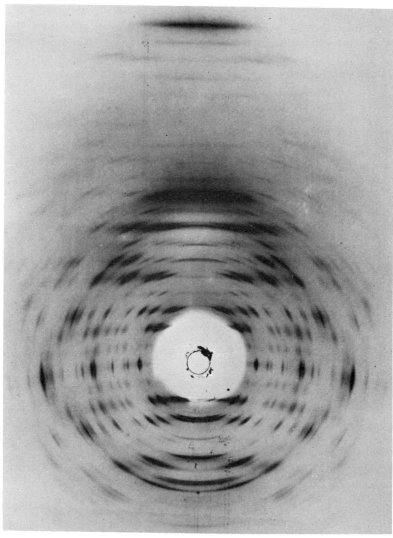

Figure 8.1m
Diffraction pattern of an orientated fibre of microcrystalline DNA.

Figure 8.1m shows a diffraction pattern produced by an orientated fibre of microcrystalline DNA. DNA is a chemical compound which occurs in chromosomes and is the substance that stores and passes on hereditary characteristics. As is to be expected of a substance which is able to perform such a complex biological function, it is of great chemical and structural complexity. By means .

Figure 8.1n
An electron density map of part of a molecule of DNA, constructed from a photograph such as 8.1m.

of X-ray photographs such as that shown in figure 8.1m, however, Professors F. H. C. Crick, M. F. H. Wilkins, and J. D. Watson were able to work out the molecular structure of DNA. For this they were jointly awarded the Nobel Prize for medicine in 1962.

Figure 8.1n shows an *electron density contour map* of part of a molecule of DNA. A scale model of part of the double-spiral structure of this molecule is shown in figure 8.1p, and a coloured photograph of this model is given on the inside of the back cover.

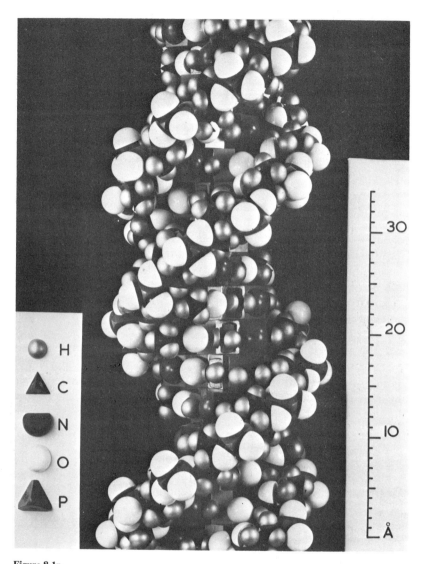

Figure 8.1p
A scale model of part of a molecule of DNA.
Photos 8.1m to 8.1p, Professor M. F. H. Wilkins, Biophysics Department, King's College, London

Computers and X-ray diffraction studies

Even for a relatively simple structure the calculations which are involved in translating a diffraction photograph into a crystal structure model can be very complex. For structures such as those of proteins and of DNA the quantity of calculation involved is so immense that it could not be achieved without the aid of computers. The development of computers has enabled X-ray diffraction studies to be extended to more and more complex structures, and in particular it is leading to a growth in our knowledge of the structures of the immensely complicated molecules upon which the processes of life on this planet depend.

Information from X-ray diffraction studies: a summary

Before discussing some other methods of obtaining structural information we should consider where X-ray methods may be used, and what information they can and cannot give. X-ray diffraction studies are limited to solids; and the most informative results are obtained by using single crystals. But useful information can be obtained if the specimen is in the form of a powder (many randomly-orientated minute crystals), or if it is in the form of a pulled thread (a method suitable for materials consisting of very long molecules).

The method detects regions of high electron density. It can give the arrangement of the units (individual molecules or ions) in the crystal, and it can give the positions of the individual atoms within those units, with the exception of hydrogen atoms, which do not have a sufficient electron density to be precisely located.

Bond lengths may be determined to better than ± 0.001 nm.* The carbon-carbon bond length in diamond is 0.154 nm, and in hexane and hexanol (as with other members of the series) it is also 0.154 nm. The knowledge of this bond-length is therefore better than ± 1 per cent (nm = nanometre, or 10^{-9} m).

X-ray diffraction is the most precise and versatile method available for the determination of structure in solids.

Electron diffraction

If a beam of electrons is passed through a gas or vapour and then allowed to fall on to a photographic plate a series of concentric darkened patches is produced on the plate. The electron beam is diffracted by the atoms in the

---

* Bond lengths and similar measurements are given in this book in nanometres, nm; 1 nm = 10^{-9} m. The quantity 10^{-10} m has for many years been known as the Ångstrom unit, Å, after A. J. Ångstrom, the Swedish physicist. Values in these units will be found in other books; 1 nm = 10 Å.

molecules of the gas or vapour, and the diffraction pattern produced is determined by the structure of the molecules. The rings in the diffraction patterns correspond to interatomic distances in the molecules (both those between atoms linked by chemical bonds and those between atoms not directly linked). By setting up a trial structure and calculating its diffraction pattern and then comparing it with the observed one it is possible to move nearer and nearer to a correct structure determination.

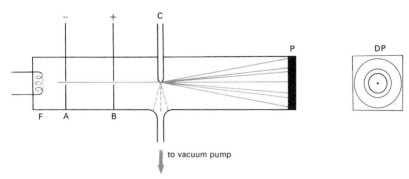

Figure 8.1q
A diagrammatic arrangement for electron diffraction.

In figure 8.1q, F is an electrically heated filament producing electrons. The electrons are collimated into a beam by the holes in plates A and B; plate B is at a positive potential with respect to A, and the electrons are thus accelerated. The gas sample is admitted through C, and the beam is diffracted by it. P is a photographic plate. DP is the developed plate, and shows a concentric circle diffraction pattern.

The method may be used for simple molecules in the gas or vapour phase. It yields information about bond angles and bond lengths. Bond lengths can be determined to ± 0.0003 nm, except for bonds to hydrogen atoms (for example C—H, N—H, O—H) where the uncertainty is usually about ± 0.001 nm. Electron diffraction locates hydrogen atoms with much greater precision than does X-ray diffraction.

Infra-red absorption spectra

Infra-red radiation has a wavelength longer than that of visible light, and it occupies the region from 2500 nm to approximately 25 000 nm. If infra-red radiation containing a wide spectrum of wavelengths is passed through a compound containing molecules it is usually found that certain wavelengths are absorbed. This is because the radiant energy is causing certain bonds in the molecule to vibrate more vigorously. Various types of vibration are possible, but the two principal ones are stretching and bending. These are shown diagrammatically in figure 8.1r, in which the water molecule is taken as an example.

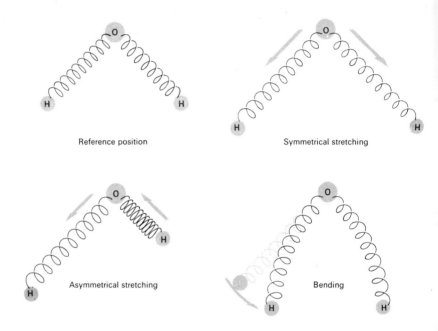

Figure 8.1r
Vibrations in the water molecule.

Since energy is quantized, the various stretching and bending modes will only absorb energy of a particular value; this value is characteristic of the bond, e.g. C—H, or N—H. Thus the wavelength at which absorption occurs is also characteristic of the bond and can be used to identify it. Some examples are given in table 8.1a.

Group	Vibration	Characteristic absorption wavelength/10^{-6} m
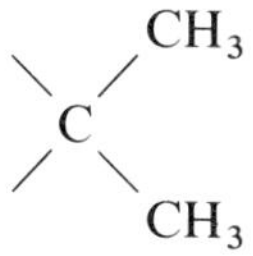 $\begin{array}{c}\diagdown/\!CH_3\\ C\\ \diagup\diagdown\\ CH_3\end{array}$	C—H bending	7.22–7.25
—CH=CH$_2$	C—H stretch	3.25–3.40
	C—H bending	10.05–10.15
Alcohols	O—H stretch	2.73–2.79
Aldehydes (saturated) (—CH$_2$CHO)	C=O stretch	5.75–5.81
Ethers (—CH$_2$—O—CH$_2$—)	C—O stretch	8.70–9.40

Table 8.1a
Some characteristic group infra-red absorptions

The sample may be in the form of a solid, liquid, or gas; a liquid (pure or solution) is most convenient. The examination of the infra-red absorption spectrum of an unknown compound is an important part in the process of identifying it, since the absorbed wavelengths give an indication of the atoms or groups of atoms which are present. Observation of the fine structure of the absorption by gases, where the energy of molecular rotation is also quantized, enables bond lengths to be determined.

8.2   Some crystal structures, and crystal properties
In this section some crystal structure models will be examined, and experiments will be done to relate the properties of some substances to their crystal structure. Some of the simplest crystal structures are those of metals: hexagonal close packing (h.c.p.); cubic close packing (c.c.p.); and body-centred cubic packing (b.c.c.). Illustrations of these are shown in figures 8.2a, b, and c.

In figure 8.2a the second layer is made by placing spheres in the hollows of the bottom layer. The third layer is made by placing spheres directly over the spheres in the bottom layer and a repeating pattern ABA is established. The fourth layer is formed by repeating the B arrangement. Thus a repeating pattern ABAB ... results; this is known as *hexagonal close packing*.

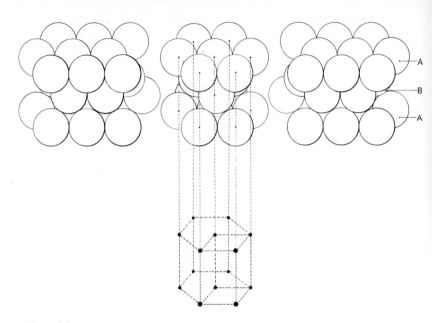

Figure 8.2a
Unit cell of hexagonal close-packed structure derived from three close-packed spheres in A : B : A : B sequence

If one considers three adjacent spheres forming a triangle in the first B layer it is possible to arrange a similar triangle in the third layer in such a way that the spheres in the third layer do not lie directly above the bottom A layer (as they do in the h.c.p. structure). This is shown in figure 8.2b. The third layer is thus a different arrangement, and could be called C. If the fourth layer is arranged to be a repeat of the bottom A layer, then the whole pattern may be repeated giving ABCABC . . . ; this is known as *cubic close packing*.

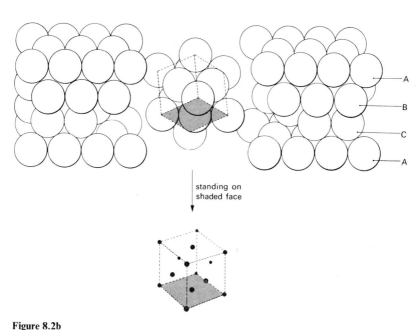

Figure 8.2b
Derivation of unit cell of face-centred cubic structures from layers of close-packed spheres in
A:B:C:A sequence.

If demountable sphere models of these structures are examined it will be found
that in a hexagonal close-packed arrangement and in a cubic close-packed
arrangement a given sphere will have 12 other spheres in contact with it. These
are the closest possible packing arrangements for equal-sized spheres. For a
given sphere there are 12 nearest neighbours, and it is said to have 12 coordination.

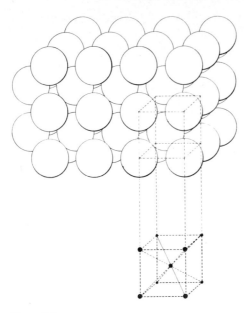

Figure 8.2c
The body-centred cubic crystal structure.

Figure 8.2c shows *body-centred cubic packing*: in this, each sphere is in contact with 8 others, and is said to have 8 coordination. Thus b.c.c. is not so close a packing arrangement as h.c.p. and c.c.p. In h.c.p. and c.c.p. structures there is only 26 per cent of empty space, but in the b.c.c. arrangement there is 32 per cent of empty space.

Almost all of the metals crystallize into one or more of these systems. Some examples are:

Hexagonal close packed: magnesium, zinc, nickel
Cubic close packed: copper, silver, gold, aluminium
Body-centred cubic: the alkali metals

The close-packed systems account for about fifty metals, and the body-centred cubic systems for about twenty metals.

There is no obvious relation between an element's structural type and its position in the Periodic Table.

Examine ball-and-spoke models of several crystal structures, for instance, sodium chloride, caesium chloride, calcium fluoride (fluorite), diamond, zinc sulphide (zinc blende), and graphite. In each work out the coordination number (number of nearest neighbours) for a given ion or atom and look for the presence of one of the three lattice arrangements which have been described (these will not always be present). Coloured photographs of models of some typical structures are given on the inside of the back cover.

A unit cell

Crystallographic studies are aided by considering a crystal to be made up of many adjacent identical 'unit cells'. A unit cell of sodium chloride was shown in figure 8.1b. It consists of three ions in each edge of the cube. By convention each edge is chosen to contain two sodium ions, but this is only a convention; two chloride ions and one sodium ion could equally well be chosen.

Does the unit cell have an empirical formula NaCl? One may find an answer to this question by considering the extent to which the various ions in the unit cell are shared.

An ion at a *corner* is shared by 8 cells, giving $\frac{1}{8}$ ion per cell.
An ion on an *edge* is shared by 4 cells, giving $\frac{1}{4}$ ion per cell.
An ion on a *face* is shared by 2 cells, giving $\frac{1}{2}$ ion per cell.
An ion *inside* the cell is not shared, giving 1 ion per cell.

Thus in a unit cell of sodium chloride we have:

	Na^+	Cl^-
At the *corners*, 8 ions, $\frac{1}{8}$ charge each	1	
On the *edges*, 12 ions, $\frac{1}{4}$ charge each		3
On the *faces*, 6 ions, $\frac{1}{2}$ charge each	3	
Inside the cell, 1 ion		1
	4	4

The unit cell therefore contains the equivalent of four sodium ions and four chloride ions, giving an empirical formula of NaCl.

An examination of the unit cells of the two cubic arrangements of metals mentioned above, cubic close packing (c.c.p.) and body-centred cubic (b.c.c.) shows that in the close-packed arrangement eight atoms are situated at the corners of a cube, and six others are situated one at the centre of each of the six faces of the cube. This arrangement is sometimes known as a *face-centred cubic structure*, and the unit cell contains the equivalent of four atoms ($8 \times \frac{1}{8} + 6 \times \frac{1}{2}$). The *body-centred cubic structure* has eight atoms at the corners of the cube, one atom at the centre of the cube, but no others on the faces. This unit cell contains the equivalent of two atoms.

These two unit cells are shown in figure 8.2d.

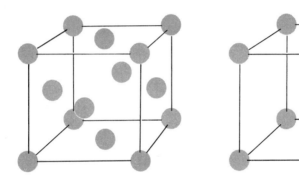

Figure 8.2d
Face-centred and body-centred unit cells.

The Avogadro constant from X-ray evidence

X-ray measurements on crystals can be used to obtain a value for the Avogadro constant, L. For this, the dimensions of the unit cell, and the number of particles that it contains, are required, and each of these quantities can be determined accurately for many crystals. The number of particles which occupy the molar volume of the substance can then be calculated, as shown in the following example.

An end view of the unit cell of sodium chloride is shown in figure 8.2e. From X-ray diffraction evidence the width of this unit cell is 0.5641 nm.

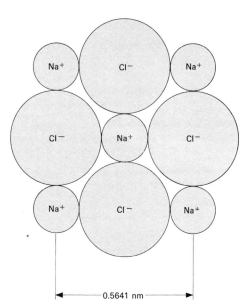

Figure 8.2e
The unit cell of sodium chloride.

The unit cube contains 4NaCl, as explained on page 225.

The formula weight of NaCl = 58.44.
The density of sodium chloride = 2.165 g cm^{-3}.

Therefore, the total volume of one mole of NaCl, including both ions and empty space, is

$$\frac{58.44}{2.165} \text{ cm}^3.$$

This contains L ion pairs of Na$^+$Cl$^-$. But the total volume share of one ion pair, that is, the volume of the two ions + their share of empty space, is

$$\frac{(0.5641 \times 10^{-7})^3}{4} \text{ cm}^3$$

$$\therefore \quad L = \frac{58.44}{2.165} \div \frac{0.5641^3 \times 10^{-21}}{4}$$

$$= 6.01 \times 10^{23}$$

This method is one of the most accurate for finding L, and has enabled its determination to be carried out to an accuracy of ± 0.01 per cent. The value of L on the ^{12}C scale is 6.02296×10^{23} mole^{-1}.

The physical properties and crystalline structures

In the following experiments we shall investigate the relationship between the physical properties of some crystals and their structures.

Experiment 8.2a
The cleavage of graphite and calcite

1   Use a magnifying glass to examine a small quantity of graphite powder. Note the flat, plate-like nature of the crystals.

Rub a little graphite powder between two fingers, and note the feel. The flat crystals slide readily over one another, markedly reducing the friction between the fingers. Graphite either on its own or suspended in oil is used as a lubricant.

Relate the appearance of the crystals to a model of the crystal structure.

2   Take a piece of calcite, and examine its shape and the cracks in it. What do you notice about the direction of the cracks?

Tap the crystal gently with a small hammer or the back of a closed penknife, or some similar instrument. What do you notice about the shape of the fragments?

Relate the shape of the fragments, and the directions of the cleavage cracks, to a model of the crystal structure.

Experiment 8.2b
Investigating the behaviour of some substances between crossed polaroids

Light waves from an electric lamp vibrate in an infinite number of different planes, at right-angles to the line of propagation. A piece of polaroid has the property of transmitting only waves vibrating in one plane.

You are supplied with two pieces of polaroid mounted at right-angles to each other, on a wooden bar. Support the polaroid assembly in a clamp stand so that the polaroid sheets are horizontal, and leave sufficient room below the sheets to place a torch bulb and holder there.

1   Place a lighted bulb beneath part of the assembly where there is only *one* piece of polaroid when viewed from above, that is where the two sheets do not overlap. Now look at the bulb and polaroid through a loose piece of polaroid. Turn the loose polaroid round and round while you watch the bulb. Interpret what you observe.

2   Examine a range of different materials as follows. Place the bulb under the centre of the crossed polaroids. Place a small sample (about $2 \times 2 \times 0.5$ mm) on a microscope slide. Insert the specimen between the crossed polaroids, until it is over the bulb. Rotate the specimen slowly, watching it carefully through the upper polaroid, and note whether any effect is produced or not.

Examine the following materials: sodium chloride, calcite (a clear, transparent fragment), sodium iodide, potassium iodide, potassium thiocyanate, fluorite (calcium fluoride), quartz, and a 1–2 cm length of human hair.

Potassium thiocyanate, K^+NCS^-, is a substance whose structure you have probably not yet met. Its structure is shown in figure 8.2f.

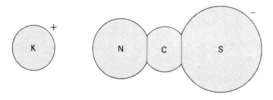

Figure 8.2f
The structure of potassium thiocyanate.

Draw four columns as shown below, and record your observations:

1	2	3	4
No effect, or almost no effect, whatever the orientation	Structure of substance listed in column 1	An effect which depends upon orientation	Structure of substance listed in column 3

From your knowledge of the shapes of the particles concerned, fill in columns 2 and 4. What do you notice? What conclusions can you draw about the types of structures which produce no effect, and the types which do produce an effect?

What do you think might be the shape of the molecular units which make up human hair?

Isotropy and anisotropy

An isotropic substance is one whose properties are the same in whatever direction they are measured. For instance, the refractive index of sodium chloride is the same whether light is passed through a crystal from side to side, top to bottom, or back to front; also the effect of sodium chloride upon polarized light is the same whatever the orientation of the crystal.

An anisotropic substance is one whose properties depend upon the direction in which they are measured. For instance, calcite has refractive indices varying from 1.49 to 1.66 depending upon the direction along which the light traverses the crystal. The thermal conductivity of calcite also is different in different crystallographic directions.

Isotropy can occur when the units which comprise a crystal are spherically symmetrical. Sodium ions and chloride ions are spherically symmetrical, and light waves or thermal vibrations are not affected differently because they approach a given ion from different directions. The carbonate ion is not spherically symmetrical, and both light waves and thermal vibrations will experience different effects on approaching a given carbonate ion from different directions.

All crystals which adopt the cubic system are isotropic with respect to polarized light, the velocity of light in them (refractive index), thermal conductivity, and electrical conductivity; they are not isotropic to some phenomena, for instance the velocity of sound, and mechanical stretching (elasticity).

All substances which crystallize in other crystal systems are anisotropic.

These statements are consequences of the relation between spherical symmetry of the packing units and isotropy, for if a packing unit departs from spherical symmetry it cannot pack in a cubic system; the distortion produces some other system.

The two extremes of departure from spherical symmetry are planar units, and linear, rod-like units. The former are exemplified by the carbonate and nitrate ions, and by giant structures consisting of layers such as graphite, cadmium iodide, iron(III) chloride, and the micas. The linear units are exemplified by the thiocyanate ion (NCS^-), by normal alkanes and normal alcohols, and by fibrous structures such as asbestos, cotton, and hair.

The anisotropy of graphite

Look at a model of the structure of graphite. Is it symmetric or asymmetric? Would you expect graphite to be isotropic or anisotropic?

Because of its anisotropic properties much research has been conducted in attempts to produce large single pieces of graphite. One of the methods developed is a hydrocarbon gas-cracking process conducted above 2000 °C. It slowly produces a material called pyrographite, a dense metallic-looking substance in which all the graphite crystallites are similarly oriented. In 1965 a piece of pyrographite $1 \times 1 \times \frac{1}{8}$ inches cost £2–£3.

Table 8.2a lists some of the physical properties of pyrographite.

Property	Along the crystal planes	At right-angles to the crystal planes
Tensile strength/N m^{-2}	124×10^6	34.5×10^6
Thermal conductivity/W cm^{-1} K^{-1}	2.0	0.025
Electrical resistivity/Ω cm	2×10^{-4}	2500×10^{-4}
Thermal expansion/K^{-1}	0.66×10^{-6}	20×10^{-6}

Table 8.2a
Physical properties of pyrographite

From these figures it will be seen that pyrographite is a highly anisotropic substance. The thermal conductivity along the crystal planes is almost 100 times greater than at right-angles to the planes. The electrical conductivity is more than 1000 times greater along the planes than at right-angles to them.

The highly anisotropic thermal conductivity has led to the use of pyrographite in rocket nozzles and for re-entry nose-cones. A layer of pyrographite enables heat to be conducted rapidly away from a zone at extremely high temperatures to low temperature areas; at the same time its poor conducting qualities at right-angles provide protection to the rocket, or re-entry capsule.

If an electric arc is struck between carbon rods under certain special conditions, it is found that the rods grow whiskers. Whiskers up to 3 cm in length have been produced. These whiskers are long single crystals of graphite, almost entirely free from defects; and they have an unusually high tensile strength of about 1000 tonnes per square inch.

As the whiskers are easily damaged they are of little use by themselves; but if they could be placed in a molten metal and the metal allowed to solidify around them, then the whiskers would give great strength to the metal and the metal would give protection to the whiskers. In this way there are possibilities of producing engineering materials of unprecedented strength, and much research is being conducted to achieve this.

As the temperature is raised, the tensile strength of most materials decreases; but the tensile strength of graphite increases with rise in temperature, and at 1500 °C it is one of the strongest substances known. If the difficulties in using graphite can be overcome it may be possible to exploit its unusual properties at high temperatures where the strength of most other materials is beginning to fail.

8.3 Structure and the Periodic Table

You will recall the trends in structure across the Periodic Table which were noticed earlier in the course during study of the elements of the period sodium to argon, and of their chlorides and oxides.

For convenience, some of the information is summarized below.

Element	Na	Mg	Al	Si	P(white)	S	Cl	Ar
Structural type	← giant lattices →				← molecules →			
Melting point/°C	98	650	660	1400	44	120	−100	−190
Latent heat of fusion/kJ mol^{-1}	2.6	8.9	11	46	0.63	1.4	3.2	1.2
Boiling point/°C	890	1100	2400	2700	280	440	−34	−190
Latent heat of vaporization/kJ mol^{-1}	89	130	290	380	12	9.6	10	6.5

⌊—metals—⌋   ⌊————non-metals————⌋

⌊ metallic bonding ⌋   ⌊—covalent bonding—⌋

Table 8.3a
Some properties of the elements sodium to argon (Numerical values are correct to *two significant figures*)

With melting points and latent heats of fusion there is a sharp break between silicon and phosphorus, reflecting a change from giant lattices to molecules.

An almost identical trend is seen with the boiling points and latent heats of vaporization, and again the break and contrast are striking.

Chlorides

NaCl	MgCl$_2$	AlCl$_3$	SiCl$_4$	PCl$_5$(g)	S$_2$Cl$_2$	Cl$_2$

└─────giant lattices─────┘   └─────────molecules─────────┘

└─────ions─────┘   └─────────covalent bonds─────────┘

Oxides

Na$_2$O	MgO	Al$_2$O$_3$	SiO$_2$	P$_2$O$_5$(g)	SO$_3$	Cl$_2$O$_7$

└─────────giant lattices─────────┘   └─────────molecules─────────┘

└─────ions─────┘   └─────────covalent bonds─────────┘

Table 8.3b
Structures of the chlorides and oxides of the elements sodium to argon

In these compounds there is a transition from ionic bonds on the lefthand side of the table, through a type of bond intermediate between ionic and covalent bonds, to covalent bonds on the righthand side of the table.

This transition raises a number of questions. What are the natures of an ionic bond and of a covalent bond? How are the particles held together? Why does the transition occur? Answers to these questions may be found in terms of the electronic structures of the atoms, and in terms of electron rearrangements.

8.4 Electron arrangements in ions and molecules

Any theory of bonding which is to be of use must provide at least a reasonably satisfactory means of accounting for the formulae of compounds, for their structure, and for the nature of the forces which hold them together. Many of the properties of materials can be interpreted in terms of simple models of how the electrons are distributed in the compounds. On the other hand there are properties, such as the electrical conductivity of metals, which can only be interpreted in terms of more sophisticated models; and there are some properties for which there is as yet no adequate explanation.

In this section we shall examine simple models of electron distribution in compounds, and the ways in which they can account for the formulae and structure of compounds, and the forces which hold them together.

1 Ions

Try to write down the evidence which you have encountered for the existence of ions. Can you also remember the names of any compounds that you have made that contain ions?

Typical compounds containing ions are lithium chloride, lithium oxide, magnesium fluoride, and magnesium oxide.

To obtain ions from neutral atoms, electrons must be transferred; and the model for ion formation is that electrons are transferred until the outer electron shell is identical with that of the nearest inert gas. The lack of chemical reactivity of the inert gases of low atomic weight is an indication of the great stability of their electronic structure; and ions with such electronic structures are also very stable.

The ions are held together in crystals by the forces of electrostatic attraction between oppositely charged (+ and −) ions. The lattice energy (see Topic 7) of the crystalline compound is a measure of the strength of this attractive force.

The diagrams opposite (figure 8.4a) show how the transfer of electrons from one atom to another gives ions. In these diagrams the nucleus of each atom is represented by its symbol, and the shells of electrons are represented by groups of dots and crosses around the nucleus. The shell of lowest energy is nearest to the nucleus, and successively higher energy levels are shown at increasing distances. The electrons in different atoms are represented by 'dots' and 'crosses'. It should not be thought that these electrons are distinguishable; this is merely a device used in the diagrams to enable their movements to be followed.

It will be seen from the first two diagrams that the electronic structures of the ions that are formed are identical to those of an inert gas. These diagrams also show the formation of lithium chloride and lithium oxide. Study these diagrams, and then copy the last two, unfinished, diagrams into your notebook. Complete the diagrams showing the electron transfers for these other two compounds, and write in the names of the inert gases whose structures are formed in these two cases.

You may like to draw similar diagrams to show how the following compounds might be formed by electron transfer: potassium oxide, and lithium hydride.

You will notice that making up the inert-gas structure leads to the experimentally-determined empirical formula.

It should be remembered when writing 'dot-and-cross' diagrams that the dots and crosses are a means of counting electrons, and showing the number present; they do not show the positions of the electrons. The electrons are distributed in space as diffuse negative charge-clouds.

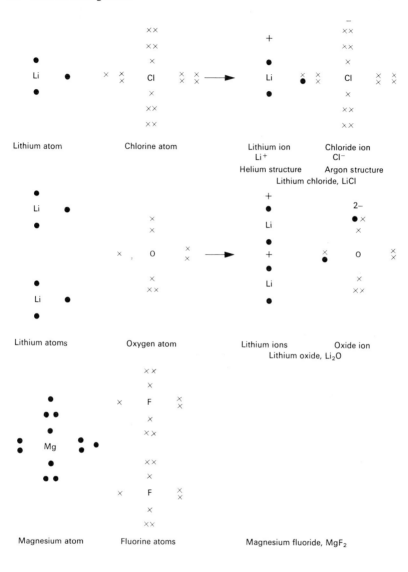

Figure 8.4a

The formation of ions.

It should also be noted that not all ions have an inert-gas structure. The ions formed by d-block elements are cases in point. The Cu^{2+} ion for example has an electronic configuration $(1s^2)(2s^22p^6)(3s^23p^63d^9)$.

What elements form ions?

In order to form positive ions the element must be ionized, that is, electrons must be removed from its atoms, and this requires energy. For a resulting compound to be stable the lattice energy must be large enough to compensate for the energy involved in ionization. This is commonly achieved for M^+ and M^{2+} compounds, but not for M^{3+} or M^{4+}. For example, the first and second ionization energies of magnesium are 736 and 1448 kJ mol^{-1} respectively, and magnesium will form ionic compounds. But the successive ionization energies of carbon are 1088, 2350, 4644, and 6192 kJ mol^{-1}, and the very large amount of energy involved in ionizing carbon cannot be recovered in any lattice energy. Consequently carbon does not form C^{4+} ions; it does not in fact form stable positive ions at all but forms bonds by another method.

In order to form negative ions, atoms must gain electrons. The formation of X^- ions from atoms (for example $Cl(g) + e^- \rightarrow Cl^-(g)$, and $Br(g) + e^- \rightarrow Br^-(g)$) is exothermic, but the formation of O^{2-} is slightly endothermic; the formation of X^{3-} ions is endothermic.

Can some ions be regarded as spheres?

When building models of ionic structures, those ions which are formed from single atoms are usually represented as spheres. It might be argued that as ions possess a complete outer electron shell it would be reasonable to suppose that the electron distribution is spherical. There is some experimental evidence for this view, based on a study of electron density maps, obtained by X-ray diffraction measurements. Figure 8.4b shows such electron density maps for sodium chloride and calcium fluoride.

Do the maps suggest that these ions are discrete entities, and if so, are the ions spherical?

How large are ions?

From figure 8.4b try to find the radius of a sodium ion and of a chloride ion. What difficulty is involved? Suggest one method for overcoming it. Comment on the accuracy of obtaining ionic radii from electron density maps. What distance does a map of this nature give accurately?

(i) sodium chloride

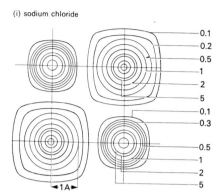

(ii) calcium fluoride

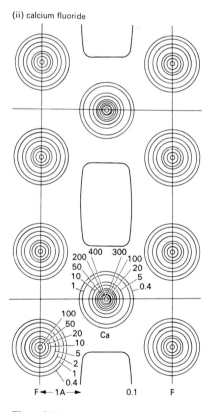

Figure 8.4b
Electron density maps for (*i*) sodium chloride and (*ii*) calcium fluoride. Contours in electrons Å^{-3}.
After, Witte, H. and Wolfel, E., (1958) Reviews of Modern Physics, **30**, 51–55

In order to compile a table of ionic radii, one ionic radius has to be fixed arbitrarily; and then other radii obtained. In addition to this problem, the size of the ion of an element varies slightly depending upon the compound it is part of; ions are slightly soft, compressible, and deformable. For these reasons, tables of ionic radii compiled by different authorities do not always agree with each other. In spite of these uncertainties, the concept of ionic radius is a useful one.

Conclusion

The electron transfer model leading to the formation of ions with electronic structures the same as those of inert gases therefore accounts for the formulae of ionic compounds, the charges on them, and thus the non-directional electrostatic forces which hold them together in giant lattices.

2   Covalent bonds

Any model of the bonding in molecules must be able to account for the formula of a molecule, its structure, and the forces which hold the atoms together.

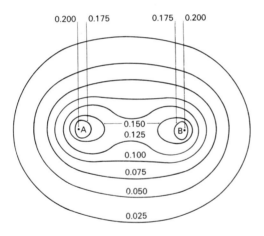

Figure 8.4c
Electron density map for the H_2^+ molecule ion. Contours in electrons Å^{-3}.
After, Coulson, C. A., (1938) Proceedings of the Cambridge Philosophical Society, **34**, 210

Figure 8.4c shows an electron density map which was calculated from theory by C. A. Coulson for the simplest possible covalent structure, the H_2^+ molecule ion. This consists of two hydrogen nuclei but only one electron; it thus has a net positive charge. The molecule H_2^+ was chosen for the calculations because of its simplicity.

What do you notice about the contours that is different from the contours in ionic compounds?

What does this tell us about the electron density between the nuclei of the atoms in the molecule?

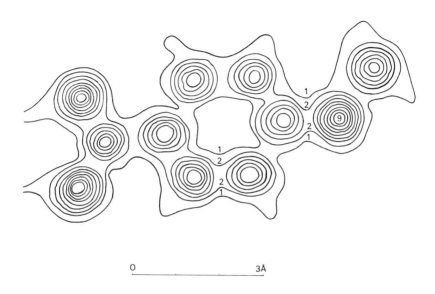

O ———————————————————— 3Å

Figure 8.4d
Electron density map of anisic acid (*p*-methoxybenzoic acid). Contours in electrons Å^{-3}.
Dr J. P. G. Richards, Department of Physics, University College, Cardiff

Figure 8.4d shows an electron density map, determined by X-ray diffraction, for crystals of anisic acid (*p*-methoxybenzoic acid). What can be said about the electron density between adjacent atoms in this molecule?

The two sets of maps (figures 8.4b, c, and d) show that in structures consisting of ions the electron density drops to zero between the ions, and the ions are discrete entities; but in molecules there is a substantial electron density at all points along the line joining the centres of two bonded atoms. Thus it seems that in bonds in molecules the electrons are *shared*.

Figure 8.4e
Charge density distribution for the H_2^+ molecule.

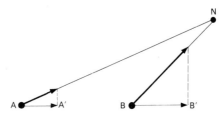

Figure 8.4f
Forces on the nuclei due to a negative charge N not between the nuclei.

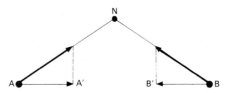

Figure 8.4g
Forces on the nuclei due to a negative charge N between the nuclei.

What is it that holds the atoms together in molecules?

Figure 8.4e shows the electron density distribution for the electron cloud in the H_2^+ molecule. Consider planes through the two nuclei such that the planes are at right-angles to the line joining the nuclei, that is, to A and B in figure 8.4f. If a portion, N, of negative charge-cloud is *outside* these planes, as in figure 8.4f, it attracts the nearer nucleus more strongly than the farther one, and tends to separate them. If a portion, N, of negative charge-cloud is *between* the two planes, as in figure 8.4g, it attracts the nuclei towards itself, and thus towards each other (the resolved parts AA′ and BB′ in figure 8.4g). Since there is very much more negative charge-cloud between the planes than there is outside them, there is a strong net attractive force holding the nuclei together.

The overall effect, therefore, is that the two positive nuclei are bound together by sharing the negative charge-cloud. This arrangement leads to a lower potential energy than if the electron charge-cloud were not shared; and a lower potential energy results in greater stability.

A similar situation exists for the neutral hydrogen molecule, H_2; in this instance, however, the two electrons are shared, and the electron density between the nuclei is greater. The binding is thus stronger than in the instance of H_2^+.

These effects are seen in the separation of the nuclei, and in the bond dissociation energies for H_2^+ and H_2, both of which are given in table 8.4a.

	Structure	Internuclear distance/nm	Bond dissociation energy/kJ mol^{-1}
Hydrogen molecule ion, H_2^+	$H \cdot H^+$	0.104	257
Hydrogen molecule, H_2	$H:H$	0.074	435

Table 8.4a
Data for the hydrogen molecule ion, and the hydrogen molecule

Figure 8.4d should now be examined again. It will be seen that between adjacent atoms there is a substantial electron density, amounting to between 1 and 2 electrons per cubic Ångstrom,* and in some instances between 2 and 3 electrons per cubic Ångstrom. These adjacent atoms are bound together by the attraction which the shared electron charge-cloud exerts upon the nuclei on either side of it.

Stoichiometry, and electron sharing

Covalent bonding exists between atoms when electrons are shared, usually in pairs. In general, the number of atoms involved is such as to enable an inert-gas electron structure to be built up around each atom. Figure 8.4h shows how this is done in the case of methane, CH_4.

The hydrogen atoms have a share in the electrons from the carbon atom, thus acquiring helium structures; and the carbon atom has acquired the neon structure by sharing electrons from the hydrogen atoms.

* 1 Ångstrom $= 10^{-10}$ m.

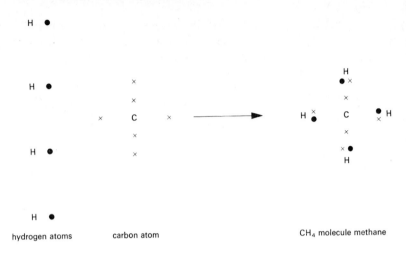

hydrogen atoms   carbon atom CH_4 molecule methane

Figure 8.4h
The formation of covalent bonds.

Figure 8.4j shows the formula and shape of a number of molecules in the period lithium to neon. The shapes of the molecules are known from electron diffraction and other physical studies. Copy these into your notebook, and then below the diagram showing the shape of each molecule draw a dot-and-cross diagram of its electronic configuration. Beryllium chloride and boron trifluoride do not follow the inert-gas general rule as far as the central atoms are concerned, but the remainder do.

When atoms of non-metals are joined together it is in general by covalent bonds.

Now try to draw dot-and-cross diagrams showing the electronic configurations in the molecules of hydrogen, H_2; chlorine, Cl_2; hydrogen chloride, HCl; chloromethane, CH_3Cl; methanol, CH_3OH; ethane, CH_3CH_3; ethanol, CH_3CH_2OH; ethene, $CH_2=CH_2$; ethyne, $CH\equiv CH$; oxygen, O_2; propane, C_3H_8; and hydrogen cyanide, HCN.

Molecular shapes, and electron distributions
Single bonds
Refer to the shape of the molecules shown in figure 8.4j and to their electron configurations. Remembering that a covalent bond consists of an internuclear negative charge-cloud. Why do you think that the molecules have the shapes that they do?

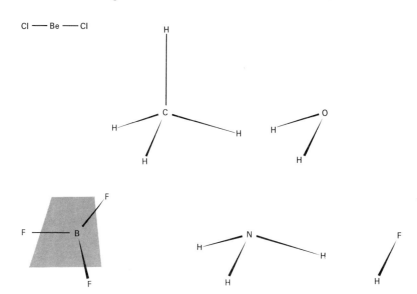

Figure 8.4j
The shapes of some molecules in the period lithium to neon.

What do you notice about the spatial arrangements of the bonds in the molecules of $BeCl_2$, BF_3, and CH_4? What does this suggest about the interaction between electron charge-clouds?

Why are the molecules of ammonia and water not planar and linear respectively?

The bond angles in methane, ammonia, and water molecules are given in table 8.4b.

	Methane, CH_4	Ammonia, NH_3	Water, H_2O
Bond angle	109.5°	107.0°	104.5°
Change in bond angle	←———2.2°———→		
	←———————————5°————————————→		

Table 8.4b
Bond angles in some hydrides

What do these figures suggest?

Multiple bonds

From your consideration of the shapes of molecules containing single bonds, and the changes in the bond angles from one hydride to another as shown in table 8.4b, you should have come to the following conclusions:

a   Pairs of electrons try to get as far away from each other as they can; this results in a tetrahedral distribution of electron pairs.

b   'Lone' pairs of electrons, that is, pairs of electrons not shared between two atoms, repel one another more strongly than shared pairs do. This causes some distortion of the tetrahedral arrangement.

If you did not come to these conclusions you may like to go back over the evidence to see how well they account for the observations.

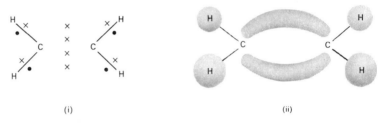

(i)   (ii)

Figure 8.4k
The electron density distribution in the molecule of ethene.

On the basis of these ideas, which were developed for single bonds, can you suggest a shape for the molecule of ethene, $CH_2=CH_2$? Figure 8.4k shows the electron density distribution in this molecule. Compare your suggestion with this.

Suggest a shape for the molecule of ethyne, $CH\equiv CH$.

The ideas developed above do enable one to predict correctly the shapes of a surprisingly large number of molecules and other structures; but they are subject to some limitations. Some of these will be discussed later.

Bond lengths, and bond energy terms

A double bond consists of two pairs of shared electrons, and therefore there might be thought to be a greater electron density between the nuclei than in the corresponding single bond. If this were the case, it should cause a greater force of attraction between the nuclei, and be reflected in a shorter bond length and a greater bond energy term.

Examine the figures given in table 8.4c, and see whether this is so.

Bond	Compound(s)	Bond length/nm	Bond energy term/kJ mol^{-1}
C—C	hydrocarbons	0.154	346
C=C	ethene	0.134	598
C≡C	ethyne	0.121	813
C—N	amines	0.147	305
C=N	oximes	0.132	615
C≡N	hydrogen cyanide	0.116	866
C—O	ethers	0.143	358
C=O	ketones	0.122	749
C≡O	carbon monoxide	0.113	1070 (bond dissociation energy)

Table 8.4c
Bond lengths and bond energy terms

Dative covalency

The two electrons which form a covalent bond between two atoms do not necessarily have to come one from each atom; both may originate from one of the atoms.

Look back at the electron configuration which you drew for the compound BF_3. The electrons around the boron atom do not equal the number corresponding to an inert gas; there are two short. Refer also to the electron structure of ammonia; of the eight electrons around the nitrogen atom, two are not shared with any other atom.

Ammonia gas and boron trifluoride gas react readily to give a white solid with the composition NH_3BF_3. This can be interpreted in terms of electron sharing as follows:

or, using single lines to represent pairs of shared electrons:

Since one pair of the shared electrons has come from one atom the bonding is sometimes known as dative covalency, and the bond is indicated by an arrow →. But the bonding is still covalent and can be written

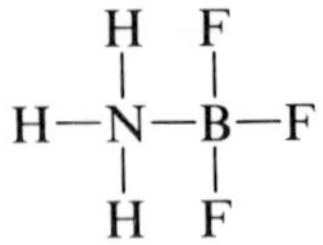

$$
\begin{array}{c c}
\text{H} & \text{F} \\
| & | \\
\text{H}-\text{N}-\text{B}-\text{F} \\
| & | \\
\text{H} & \text{F}
\end{array}
$$

Another example of dative covalency occurs in the ammonium ion, NH_4^+. This is formed by the combination of a hydrogen ion with an ammonia molecule.

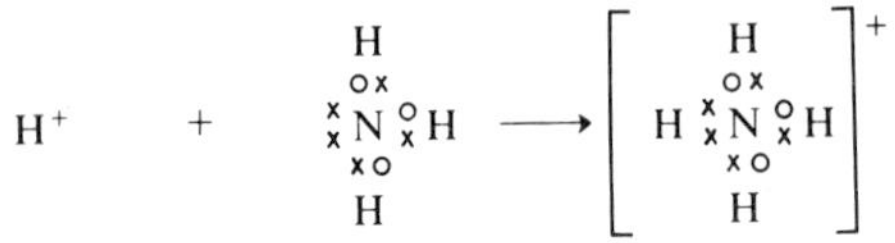

$$
\text{H}^+ \quad + \quad
\begin{array}{c}
\text{H} \\
\text{O X} \\
\text{X N }^\text{O}_\text{X}\text{ H} \\
\text{X O} \\
\text{H}
\end{array}
\longrightarrow
\left[
\begin{array}{c}
\text{H} \\
\text{O X} \\
\text{H }^\text{X}_\text{X}\text{ N }^\text{O}_\text{X}\text{ H} \\
\text{X O} \\
\text{H}
\end{array}
\right]^+
$$

hydrogen ion   ammonia   ammonium ion
(no electrons)   molecule

In the ammonium ion the hydrogen atoms each have a share in 2 electrons, giving a helium structure; and the nitrogen atom has a share in 8 electrons, giving a neon structure. The ion has an overall charge of $+1$, originating from the hydrogen ion; it is distributed all over the ion, and is not located on any particular atom.

The sequence shown above leads to a representation of the ammonium ion as in *a* below:

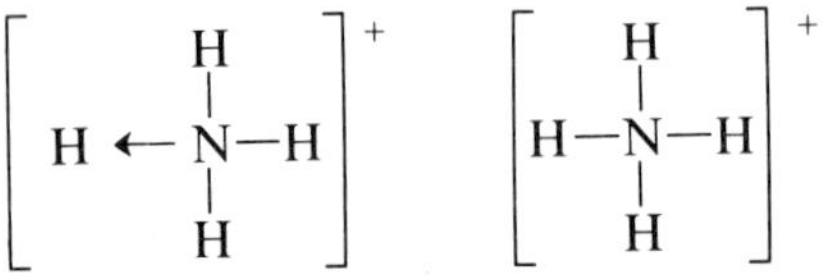

$$
\left[
\begin{array}{c}
\text{H} \\
| \\
\text{H} \leftarrow \text{N}-\text{H} \\
| \\
\text{H}
\end{array}
\right]^+
\qquad
\left[
\begin{array}{c}
\text{H} \\
| \\
\text{H}-\text{N}-\text{H} \\
| \\
\text{H}
\end{array}
\right]^+
$$

a   *b*

But as all the N—H bonds have the same length, and the hydrogen atoms are indistinguishable, a better representation is that in *b*.

The nitric acid molecule and the carbon monoxide molecule contain dative covalent bonds.

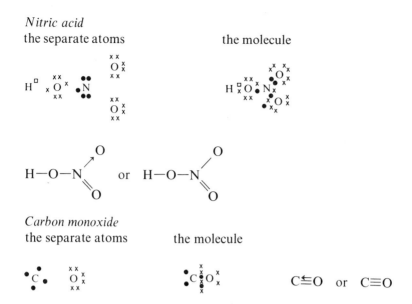

Nitric acid
the separate atoms   the molecule

H—O—N⟨O / O   or   H—O—N⟨O / O

Carbon monoxide
the separate atoms   the molecule

C≚O   or   C≡O

How is an oxonium ion, H_3O^+, formed from a water molecule? Draw an electronic structure for the ion. How do you think the positive charge is distributed?

Would it be possible, in terms of electrons, to form an ion H_4O^{2+}? Try to draw such a structure. Have you ever heard of such an ion? Suggest a reason for your answer.

Draw a likely shape for the molecule of nitric acid, and for the ammonium ion.

3 Intermediate types of bonds

Polarization of ions

Are bonds either ionic or covalent with no intermediate situation; or are pure ionic and pure covalent bonds the extreme types, with a complete range of intermediate situations existing in between?

Refer back to Topic 7, table 7.4, where theoretical values of lattice energies were compared with experimentally determined values. The theoretical values were obtained on the assumption that the ions were spheres, and that transfer of charge had taken place by complete units (i.e. electrons). The agreement for the sodium and potassium halides was within 1 per cent, and therefore it appears that the assumption was a reasonable one in these instances.

Table 8.4d gives the values for some other compounds.

Compound	Structure	Lattice energy/kJ mol^{-1}		
		Calculated	Experimental	Difference
AgF	NaCl	−869.9	−966	96
AgCl	NaCl	−768.6	−916	147
AgBr	NaCl	−758.5	−908	150
AgI	zinc blende	−735.9	−865	129
ZnO	wurtzite	−4088	−4033	−55
ZnS	zinc blende	−3427	−3615	188
ZnS	wurtzite	−3414	−3602	188
ZnSe	zinc blende	−3305	−3611	306

Table 8.4d
Calculated (Born–Mayer) and experimental (Born–Haber cycle) lattice energies for some compounds

Work out two or three discrepancies as approximate percentages of the experimental values. Do you think that the pure ionic model holds in these instances?

Spectroscopic studies of the vapours of the alkali metal halides show these to contain diatomic molecules, MX, and they reveal that the internuclear distance in these molecules is less than in the corresponding ionic solid. The examples of lithium bromide and lithium iodide illustrate this, and values for these compounds are given in table 8.4e.

	Internuclear separation/nm	
Halide	Crystal	Vapour
LiBr	0.275	0.217
LiI	0.300	0.239

Table 8.4e
Internuclear distances in lithium halides

The shortening of the separation implies stronger bonding than in the crystal, and this can only be achieved by a higher concentration of electrons between the two nuclei. This implies a distortion of the electron cloud of one ion, or both, from a spherical distribution.

Figure 8.4l illustrates the effect.

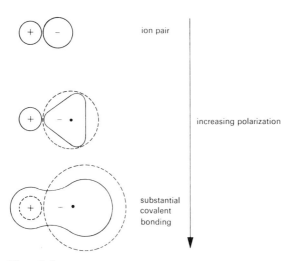

Figure 8.4l
Increasing polarization of a negative ion by a positive ion.

The polarization of the ion represents the start of transition from ionic bonding to covalent bonding.

What factors might affect the extent to which an electron charge-cloud around an ion is distorted?

What type of ion would be best at causing distortion? What type of ion would be most easily distorted? Consider the sizes of ions, and the number of charges on ions. Make a list of the various situations.

Is there any evidence in table 8.4d to support your ideas?

Bond polarization, and electronegativity

Two electrons shared between two atoms constitute a covalent bond between these atoms. It is reasonable to ask whether the electrons are always shared equally between the two atoms, or whether some elements are more 'electron-attractive' than others. It is found that elements do differ considerably in their electron-attractiveness. The term used for electron-attractiveness is *electronegativity*.

The electronegativity of an atom represents the power of an atom in a molecule to attract electrons to itself.

Many attempts have been made to allot numerical values for the electro-negativities of the elements, but so far no wholly satisfactory method has been devised; each method suffers from shortcomings. But whatever numerical scale is used, the trends in the values of the electronegativities of the elements in the Periodic Table are clear.

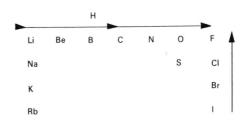

Figure 8.4m
Trends in electronegativity in the Periodic Table.

It can be seen from the trends, that the most electronegative element is fluorine; chlorine, oxygen, and nitrogen are also very electronegative.

If two atoms bonded covalently are atoms of the same element, then the attractions of their nuclei for the bonding electrons are the same, and the bonding electrons will be shared equally between them. But if the atoms are not of the same element, the two nuclei exert different degrees of attractive force on the bonding electrons, and these electrons are displaced towards one atom.

equal sharing   unequal sharing

This unequal sharing of electrons is known as *bond polarization*. It represents the departure of the bond from being purely covalent, and it introduces some ionic character into the bond.

Thus the polarization of ions represents the existence of some covalent character in the ionic bonding; and the polarization of a covalent bond represents the existence of some ionic character in the covalent bond.

One important conclusion from the last few sections is that wholly ionic and wholly covalent bonds are extreme types, and examples occur over the whole range of intermediate types: bonds can be partially ionic and partially covalent in character.

Electronegativity and polar molecules

Using the dot-and-cross diagrams above for the HCl molecule and the CH_3CH_2Cl molecule as a starting-point, draw representations of the electron charge-cloud distributions in the two molecules. Do this both for the polarized bonds and the lone pairs of electrons. Superimpose on these drawings a series of + signs to indicate the positions of the atomic nuclei. Now consider whether you think the 'centre of gravity' of the positive charges, and the 'centre of gravity' of the negative charges coincide.

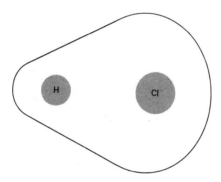

Figure 8.4n
The shape of the HCl molecule.

Figure 8.4n represents schematically the shape of the HCl molecule. Copy the figure into your notebook, and draw in by means of a + and a − the relative positions of the centre of positive charge and the centre of negative charge.

What implications do you think this has for the properties of the molecule?

Suggest one or two experiments which might be conducted to obtain evidence for the reality or otherwise of your suggestions, either for the HCl molecule or for some more readily handled material such as the liquid 1-chlorobutane, $CH_3CH_2CH_2CH_2Cl$ (for which the same general arguments apply).

Suggest relative positions for the centre of positive charge and the centre of negative charge in each of the molecules shown in figure 8.4p.

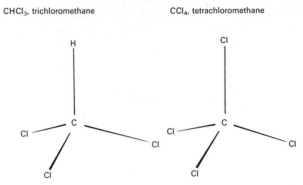

CHCl₃, trichloromethane   CCl₄, tetrachloromethane

Figure 8.4p
The molecules of $CHCl_3$ and CCl_4.

What elements in addition to chlorine would be likely to produce these effects?

In asymmetric molecules such as HCl and CH_3CH_2Cl the centre of positive charge does not coincide with the centre of negative charge, and a permanent dipole results. Such molecules are said to be *polar*. Highly electronegative elements such as F, Cl, O, and N cause polarity in molecules. They do so partly by virtue of the bond polarization which they produce, and partly by virtue of their lone pairs of electrons.

Polarity in molecules has important effects on the physical and chemical properties of the substances, and on the mechanisms by which they undergo reaction. These matters will be discussed more fully elsewhere.

Look back to the charge-cloud diagrams which you drew in your notebook for the molecules shown in figure 8.4j. Decide which of these molecules are polar and which are non-polar, and label them as such in your drawing.

4   Delocalization of electrons
Do single, and double, (and triple) covalent bonds always represent the electron distribution adequately when electrons are shared?

We shall examine structural evidence and thermochemical evidence for some compounds, with a view to obtaining an answer to this question.

Benzene
Structural studies
Benzene molecules have the formula C_6H_6 and the molecules are known to be ring structures.

Draw a molecular structure for benzene, showing the bonds present. The co-valency of carbon is always four. Use a line — to represent one covalent bond.

Having drawn a bond structure diagram, look up the bond lengths given in table 8.4c. What would you expect the bond lengths in the benzene molecule to be? Would you expect the shape of the molecule to be symmetrical? Sketch the shape you would expect.

Is there any other arrangement of bonds which would do equally well?

X-ray diffraction studies of benzene, and electron diffraction studies show that the molecule is symmetrical and planar. They also show that the internuclear distance, from carbon to carbon, is 0.140 nm.

What do you conclude about the structure which you proposed for benzene?

Thermochemical data
   a Heats of hydrogenation

cyclohexene   cyclohexane

The compound cyclohexene contains a carbon-to-carbon double bond, and in the gas phase, in the presence of a catalyst, it can be reduced to cyclohexane. In doing so, heat is evolved. The heat of hydrogenation is $-120\,kJ\,mol^{-1}$.

Supposing benzene had the structure which you proposed for it above, calculate what the gas-phase heat of hydrogenation of benzene might be.

$$C_6H_6(g) + 3H_2(g) \longrightarrow$$

benzene   cyclohexane

It has been found experimentally that the heat of hydrogenation of benzene in the gas phase is $-208\,kJ\,mol^{-1}$.

Is benzene more stable or less stable than your calculation suggested, as far as hydrogenation is concerned? By how many $kJ\,mol^{-1}$ is it more stable or less stable?

b Standard heat of formation from atoms

Use the table of bond-energy terms in the *Book of Data* to obtain a value for the standard heat of formation of gaseous benzene from atoms, assuming the structure which you put forward earlier for benzene.

The value found experimentally, from the heat of combustion of benzene and the appropriate energy cycles, is $-5514\,kJ\,mol^{-1}$.

Is benzene more stable or less stable than the structure which you proposed for it? By how many $kJ\,mol^{-1}$ is it so? How does this value compare with the value from hydrogenation measurements? Do you think there is good or poor agreement? Why?

What do the structural determinations and the thermochemical measurements seem to indicate about the bond situation in benzene?

Nitric acid, and the nitrate ion

Refer back to the electron diagram which you drew earlier for the structure of nitric acid.

Would any other arrangement of the bonds have done equally well?

In a range of other compounds the average $N=O$ bond length is 0.114 nm and the average $N-O$ bond length is 0.136 nm. On the basis of your diagram what would you expect the bond lengths to be in nitric acid?

Electron diffraction studies of nitric acid vapour show the molecule to have the structure shown below.

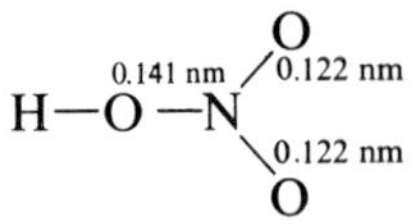

bond lengths 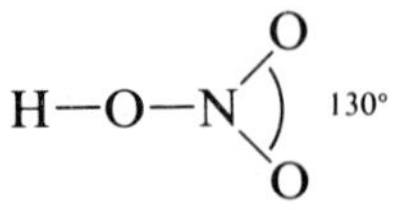   bond angles

Are the bond lengths what you expected? If not, does the situation bear any resemblance to the situation found in benzene?

Draw an electron diagram for the nitrate ion, NO_3^- ; and then draw a bond diagram using a line —— to represent a single covalent bond and an arrow ——→ to represent a dative covalency. The nitrate ion possesses one electron which has been transferred to it from an atom which is now a positive ion.

Is there any other way in which the bonds could be arranged?

What would you expect the bond lengths in the nitrate ion to be?

The structure of the nitrate ion has been found to be as shown below.

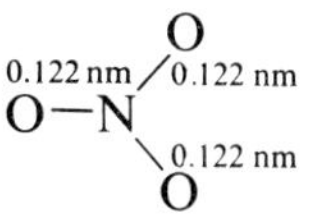

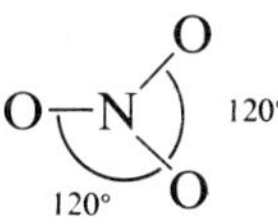

What can be concluded about the nature of the bonds in the nitrate ion?

Using the evidence from benzene, nitric acid, and the nitrate ion, what can be said about the likely bond lengths in bond systems where single and double bonds may be represented diagrammatically by two or more interchangeable arrangements?

Formic acid, and the formate ion
The structural formula of formic acid is

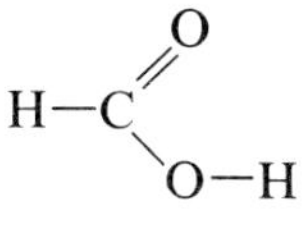

An electron diffraction study on formic acid vapour shows it to have the structure as now shown.

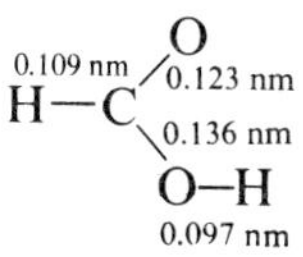

bond lengths 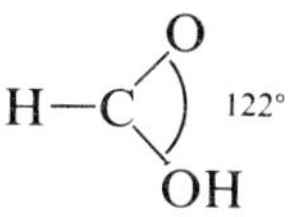   bond angles

Earlier in this topic it was seen that the shapes of molecules could be explained by supposing that pairs of electrons repel one another, and thus bonds, and lone pairs of electrons, tend to get as far away from one another as possible.

Are the bond angles in formic acid what you would expect from this theory? Would a small departure from the simple predicted angle seem a likely situation?

Sodium formate has the formula $HCO_2^- Na^+$, and that of the formate ion is HCO_2^-. Draw an electron structure for the formate ion; and then draw a diagram using — for a single covalent bond and $=$ for a double covalent bond.

Are alternative diagrammatic arrangements of the bonds possible? What sort of length would you expect the C—O distance to be?

X-ray diffraction studies of sodium formate show the formate ion to have the structure given below.

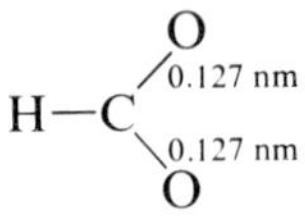

bond lengths 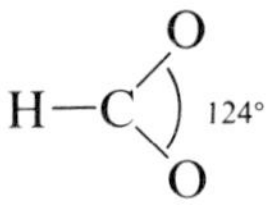   bond angles

Are the bond lengths within the limits which you expect?

What can you say about the nature of the C—O bonds in the formate ion?

Can the structures of benzene, nitric acid, the nitrate ion, and the formate ion be described adequately in terms of the types of bonds which you have so far studied?

Where equivalent alternative bond structures can be drawn for a compound the actual situation is neither of these; the bond lengths in these situations often prove to be equal, and the available electrons must therefore be distributed equally among the atoms concerned. It is believed that each atom is bonded to the next by two electrons between the nuclei, forming a single covalent bond, and that the remaining available electrons are distributed as charge-clouds above and below all the atoms concerned. These electron charge-clouds are not associated with any particular atom but are mobile over the whole atomic system; they are thus known as *delocalized electrons*. Representations of the situation in benzene, nitric acid, and the nitrate ion are given in figure 8.4q.

Draw a representation of the electron distribution in the formate ion.

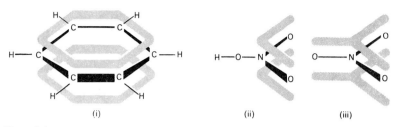

Figure 8.4q
Delocalization of electrons in (*i*) benzene, (*ii*) nitric acid, and (*iii*) the nitrate ion.

The delocalization of electrons in the benzene molecule has important implications for the chemisty of benzene, and this is developed in the topics on carbon chemistry. In order to emphasize the presence of delocalized electrons in the benzene ring a symbol used commonly for benzene is a hexagon with a circle inside it:

a Kekulé structure

But for some purposes the hexagon with a ring is not so convenient a symbol as a hexagon with alternate single and double bonds (known as a Kekulé structure, after the German chemist who proposed it). Where structures are being discussed the former is very satisfactory; where reaction mechanisms are being discussed the latter is often more useful.

Refer back to the structure and properties of graphite, and in particular its anisotropic conduction of electricity. How is it that graphite conducts electricity at all, do you suppose, and why should it conduct better in one direction than in another?

5   Metallic bonds

Three significant properties of metals are their high melting points (as contrasted with most non-metals), their high electrical conductivity, and their high thermal conductivity. Any model of the nature of the bonding in metals must be able to account for these properties. Figure 8.4r shows a representation of a simple model of bonding in a solid metal; it consists of metal ions surrounded by a sea of mobile valency electrons.

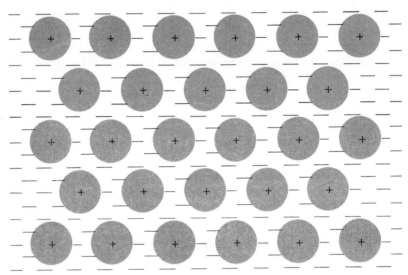

Figure 8.4r
A simple representation of bonding in a metal lattice.

The shared electron sea bonds the metal ions tightly into the lattice and confers a relatively high melting point, while the mobile electrons provide a means of conducting electricity and heat. The mobile electrons are another example of delocalization.

This model is an oversimplification, and it is unable to account for all of the properties of metals; the more sophisticated models are, however, too advanced to consider in this course.

Figure 8.4s shows an electron density map for aluminium. Are the ions spherical?

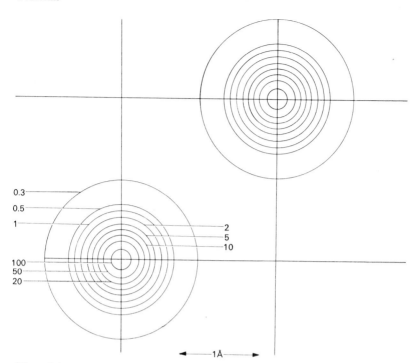

Figure 8.4s

Electron density map of aluminium. Contours in electrons Å^{-3}. (The average density of electrons between the ions is found to be 0.21 electrons Å^{-3}.)

After, Witte, H. and Wolfel, E. (1958) Reviews of Modern Physics, **30**, 51–55

Problems

* Indicates that the *Book of Data* is required.

1   The following are data relating to crystalline potassium bromide, K^+Br^-, which has a structure similar to that of sodium chloride.

Internuclear distance K^+ to Br^- 0.3285 nm
Density   2.75 g cm^{-3}
Formula weight   119.01

i   Select a convenient crystal unit cell and *either* make a space-filling model of it *or* draw a sketch of it.
ii   What are the dimensions of the unit cell?
iii   How many ion pairs does the unit cell contain?
iv   Calculate the total volume share of one ion pair.
v   Using the density and formula weight of K^+Br^- calculate the total volume (ions plus empty space) of one mole of K^+Br^-.

vi   Using the total volume share of one ion pair (from part *iv*) and the total volume share of *L* ion pairs (from part *v*) calculate a value for the Avogadro constant.

vii   In calculating the Avogadro constant, which of the following limits most severely the accuracy of your estimate?

The accuracy of

    A   the K^+ to Br^- internuclear distance

    B   the density of potassium bromide

    C   your method of computation (state whether by slide rule, logarithm tables etc.)

    D   both A and C

    E   both B and C

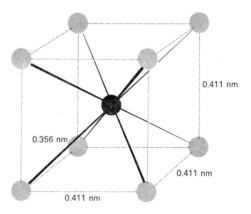

Figure 8.5
A unit cell of caesium chloride.

2   Figure 8.5 shows a unit cell of caesium chloride with a caesium ion at its centre.

i   Which of the following most precisely describes this type of crystal structure?

    A   giant lattice

    B   double face-centred cubic

    C   body-centred cubic

    D   hexagonal closed-packed

    E   double simple cubic

ii   Using only the information supplied, deduce the empirical formula of caesium chloride. Is your answer consistent with the positions of caesium and chlorine in the Periodic Table?

iii   What is the coordination number of (a) the caesium ions and (b) the chloride ions?

iv   The ratio of the radii of caesium and chloride ions is 0.93. That of sodium and chlorine in sodium chloride is 0.52. In the light of this, is the difference in coordination number of the ions in caesium chloride from that in sodium chloride to be expected or not? Explain how you arrive at your answer.

v   Given the following additional data on caesium chlorides, calculate a value for the Avogadro constant.

Density 3.988 g cm^{-3}
Formula weight   168.36

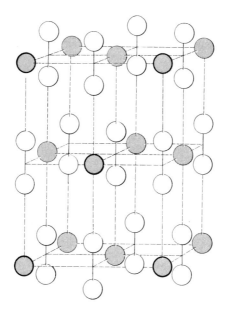

Figure 8.6
The crystal structure of calcium carbide.

3   Figure 8.6 shows the crystal structure of calcium carbide, $Ca^{2+}C_2^{2-}$.
i   Which of the following structures does this crystal *most closely* resemble?
    A   graphite type
    B   simple cubic
    C   body-centred cubic
    D   hexagonal close packed
    E   sodium chloride type
ii   Would you expect this crystal to show isotropy or anisotropy? Explain why you answer as you do.

iii   The C_2^{2-} ion has a carbon-carbon internuclear distance of 0.120 nm. Compare this with values for carbon-carbon bond lengths given in table 8.4c (page 245) and hence draw a dot-and-cross diagram for the probable structure of this ion. Name a molecule that is isoelectronic with this ion (that is, a molecule which has the same number of electrons).

iv   Compound $\Delta H_{f,298}/\text{kJ mol}^{-1}$

    $CaC_2(s)$ -63
    $Ca(OH)_2(aq)$ -1004
    $H_2O(l)$ -285
    $C_2H_2(g)$ $+226$

Use these data to calculate the standard enthalpy change for the reaction between calcium carbide and water, producing acetylene gas and calcium hydroxide. Is the reaction exothermic or endothermic?

v   Anhydrous barium peroxide $Ba^{2+}O_2^{2-}$ has a similar crystal structure to that of $Ca^{2+}C_2^{2-}$ and its reaction with water is exactly parallel also. Using this information describe its reaction with water by means of an equation.

4   The following is an extract from a textbook of structural chemistry:
'In all its compounds, nitrogen has four pairs of electrons in its valency shell. According to the numbers of the lone pairs, there are the five possibilities exemplified by the series

A	B	C	D	E
NH_4^+	NH_3	NH_2^-	NH^{2-}	N^{3-}
ammonium ion	ammonia	amide ion	imide ion	nitride ion

The last three, the NH_2^-, NH^{2-} and N^{3-} ions, are found in the salt-like amides, imides, and nitrides of the most electropositive metals.'

i   Draw dot-and-cross diagrams of the structures A to E.

ii   Sketch the shapes you would expect NH_4^+ ions and NH_3 molecules to have. Explain the differences, if any, in the HNH bond angles in NH_4^+ and NH_3.

iii   Show by means of a sketch the shape you would expect the amide ion to have. Make an estimate of the likely values of the bond angles.

iv   The ammonia molecule, NH_3, can form the positive ion NH_4^+. Would you expect methane, CH_4, to form an ion CH_5^+? Give reasons for your answer.

5   The carbonate ion, CO_3^{2-}, has a planar structure as shown in the diagram below.

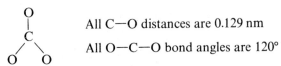

All C—O distances are 0.129 nm

All O—C—O bond angles are 120°

The diagram shows only internuclear separations and shape, not bonding

i What structures can you draw for this ion using the dot-and-cross method?

*ii Select any *one* of the structures you have drawn. What bond lengths does it suggest the ion should have? Do these agree with the observed internuclear separations? If not, how do they differ?

iii Suggest a way of explaining the nature of the bonds in the carbonate ion.

iv It has been suggested that molecules of carbonic acid, H_2CO_3 are present in low concentration in aqueous solutions of carbon dioxide. How, if at all, would you expect the bond lengths and angles between carbon and oxygen atoms in the carbonic acid molecule to differ from those in the CO_3^{2-} ion?

6 *Data for the question*

Bond			Bond length /nm	Bond energy terms/kJ mol^{-1}
N—N	hydrazine	N_2H_4	0.147	163
N=N	azomethane	CH_3—N=N—CH_3	0.120	410 ⎰ bond dis-
N≡N	nitrogen	N_2	0.110	945 ⎱ sociation energy

i Do these figures support the statement in the first paragraph of the section headed 'Bond lengths and bond energy terms' on page 244?

ii Compare the bond energy term for N—N with that for C—C in table 8.4c, page 245. Does this help to explain why large molecules based on carbon are known while those based on nitrogen are not? Give reasons for your answer.

iii Compare by means of dot-and-cross diagrams the isoelectronic molecules CO and N_2 (isoelectronic means having the same number of electrons). In view of what you know about their relative tendencies to react chemically, is there anything about their bond dissociation energies which surprises you? If so, what?

iv Draw (a) a dot-and-cross diagram for the structure of hydrazine, and (b) a sketch of a hydrazine molecule showing the bond angles you would expect to find in the molecule.

v Draw (a) a dot-and-cross diagram for the structure of azomethane, and (b) a sketch of one structure of azomethane showing the bond angles you would expect to find in the molecule. Can you then sketch a second structure showing an azomethane molecule with the same bonding but a different shape?

7 Sulphur forms a chloride, SCl_2, in the gas phase. Draw a diagram of the molecule showing the shape you would expect it to have, indicating the approximate value you would expect the bond angle to have.

8 Use the general trends in electronegativity evident from figure 8.4m, to explain the following:

i Sodium hydride has a structure which contains the ions Na^+ and H^-.

ii Carbon hydride (methane) has a covalent molecular structure and the electrons are evenly shared in the bonds between the carbon and hydrogen atoms in the methane molecules.

iii Chlorine hydride (hydrogen chloride) gas has a covalent molecular structure but the molecule has a dipole

H—Cl

$\delta+$ $\delta-$

iv Lithium forms a crystalline fluoride Li^+F^- whereas oxygen forms a gaseous fluoride OF_2.

Topic 9
Carbon chemistry, part I

The purpose of this topic and of Topic 13 is to provide an introduction to the chemistry of carbon compounds. This is a very large branch of chemistry, and is often referred to as *organic chemistry*.

We shall begin by finding out how the structures of molecules can be determined from a study of their properties. We shall then see something of the range of structures possessed by carbon compounds, and in later work concentrate our attention on a few of these.

In addition to investigating the chemical reactions and properties of a few of the types of carbon compounds, Topic 9 has several other objectives.

By comparing the behaviour of certain atoms in different molecular environments, for example the chlorine atoms in the two molecules shown in figure 9.1a, we shall aim to find out how the structures of molecules influence their properties.

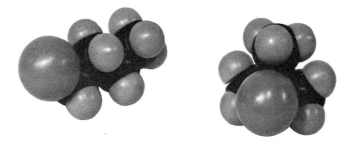

Figure 9.1a
Space-filling models showing the chlorine atom in different molecular environments.

We shall try to explain the course of some of the reactions that we meet in terms of the probable behaviour of the electrons in the molecules.

We shall also see something of the industrial importance of the materials that are being studied.

When you have understood the principles involved in this relatively simple part of carbon chemistry you will be in a good position to begin some investigations into the more complicated materials, whether they are found in living organisms or manufactured synthetically. Some carbon compounds of this more complicated type will be dealt with in Topic 18, and others in the Special Studies.

9.1   The structure of carbon compounds

The molecular structures of carbon compounds can be most astonishingly complex. Some of the more complicated carbon compounds include drugs, dyes, insecticides, vitamins, plastics, and a host of other vital materials. Complex carbon compounds also provide all the fats, proteins, carbohydrates, nucleic acids, and other materials from which living materials are made. You will appreciate that to understand the behaviour of these substances the essential thing to know is their molecular structure. The procedure is outlined in figure 9.1b.

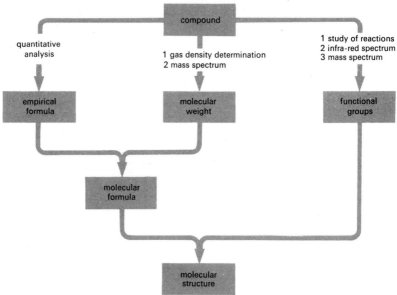

Figure 9.1b
Determination of molecular structure.

Consider the example of dodecane. Quantitative analysis will give the relative numbers of atoms of each element present in a compound. For dodecane there are thirteen hydrogen atoms to every six carbon atoms, and this result can be expressed as an *empirical formula*, C_6H_{13}. The empirical formula shows the *ratio* of the numbers of atoms of each element present in one molecule of a compound.

The next stage of the investigation is a determination of the relative molecular mass of the compound. This is needed in order to obtain the *molecular formula*.

The molecular formula is the *actual* number of atoms of each element present in one molecule of a compound. Thus the empirical formula of dodecane was C_6H_{13} but the molecular formula is $C_{12}H_{26}$.

Relative molecular masses (molecular weights) can be found by the determination of the gas density of the compound, as explained in Topic 3. At the present time determinations of the molecular weight of a carbon compound are often carried out in industrial and other research laboratories using a mass spectrometer, a method suitable for a wider range of compounds than the volatile liquids needed for the method described in Topic 3.

The use of the mass spectrometer for the accurate determination of relative atomic masses (atomic weights) has already been described in Topic 4. The instrument can also be used to determine molecular weights accurately, and this method is particularly suitable in the case of carbon compounds.

For the molecular weight determination, the compound under investigation is injected into the instrument as a vapour. (It must, of course, be stable at whatever temperature is needed to turn it to a vapour at about 10^{-6} mmHg, the pressure inside the instrument.) High velocity electrons then bombard the molecules, and produce a variety of positively-charged ions.

In the case of dodecane, $C_{12}H_{26}$, if one electron is knocked out of the molecule by the bombardment, the $C_{12}H_{26}^{+}$ ion will be formed, and the detector will show the presence of an ion of mass number 170. The electron bombardment, however, not only has the effect of knocking out electrons from the molecules; it may also break the molecules into smaller pieces, such as the $C_6H_{13}^{+}$, $C_5H_{11}^{+}$, $C_4H_9^{+}$ and other ions (see figure 9.1c).

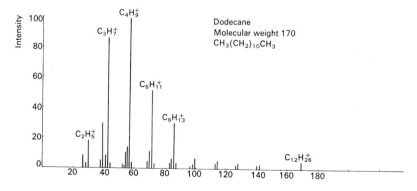

Figure 9.1c
Mass spectrum of dodecane.

Mass/charge ratio

The ion detected to have the highest mass, the 'parent ion', indicates the molecular weight of the compound. From this, and from a knowledge of the elements present, some idea of the molecular formula can be obtained by reference to tables of masses which have been compiled for the purpose. For example, if the mass of the parent ion was 200, and the compound contained C, H, and O only, possible molecular formulae would include $C_{10}H_{16}O_4$, $C_{11}H_4O_4$, $C_{11}H_{20}O_3$, and six others. Which one is correct can usually be found using a high-resolution instrument giving a higher degree of accuracy, for there are small variations in molecular weight between each of the possible examples. For example, using the values of the most abundant isotopes,

$$^{16}O = 15.99491$$

$$^{12}C = 12.00000$$

$$^{1}H = 1.007829$$

the molecular weights of the formulae given above are

$$C_{10}H_{16}O_4 = 200.1049$$

$$C_{11}H_4O_4 = 200.0110$$

$$C_{11}H_{20}O_3 = 200.1413$$

for molecules made up of atoms of the stated isotopes. The presence of other isotopes will of course lead to small amounts of molecules having slightly different masses.

Having found the molecular formula, some idea of the structure of the compound can be obtained from the ions of smaller mass, caused by the break-up of some of the original molecules under high velocity electron bombardment.

Alternatively, as the nature and proportions of these different fragments are characteristic of the original compound, the spectrum obtained can be used as a sort of 'fingerprint' for identification purposes, by comparison with the spectrum obtained from a sample of the authentic compound.

Once the molecular formula has been found, the molecular structure can be determined. This can be done by determination of, for example, the infra-red spectrum and the X-ray diffraction pattern as explained in Topic 8. Organic chemists were, however, determining molecular structures long before these techniques were invented. Their methods, still of use at the present time, involve a study of the chemical reactions of the substances being investigated. You will be following in class one or more examples of how this is done. Write a short account of the arguments used in each case.

The scope of molecular structures in simple carbon compounds
On the following three pages are the structural formulae of a number of representative carbon compounds. Have a careful look at them. Can you see any patterns that may recur, that is, any groups of atoms that appear in several different structures? If so, what are they, and in what compounds do they appear? Have these compounds similar names, and if so, in what way are they similar?

Names and structures of some representative carbon compounds
Compounds containing C and H only

```
     H              H  H              H  H  H
     |              |  |              |  |  |
  H—C—H          H—C—C—H          H—C—C—C—H
     |              |  |              |  |  |
     H              H  H              H  H  H

  methane          ethane              propane
```

```
    H  H  H  H              H  H  H
    |  |  |  |              |  |  |
 H—C—C—C—C—H            H—C—C—C—H
    |  |  |  |              |  |  |
    H  H  H  H              H     H
                               |
                            H—C—H
                               |
                               H

     butane              2-methylpropane
```

```
                                                        H
                                                        |
                                                     H—C—H
   H  H  H  H  H          H  H  H  H           H     H
   |  |  |  |  |          |  |  |  |            |     |
H—C—C—C—C—C—H        H—C—C—C—C—H          H—C—C—C—H
   |  |  |  |  |          |  |  |  |            |     |
   H  H  H  H  H          H     H  H           H     H
                             |                    |
                          H—C—H                H—C—H
                             |                    |
                             H                    H

     pentane             2-methylbutane        2,2-dimethyl-
                                                  propane
```

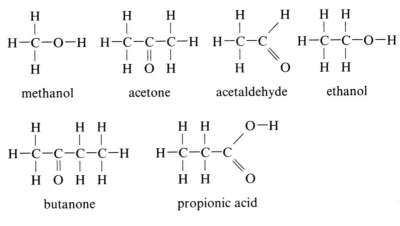

ethylene   propylene   acetylene

cyclohexane   cyclohexene

benzene   styrene

Compounds containing C, H, *and* O *only*

methanol   acetone   acetaldehyde   ethanol

butanone   propionic acid

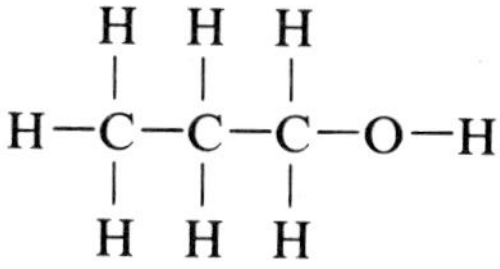

propan-1-ol

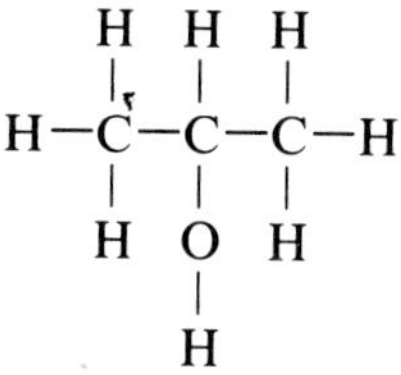

propan-2-ol

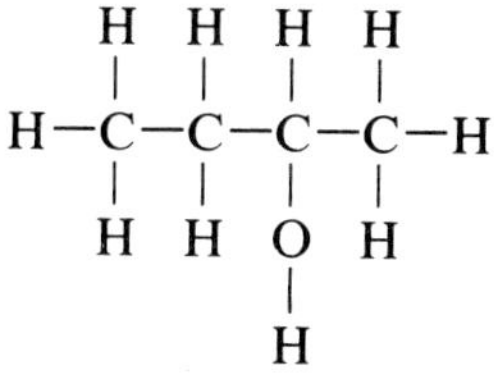

butanol-2-ol

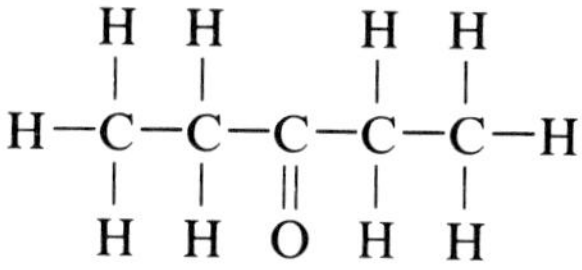

pentan-3-one

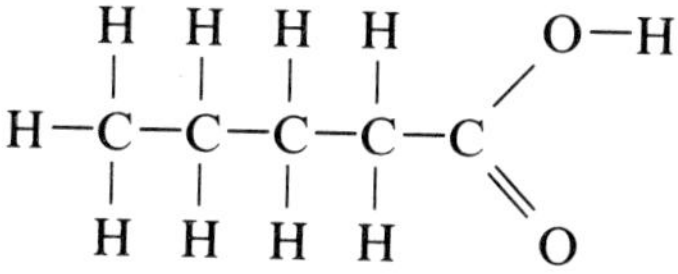

pentanoic acid

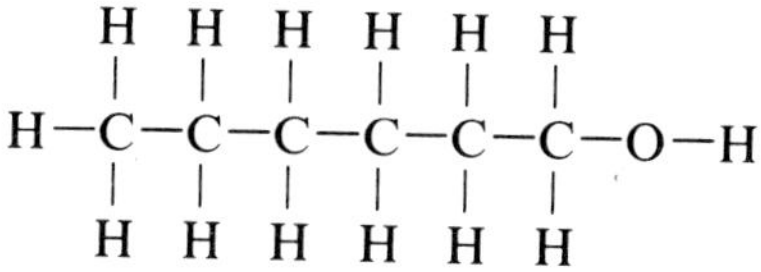

hexan-1-ol

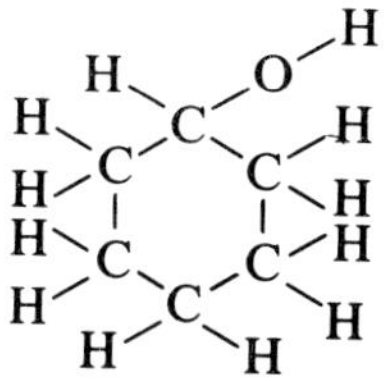

cyclohexanol

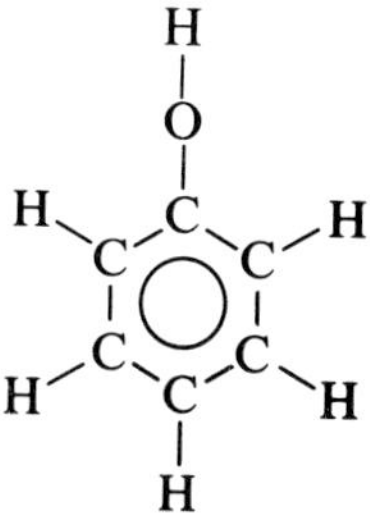

phenol

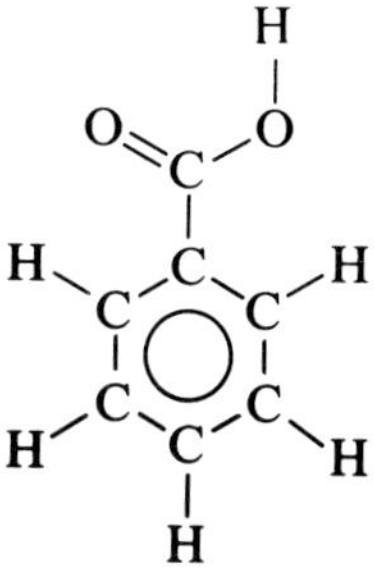

benzoic acid

Compounds containing other elements

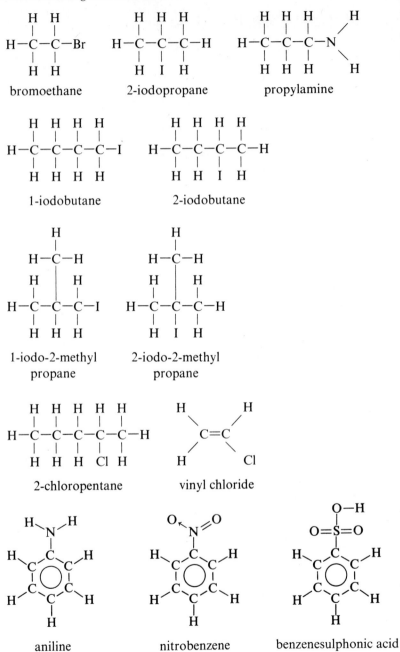

bromoethane

2-iodopropane

propylamine

1-iodobutane

2-iodobutane

1-iodo-2-methyl
propane

2-iodo-2-methyl
propane

2-chloropentane

vinyl chloride

aniline

nitrobenzene

benzenesulphonic acid

9.2   An investigation of hydrocarbons

We shall begin our experimental investigations with a study of some hydrocarbons. These materials are all extracted from naturally-occurring crude oil, and are raw materials for organic chemical synthesis.

Look up some physical data for some hydrocarbons in the *Book of Data*. Can you connect the structures of the various compounds with their names? You will probably need some help in this. Some rules of nomenclature are given in an appendix at the end of this topic.

What is the influence of structure on boiling point and melting point? Do 'branches' on the carbon chains make any *systematic* difference? Are compounds with 'rings' rather than 'chains' of carbon atoms noticeably different? This is discussed further in Topic 10.

If you detect any systematic differences, can you suggest reasons why these particular structures have the effect that they do? If samples of hydrocarbons are sealed in tubes, together with a small air bubble, and the tubes are inverted, their rates of flow can be compared (experiment 9.2a). Alternatively, viscosities can be found from the *Book of Data*. Can you see any reason for the differences that are to be found?

Having had a look at some physical properties, go on to experiment 9.2b. When you have done the experiments, the results will need to be carefully discussed, to find out what changes were taking place.

Experiment 9.2b

To compare the properties of some hydrocarbons

You are now going to compare some of the chemical properties of three hydrocarbons, hexane, cyclohexene, and benzene. Benzene vapour is poisonous and this substance should be handled in a fume cupboard. Before you begin, make sure you know and understand the structural formulae of the substances you are going to use. A good idea is to build ball-and-spoke models of the hydrocarbons.

Procedure

Carry out the following tests on a sample of each of the three hydrocarbons. Note your observations carefully in tabular form and use them to compare the behaviour of the compounds.

1 *Combustion.* Place 3 or 4 drops only of the hydrocarbon onto a hard-glass watchglass and remove the bottle to a safe place well away from flame. Set fire to the hydrocarbon and examine the luminosity and sootiness of the flame.

2 *Oxidation.* Add 2 to 3 cm^3 of a mixture of dilute potassium permanganate solution and dilute sulphuric acid to a few drops of hydrocarbon in a 150×16 mm test-tube. Shake the contents and try to tell from the colour of the permanganate if it oxidizes the hydrocarbon either at room temperature or on warming.

3 *Action of bromine.* Add a few drops of bromine dissolved in tetrachloromethane to 1–2 cm^3 of the hydrocarbon in a 150×16 mm test-tube. What happens to the colour of the bromine?

4 *Action of sulphuric acid.* Place 1–2 cm^3 of concentrated sulphuric acid in a 150×16 mm test-tube held in a test-tube rack and add drop by drop about 10 drops of the hydrocarbon. Do the substances mix or are there two separate layers in the test-tube?

Hexane is one of a type of hydrocarbons known as *alkanes*; cyclohexene is an *alkene*, and benzene is an *arene*. On the basis of your experiment, do you think you could distinguish between these types?

You should now obtain an 'unknown' hydrocarbon to examine. Try to decide if it is an alkane like hexane, an alkene like cyclohexene, or an arene like benzene.

Summarize your experimental results in the form of a table. You will be discussing in class the significance of the reactions you have investigated, especially the reaction of alkenes with bromine.

The addition reaction of alkenes with bromine

The reaction of bromine with alkenes is called an *addition reaction* because two molecules combine, or 'add' together, to produce one molecule:

$$CH_2{=}CH_2 + Br_2 \longrightarrow CH_2BrCH_2Br$$

and also

$$CH_2{=}CH_2 + Cl_2 \longrightarrow CH_2ClCH_2Cl$$

The chlorination and bromination of ethylene are industrially important. Both reactions occur readily at ordinary pressure and temperature.

In fact, addition reactions are the dominant feature of alkene chemistry (see section 13.1). We are going to look at this in a little more detail because it will lead us to an important aspect of the reactions of carbon compounds.

Some significant experimental results were reported by A. W. Francis in the *Journal of the American Chemical Society* in 1925, and part of the article now follows:

> A 1-litre pressure bottle was filled two-thirds full with a saturated salt solution, and sufficient halogen was added to saturate it. The bottle was closed with a cap containing a bicycle valve, and a moderate pressure (4 or 5 atmospheres) of ethylene was added from a cylinder. The bottle was well shaken for one minute or until the pressure had become practically atmospheric, and more ethylene was added. When the solution had become colourless, more halogen and ethylene were introduced. The process was continued until a sufficient amount of oil had been accumulated. This was separated from the aqueous solution, washed with water, dried with calcium chloride and examined by determination of density, refractive index or boiling point. In each case a mixture was obtained, and a partial separation was made by fractional distillation.
>
> From ethylene, bromine, and sodium chloride solution a mixture of 1,2-dibromoethane and 1-chloro-2-bromoethane was obtained. The product mixture, an oil, contained about 46 per cent of C_2H_4ClBr as estimated from the index of refraction, 1.51 (the value for $C_2H_4Br_2$ in the literature is 1.53; for $C_2H_4Cl_2$, 1.44). A portion was separated by fractional distillation with the following properties: b.p., 106–8°: d. 1.70 (the values for C_2H_4ClBr in the literature are as follows: b.p., 106°; d. 1.69; for $C_2H_4Br_2$, b.p., 131°; d. 2.19).
>
> 1-bromo-2-nitratoethane was obtained from ethylene, bromine, and sodium nitrate solution. Before distillation the product mixture was washed with sodium bicarbonate solution to remove any trace of nitric acid. The oil began to boil at 132° ($C_2H_4Br_2$) but a portion boiled at 163–5°; d. 1.78 (the values in the literature for $C_2H_4BrNO_3$ are as follows: b.p., 164°; d. 1.73). A sample analysed with titanium(III) chloride required about 80 per cent of the calculated amount, while pure 1,2-dibromoethane failed to oxidize titanium(III) chloride. In the distillation the last trace exploded with evolution of brown nitrous fumes, recognized also by their odour.

An equation for one reaction carried out by Francis could be written as:

$$CH_2{=}CH_2 + Br_2 + NO_3^- \longrightarrow CH_2BrCH_2NO_3 + Br^-$$

Can you make any proposals for possible reaction pathways for this reaction?

9.3 The behaviour of the hydroxyl group in primary alcohols

In this section, we are going to examine the compound ethanol. This compound is an example of a *primary alcohol*. In its molecular structure ethanol has a functional group, the hydroxyl group $-OH$, which is attached to an ethyl group, C_2H_5-. All alcohols have the same functional group; primary alcohols have this group attached to a carbon atom which is attached directly to only one other carbon atom. In addition, methanol, CH_3OH, is also a primary alcohol.

Make sure you can write the structural formula of ethanol before you begin. Can you write the structural formulae of two or three other primary alcohols? Make up space-filling or ball-and-spoke models to be sure you understand the structures associated with the formulae you have written down.

To begin, carry out experiment 9.3a so as to become familiar with the chemical properties of ethanol. In this experiment you will be converting ethanol into several other carbon compounds. The products are either gases or volatile liquids, and require some care and special techniques to handle successfully.

Wherever possible you should verify that a change has taken place by doing some simple tests on the product, to see that it is not in fact unchanged starting material, and if possible compare it with a bottle of the authentic material taken from the laboratory supply. You need only make sufficient of each product to carry out the tests; in this way you will have more time for other experiments later.

Experiment 9.3a
Some properties of ethanol

Procedure
a   The dehydration of ethanol to form ethylene
Set up the apparatus shown in figure 9.3a(i). Put cotton wool, loosely packed, to a depth of about 2 cm in the 150×16 mm test-tube and drop 2 cm^3 of ethanol onto it with a teat pipette. Clamp the tube horizontally and place granules of pumice stone in it as shown in the figure. Attach the delivery tube and arrange the rest of the apparatus as shown. Heat the pumice stone fairly strongly (but not so strongly as to melt the glass) giving an occasional flick at the ethanol with the Bunsen burner. After air has been expelled from the apparatus, collect the gas (ethylene) which is evolved in three 150×25 mm test-tubes. Close each with a cork when full. Carry out the following tests on the gas, recording the results of each one before going on to the next.

    1   Add a little bromine water to a test-tube of ethylene and shake it. Ethylene will react with the bromine to form 1,2-dibromoethane. What do you see happen?

    2   Add a little very dilute potassium permanganate solution (about 0.01M), acidified with dilute sulphuric acid, to a test-tube of ethylene and shake. What do you see happen here? What is the interpretation of this?

    3   Using the last test-tube to be filled (so as to minimize the risk of having air mixed with the ethylene) set fire to the gas and examine the nature of the flame. Make sure the gas for this experiment is collected in a suitably wide test-tube.

Ethylene is a hydrocarbon. On the basis of your tests, is it an alkane, an alkene, or an arene?

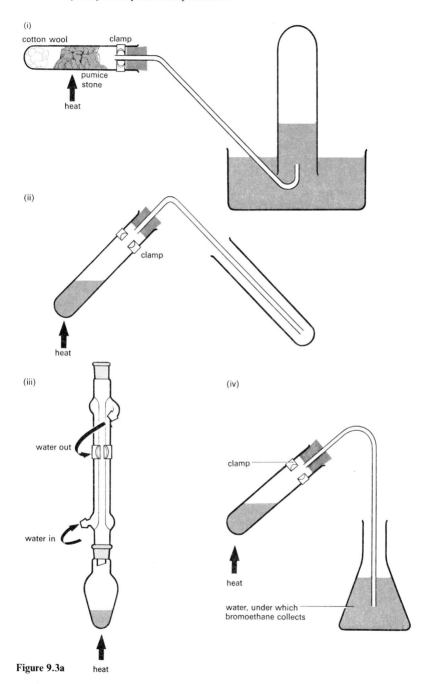

(i)
cotton wool clamp
pumice stone
heat

(ii)
clamp
heat

(iii)
water out
water in
heat

(iv)
clamp
heat
water, under which bromoethane collects

Figure 9.3a

b The oxidation of ethanol to acetaldehyde

Place about 10 cm^3 of dilute sulphuric acid in a $150 \times 25 \text{ mm}$ test-tube and add about 3 g of sodium dichromate and 2–3 anti-bumping granules. Shake the contents of the tube until solution is complete (warming if necessary but cooling afterwards). Cool the mixture and add about 1 cm^3 of ethanol in drops from a teat pipette, shaking the tube so as to mix the contents, and then assemble the apparatus as in figure 9.3a(ii). Gently distil 2–3 cm^3 of liquid into the $150 \times 16 \text{ mm}$ test-tube taking care that none of the reaction mixture splashes over. The product is an aqueous solution of acetaldehyde. Carry out the following tests on it, recording your results in tabular form. Leave a column in your table in which to record the results of each test when performed on ethanol.

  1   Notice the smell of the product.
  2   Will it neutralize an appreciable volume of sodium carbonate solution?
  3   Mix 1 cm^3 of Fehling's solution A with 1 cm^3 of Fehling's solution B, add a few drops of the product, and boil the mixture in a $150 \times 25 \text{ mm}$ test-tube. Fehling's solution contains a copper(II) compound. What might the precipitate be, and what type of reaction has taken place? What does this tell you about the nature of acetaldehyde?
  4   To 2 cm^3 of silver nitrate solution add *one drop* of bench sodium hydroxide solution, followed by sufficient bench ammonia solution just to bring the precipitate into solution. (*Caution:* Do not allow this mixture to stand as dangerously explosive silver salts are sometimes formed.)

Add a few drops of the acetaldehyde solution and warm very gently. The solution contains a silver(I) compound. What happens to it, and what type of reaction has taken place? Is this consistent with the result you obtained from test 3?

Compare these results with those obtained using ethanol (the starting material).

c The oxidation of ethanol to acetic acid

Place about 10 cm^3 of dilute sulphuric acid in a 50 cm^3 pear-shaped flask and add about 5 g of sodium dichromate and 2–3 anti-bumping granules. Shake and warm the contents of the flask until solution is complete, and then *cool thoroughly* before very carefully adding 2 cm^3 of concentrated sulphuric acid, so as to increase the concentration of the acid. *Cool again* and add about 1 cm^3 of ethanol *in drops* from a teat pipette, pointing the flask away from your face. Place a condenser in the flask for reflux, as shown in figure 9.3a(iii). Boil the contents of the flask gently, not allowing any vapour to escape, for 10 minutes. At the end of that time remove the Bunsen burner and transfer the reaction mixture to the apparatus shown in figure 9.3a(ii). Gently distil 2–3 cm^3 of liquid into the $150 \times 16 \text{ mm}$ test-tube. This is an aqueous solution of acetic acid. Carry out the tests listed for acetaldehyde. Is there a difference?

d Bromoethane from ethanol

Place about 5 cm³ of ethanol in a 150 × 25 mm test-tube and cautiously add, 0.5 cm³ at a time, about 5 cm³ of concentrated sulphuric acid. Ensure that the contents of the test-tube are well mixed, and cool them by holding the test-tube under a running cold-water tap. Quickly add 6 g of powdered potassium bromide and arrange the apparatus as shown in figure 9.3a(iv).

Heat the test-tube by means of a low Bunsen burner flame, so as to distil the bromoethane which is made, and collect this product under water so as to ensure complete condensation. *Do not remove the flame without first removing the delivery tube*, or water may be sucked up into the hot test-tube.

Next, take the conical flask, tip off the bulk of the water, and remove as much as possible of what remains using a teat pipette. The bromoethane may contain some hydrogen bromide which has been carried over during the distillation. Add approximately 20 cm³ of water to dissolve this, swirl the contents of the flask, and remove the water as before. The residual bromoethane will then be seen to be a colourless liquid somewhat milky in appearance owing to the presence of water. Carry out the following tests, comparing your results with those obtained from (a) ethanol and (b) potassium bromide solution.

   1   Put a few drops of bromoethane into some silver nitrate solution.
   2   Put a few drops of bromoethane into some sodium hydroxide solution. Shake well and warm very gently. Acidify the solution with dilute nitric acid and add silver nitrate solution.

Can you draw any conclusions concerning the type of bonding by which the bromine atom is held in bromoethane?

e Ethyl acetate from ethanol

Place 2 cm³ of ethanol in a 150 × 16 mm test-tube, add 1 cm³ of glacial acetic acid, and two or three *drops* of concentrated sulphuric acid. Warm the mixture gently over a low Bunsen burner flame for a few minutes, when ethyl acetate will be formed.

What does the product smell like? How does this compare with the smells of the starting materials? Is it like the smell of ethyl acetate from the bottle? Pour the contents of the test-tube into a small beaker of sodium carbonate solution, to neutralize any excess of acid. Stir well and smell again.

Summarize the reactions of ethanol in the form of a chart.

Now that you are familiar with the reactions of ethanol, the reactions of some other alcohols can be examined to see if the functional group has essentially the same properties no matter what alkyl group it is bonded to.

Experiment 9.3b
A comparison of other primary alcohols with ethanol
These experiments should be carried out quickly as test-tube reactions. The emphasis should be on noting signs of reaction rather than attempting to collect and examine products. If you wish, carry out the reactions with ethanol at the same time as you do them with another primary alcohol chosen from any available.

Procedure
1   To 1 cm³ of an alcohol in an evaporating basin add one small cube of sodium. Is there any sign of reaction? If so, is reaction faster or slower or much slower than with ethanol? Add ethanol to dissolve all traces of sodium before throwing away the reaction mixture.

2   To a few cm³ of dilute sulphuric acid add a few drops of potassium (or sodium) dichromate solution. Next add 2 drops of an alcohol and heat the reaction mixture until it just boils. Is there any sign of reaction? Is there any change of smell suggestive of a new organic compound?

Experiment 9.3c
The preparation of 1-bromobutane
In this experiment the object is to prepare 1-bromobutane using apparatus and techniques designed to give the maximum yield of pure product. You will learn something of the procedures necessary when handling volatile, inflammable liquids and of the importance of sound technique if good yields are to be obtained from organic reactions. You should compare this detailed procedure with the instructions given in experiment 9.3a for the preparation of a small sample of bromoethane.

In a pear-shaped 50 cm³ flask place 10 g of sodium bromide, 10 cm³ of water, and 7.5 cm³ (6 g) of butanol. The *weight* of butanol used should be determined, e.g. as the loss in weight of the measuring cylinder used to measure out the butanol. This will enable an accurate determination of yield to be made.

Set up the apparatus for 'reflux with addition' as shown in figure 9.3b(i), with the flask partially immersed in a cold water bath. Place 10 cm³ of concentrated sulphuric acid in the tap funnel and allow it to drip slowly into the reaction mixture which should be shaken gently to ensure good mixing.

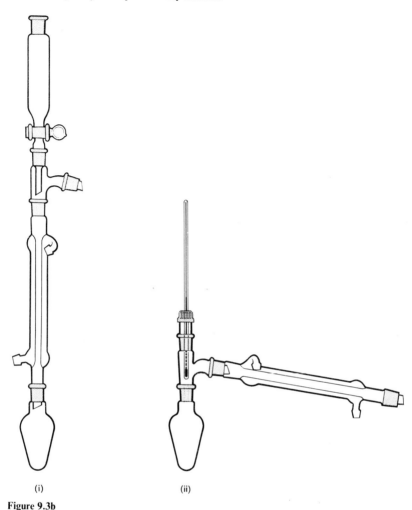

(i)   (ii)

Figure 9.3b
(*i*) Addition of sulphuric acid, followed by refluxing. (*ii*) Distillation of crude and final product.

Note the condition of the reaction mixture at this stage; it should be only slightly discoloured by bromine.

Remove the tap funnel from the top of the condenser and the water bath surrounding the flask. Support the flask on a tripod and gauze. Heat for 45 minutes with a low Bunsen burner flame so that the mixture refluxes gently. If the refluxing is too vigorous, liquid may be 'held up' in the condenser; if this happens, remove the flame temporarily or liquid may be lost from the top of the condenser.

At the end of the reflux period allow the apparatus to cool and then rearrange it for the distillation as in figure 9.3b(ii). (At this stage the thermometer is not necessary and may be replaced by a stopper.) There should be two layers in the flask; which do you imagine is the organic layer? The table of densities (table 9.3) may help you to decide. By distilling the reaction mixture all non-volatile substances will be separated from volatile ones.

Liquid	Density/g cm^{-3}
butanol	0.81
water	1.0
concentrated hydrochloric acid	1.2
1-bromobutane	1.3
sulphuric acid and water (1:1)	1.4

Table 9.3
Densities of liquid used in experiment 9.3c

Boil the reaction mixture and collect the distillate in a measuring cylinder. The bromobutane distils over together with water and the 'steam distillation' should be continued until the upper layer in the flask has disappeared and no more oily drops are seen in the condenser. About 10 cm^3 will be collected.

Dismantle your apparatus, clean it, and put it in an oven to dry.

Note the volumes of the two liquids collected. What substances are they? Mistakes can be made by trying to remember by rote rather than by understanding, which layer of two immiscible liquids is the desired product and which is to be thrown away.

Remove the upper layer of water using a teat pipette and transfer the lower organic layer to a tap funnel. Then wash the crude product with concentrated hydrochloric acid to remove any unreacted butanol.

Swill the measuring cylinder with about 10 cm^3 of concentrated hydrochloric acid to collect any traces of crude product and transfer to the tap funnel. Stopper the tap funnel and shake well for a minute (after the first shake release any excess of pressure). Allow the two layers to separate completely (which is the organic layer now?) and then run the lower layer into a clean measuring cylinder. What is the volume of product now? How do you account for any change?

Since hydrogen chloride is soluble in organic liquids and is volatile, it is now necessary to wash the bromobutane with sodium hydrogen carbonate solution (0.5 g in 10 cm^3 of water). Why do you imagine a weak base has been chosen for washing the bromobutane?

Transfer the bromobutane from the measuring cylinder to a cleaned tap funnel. Swill the measuring cylinder with the sodium hydrogen carbonate solution and add to the tap funnel. Shake the tap funnel cautiously making sure that excess pressure due to the formation of carbon dioxide is regularly released.

Allow the two layers to separate and run the lower layer, which will probably be cloudy, into a small (50 cm^3) conical flask, taking care to let none of the aqueous layer through. The bromobutane must now be dried to remove any water emulsion and also dissolved water.

Add small quantities of anhydrous sodium sulphate and swirl the conical flask after each addition until it can be seen that the bromobutane is crystal clear. Stopper the conical flask and allow to stand for 5 minutes.

Set up the cleaned apparatus for distillation as in figure 9.3b(ii) using a 0–110 °C thermometer. Transfer the dried bromobutane to the pear-shaped flask through a small filter funnel containing a small cotton wool plug to retain the desiccant. The desiccant and cotton wool should be well pressed to remove as much bromobutane as possible. Much product can be lost at this stage by poor technique.

Boil the bromobutane and collect the distillate which boils in the range 100–103 °C in a preweighed specimen bottle. Determine the weight of pure product and calculate the yield based on the butanol used:

$$CH_3CH_2CH_2CH_2OH \longrightarrow CH_3CH_2CH_2CH_2Br$$

m.w. 74.1   m.w. 137

The yield should be 50–70 per cent.

Summary of the reactions of primary alcohols

1   *Dehydration to alkenes*

Concentrated phosphoric acid is a good dehydrating agent for alcohols, or alcohol vapour can be passed over a catalyst such as pumice at about 400 °C.

$$RCH_2CH_2OH \longrightarrow RCH=CH_2 + H_2O$$

2 Oxidation to form a C=O group

When alcohols are treated with oxidizing agents such as acidified sodium dichromate, aldehydes may be obtained provided they are not exposed to further oxidation.

$$3RCH_2OH + Cr_2O_7^{2-} + 8H^+ \longrightarrow 3RCHO + 2Cr^{3+} + 7H_2O$$

Under more vigorous conditions and with an additional quantity of oxidizing agent carboxylic acids are obtained.

$$3RCH_2OH + 2Cr_2O_7^{2-} + 16H^+ \longrightarrow 3RCO_2H + 4Cr^{3+} + 11H_2O$$

3 Substitution to form halogenoalkanes

Different procedures are adopted for the different halogens, dependent on their relative reactivity and the halogen compounds available, but the overall reaction can be expressed

$$ROH + H\,Hal \rightleftharpoons R\,Hal + H_2O$$

Conditions are usually selected that will remove water from the reaction mixture, or provide an excess of halogen reagent.

$$ROH + HCl \longrightarrow RCl + H_2O$$

(and anhydrous $ZnCl_2$ to remove water)

$$ROH + HBr \longrightarrow RBr + H_2O$$

(HBr added as NaBr and H_2SO_4)

$$ROH + PI_3 \longrightarrow RI + H_3PO_3$$

(PI_3 added as iodine and red P)

Alternative reagents are bromine and red phosphorus, or potassium iodide and phosphoric acid. Consult a practical book if you wish to see how these different reagents are used.

4 Formation of esters

Reaction of an alcohol with a carboxylic acid eliminates water and forms an ester

$$RCO_2H + R'OH \rightleftharpoons RCO_2R' + H_2O.$$

An acid catalyst is usually added.

Revision question

You should now be able to deduce what reagents, organic and inorganic, are required for the synthesis in one step from a primary alcohol of the following products.

1   oct-1-ene
2   1-iodopentane
3   ethyl heptanoate
4   propanal

Background reading
        An alcohol in insects
The freezing point of an insect is the temperature at which some of its body water
can be frozen. For most insects this would normally be about $-1\,^{\circ}C$ but it
depends upon the concentration and nature of the solute in the tissues. Just as
wise motorists protect their car radiators by the addition of antifreeze in which
the important component is ethane-1,2-diol (glycol), so it appears that some
insects accumulate propane-1,2,3-triol (glycerol) in their body water during the
autumn (figure 9.3c). In this way the insects can withstand low temperatures,
and avoid the danger of cell damage by the formation of ice crystals. Additional
protection is gained because the formation of ice crystals is delayed to below the
freezing point (supercooling). If they contain 15 per cent propane-1,2,3-triol
the eggs of the moth *Alsophila pometaria* can be supercooled to $-45\,^{\circ}C$ before
ice crystals form.

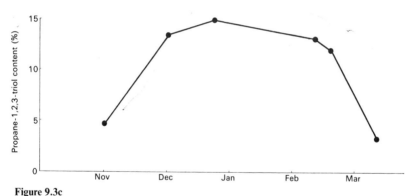

Figure 9.3c
Propane-1,2,3-triol content in the moth *Alsophila pometaria*.

9.4   **What is the effect of attaching a hydroxyl group to a benzene ring?**

In alcohols the hydroxyl group, $-OH$, is attached to an alkyl group. You are
now going to examine phenol, a compound the molecules of which have a
hydroxyl group attached to a benzene ring. Your principal objective will be to
find out if the properties of the hydroxyl group are modified by being attached
to this different type of group, but you will also see some of the characteristic
features of the chemistry of phenol.

Experiment 9.4a

To investigate the properties of phenol

Note. Phenol is corrosive, and both crystals of it and solutions containing it will cause unpleasant blisters if allowed to come in contact with your skin. *You must handle it very carefully.*

Procedure

1   Place about $5\,cm^3$ of water in a $150 \times 16\,mm$ test-tube and add a few small crystals of phenol. Notice the characteristic smell of phenol. Does phenol dissolve in water?

Test the phenol-water mixture with universal indicator. What is the pH of the solution? Compare the result with the effect of ethanol on universal indicator.

2   *Action of sodium hydroxide.* To about $5\,cm^3$ of 2M sodium hydroxide solution in a test-tube add about one-fifth of this volume of phenol crystals, and notice how much more easily they dissolve in alkali than in water. Now add about $2\text{--}3\,cm^3$ of concentrated hydrochloric acid. What do you observe and what does this tell you?

3   *Action of sodium.* Place a few crystals of phenol in a dry test-tube and warm them until they are molten. Add a small cube (2–3 mm side) of sodium and watch carefully. Do not heat the tube continuously. What do you observe? Dispose of the contents with care: do *not* pour down the sink in case some sodium remains but carefully add $1\text{--}2\,cm^3$ of ethanol to the test-tube, wait until all bubbling has ceased, and then pour away.

Did ethanol behave similarly to phenol?

4   *Action of iron(III) chloride.* To a small portion of a solution of phenol in water add a few drops of neutral iron(III) chloride solution and notice the intense violet colour that is formed. The formation of intense colours is characteristic of several compounds having hydroxyl groups attached to benzene rings. Find out if it also happens with ethanol.

5   *Combustion.* Set fire to a small crystal of phenol on a piece of porcelain or crucible lid held with a pair of tongs. What sort of flame do you see? Is it like the flame obtained when ethanol is burnt? Recollect the combustion of hydrocarbons in experiment 9.2b; does the presence of a hydroxyl group alter the result?

To explain the differences in chemical properties between ethanol and phenol their structures must be looked at more closely, and also the structures of their reaction products. Thus to explain the greater acidity of the hydroxyl group in phenol as compared to ethanol complete processes must be considered.

$CH_3CH_2OH + $ (proton acceptor) $\rightleftharpoons CH_3CH_2O^- + H^+$ (proton acceptor)

Structural studies show that the $C{-}O$ bond distance in alcohols is 0.143 nm, whereas in phenol it is only 0.136 nm. Thermochemical data indicate a stabilization energy which is 21 kJ mol^{-1} greater for phenol than for benzene itself. This suggests that there is an interaction between the hydroxyl group and the benzene ring. It is considered that one of the non-bonding pairs of electrons of the oxygen atom in phenol is delocalized, entering into the delocalized system of the benzene ring. This withdrawal of electrons from the hydroxyl group should promote the formation of a proton.

There is no tendency to draw electrons away from the hydroxyl group in ethanol nor is there any special factor to stabilize the ethoxide ion $CH_3CH_2O^-$ if formed. In contrast the phenoxide ion, O^- can be considerably stabilized

by its negative charge interacting with the delocalized electrons of the benzene ring.

In general it seems likely therefore that the behaviour of a functional group will be influenced by its possible interactions with hydrocarbon groups.

Phenol differs from ethanol in other respects. It does not react with potassium bromide and sulphuric acid to give a halogen compound as does ethanol. Halogen compounds such as chlorobenzene, C_6H_5Cl, and bromobenzene, C_6H_5Br, can be made, but different methods have to be employed. Such compounds are normally made by the action of the halogen on the hydrocarbon as mentioned at the end of this section.

Phenol does not react with acetic acid to form an ester in the easy manner that ethanol does. Esters can, however, be prepared easily if an acid chloride is used instead of an acid. (Acid chlorides are more reactive than acids.) In the next experiment you will make the ester phenyl benzoate using an acid chloride called benzoyl chloride. This compound is related to benzoic acid, as the following structural formulae show, and both compounds react with alcohols to give esters known as benzoates.

benzoic acid   benzoyl chloride   phenyl benzoate

Experiment 9.4b
The preparation of phenyl benzoate

In this experiment, you will make phenyl benzoate by the reaction of phenol with benzoyl chloride. Your aim in preparative work of this sort must be to make as pure a product as possible, in as high a yield as possible.

Procedure

Select a wide-mouthed glass-stoppered bottle of about 250 cm^3 capacity, and in it place 90 cm^3 of 2M sodium hydroxide solution and 5 g of phenol. Go to a fume cupboard and pour into the bottle with care 9 cm^3 of benzoyl chloride. (*Caution:* This liquid must be kept in the fume cupboard and not spilt on the hands or clothing. Its vapour is harmful and produces an extremely irritating effect on the eyes.) Grease the stopper of the reaction bottle lightly with Vaseline and replace it in the bottle, then, holding the stopper in place securely, shake the bottle vigorously at intervals for a period of 15 minutes.

At the end of this time collect the material obtained by suction filtration using a Buchner filtration apparatus, preferably still in the fume cupboard, for traces of benzoyl chloride may still persist. Break up any lumps on the filter-paper with a glass rod, but be careful not to puncture the paper. Wash the solid well with water and discard the filtrate. Recrystallize the solid ester from ethanol, taking care to see that just sufficient is used to prevent the ester separating as an oil at a temperature above its melting point. The phenyl benzoate is obtained as white crystals, which melt under 100 °C. Use a hot water bath; do not heat the ethanol directly with a Bunsen burner flame because it may catch fire.

Take the melting point of your sample and calculate the percentage yield you have obtained.

One form of apparatus used for the determination of melting points is illustrated in figure 9.4. It consists of a hard-glass boiling tube partly filled with dibutyl phthalate and holding a thermometer and a copper wire stirrer. Put a sample of phenyl benzoate into a small thin-walled capillary tube sealed at one end, and by gentle tapping, or rubbing with the milled edge of a coin, transfer it to the closed end. Fix the tube in the position shown in the figure by means of a rubber band. Slowly heat the boiling tube by means of a very low Bunsen burner flame and move the stirrer up and down so as to maintain an even rise of temperature. Watch the crystals in the melting point tube carefully, and the moment they melt make a note of the temperature. Repeat the process with a fresh melting point tube containing another portion of the compound, in order to obtain a more accurate value for the melting point. The temperature may now be raised rapidly to within 10° of the melting point previously obtained, but must then be raised very slowly with constant stirring (about 2° rise per minute) until the crystals melt. The temperature at which the crystals first melt and also the temperature at which melting is complete should be noted. For pure substances this melting range is narrow and the melting point is termed 'sharp'.

If the compound under examination is then recrystallized and dried, and the melting point again determined, it may be found to be a little higher than before. This is because the melting point of a pure compound is always lowered by the presence of impurities. The compound can be made completely pure by repeated recrystallization until constancy of melting point has been obtained.

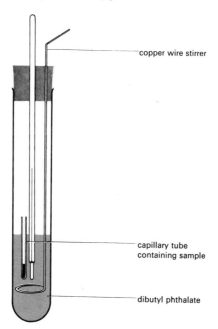

copper wire stirrer

capillary tube
containing sample

dibutyl phthalate

Figure 9.4
Melting point apparatus.

Background reading
The importance of phenol

Phenol has given its name to all compounds which have a hydroxyl group attached to a benzene ring; they are known collectively as 'the phenols'. Examples include the three isomeric *cresols*:

OH $\quad$ CH₃   OH $\quad$ CH₃   OH $\quad$ CH₃

ortho-cresol $\qquad$ *meta*-cresol $\qquad$ *para*-cresol

and the xylenols, which have two methyl groups and one hydroxyl group attached to a benzene ring. By writing out the formulae you may like to satisfy yourself that there are six possible isomeric xylenols.

Notice the prefixes given to the word *cresol* to distinguish between the three isomers formulated above. These are used for all compounds having two substituents attached to the same benzene ring; *ortho* (*o*-) if the substituents are attached to adjacent carbon atoms, *meta* (*m*-) if they are separated by one carbon atom, and *para* (*p*-) if they are separated by two carbon atoms.

Phenol itself is an industrial raw material of considerable importance. Its use as an antiseptic, first practised by Lister in the 1860s, is widely known and it has been publicised as a component of soap since the last century, e.g. Pears' 'Carbolic Soap' and Wrights' 'Coal Tar Soap'. Far more important currently is its use in the production of phenolic resins used in plastics manufacture (e.g. Bakelite). The laboratory preparation of some phenolic resins is described in Topic 18.

A number of substituted phenols have important uses. One interesting example is their use as germicides in proprietary articles such as Dettol and TCP, a modern development of Lister's discovery:

4-chloro-3,5-dimethylphenol (or Dettol)

The effectiveness of germicides can be expressed as a phenol coefficient:

$$= \frac{\text{concentration of phenol required to kill the germs}}{\text{concentration of germicide required to kill the same germs}}$$

Some values obtained using the germ *Salmonella typhosa* are given below.

Germicide	Phenol coefficient
Phenol	1
2-chlorophenol	4
2,4-dichlorophenol	13
2,4,6-trichlorophenol	23
Dettol	280

9.5 How does the hydroxyl group behave as part of the carboxyl group?

You have already seen something of the properties of two types of compounds containing the —OH group of atoms. In the next experiment you are going to examine a third type of compound, a carboxylic acid, taking as the example for study acetic acid.

In carboxylic acids the hydroxyl group is attached to an acyl group,

where R stands for any alkyl group. Once again, try to use the reactions to see how the properties of the hydroxyl group are modified by being in this different type of environment. Is there any difference in acidity? Can you offer an explanation of any differences that you may find?

Figure 9.5a
Space-filling models of acetic acid, ethanol, and phenol.

Experiment 9.5
Some properties of acetic acid

Unless otherwise stated, the name *acetic acid* refers to pure acetic acid, sometimes known as *glacial acetic acid*. Glacial acetic acid freezes at 17 °C and contains no water. To use dilute acetic acid, which does contain water, in some
reactions could be hazardous.

Procedure

1 Place a few drops of acetic acid in an evaporating basin and add water
in drops. Are the two miscible in all proportions? What is seen when universal
indicator is added to the mixture? What is the pH of the solution?

2 *Action of sodium hydroxide.* Place a few cm³ of sodium hydroxide
solution in a basin. Add acetic acid from a teat pipette, stirring the mixture with
a glass rod, until a drop withdrawn on the glass rod and placed on a piece of
universal indicator paper is found to have a pH of about 7. Evaporate the
contents of the basin over a water bath to obtain crystals of hydrated sodium
acetate. What shape are the crystals? If evaporation is slow, carry on with parts
(3) and (4) of this experiment whilst you are waiting.

3 *Action of sodium.* Place 1–2 cm³ of *glacial* acetic acid in a dry basin and
add one small cube of sodium. Do not pour this mixture away down the sink
unless you are sure that all the sodium has finished reacting. If in doubt, add a
little more acid. On no account add any water. Finally empty out the basin and
clean it.

4 *Action of iron(III) chloride.* To a few cm³ of a solution of sodium acetate
in water (so that the solution is neutral) add a few drops of neutral iron(III)
chloride solution. How does the result compare with that obtained for phenol?

5 *Action of phosphorus pentachloride.* Place 1–2 cm^3 of *glacial* acetic acid in a dry evaporating basin and add a little phosphorus pentachloride. Is there any sign of reaction? Be sure not to leave the stopper off the phosphorus penta-chloride bottle because it reacts readily with moisture.

Background reading
Naturally occurring carboxylic acids

Lipids are cell materials which are insoluble in water. They are commonly known as fats or oils and their molecules are esters of the alcohol, $CH_2OHCH(OH)CH_2OH$, with long chain carboxylic acids e.g. $CH_3(CH_2)_{16}CO_2H$. Systematic names for the alcohol and acid are propane-1,2,3-triol and octadecanoic acid while the more commonly used trivial names are glycerol and stearic acid. Other lipids of more complex structure also occur.

The important function of lipids is to act as an energy store, as they have the best calorific value of all foods; plant seeds are particularly rich in lipids for this reason. Nutmegs for example yield up to 40 per cent of fat. A feature of lipids, which is related to their metabolism, is that all the commonly occurring carboxylic acids have an even number of carbon atoms. This feature of lipid metabolism has been investigated by feeding animals synthetic arene-substituted carboxylic acids. A C_{even} alkyl chain is completely assimilated and benzoic acid is excreted:

$$\langle\bigcirc\rangle - (CH_2)_{even} - CO_2H \xrightarrow{\text{degrades to}} \langle\bigcirc\rangle - CO_2H$$

But a C_{odd} alkyl chain is degraded to phenylacetic acid, showing the animal's inability to utilize the odd carbon atom:

$$\langle\bigcirc\rangle - (CH_2)_{odd} - CO_2H \xrightarrow{\text{degrades to}} \langle\bigcirc\rangle - CH_2 - CO_2H$$

A number of fats and oils are of economic importance as they are used to make foodstuffs, such as margarine, and also soap. Because of the value of fats and oils for foodstuffs, synthetic detergents based on petrochemicals were introduced, but these original synthetic detergents caused serious pollution in rivers and proved to be toxic to fish. These problems did not arise with soap because it could be degraded by bacteria. The original synthetic detergents had branched-chain alkyl groups and could not be degraded. Biodegradable synthetic detergents have now been introduced which have straight-chain alkyl groups similar to the alkyl chains in natural fats and oils (figure 9.5b).

a soapy detergent

$$\wedge\!\wedge\!\wedge\!\wedge\!\wedge\!\wedge\!\wedge\!\wedge\!\wedge\, CO_2^- \, Na^+$$

a synthetic detergent

$$\wedge\!\wedge\!\wedge\!\wedge\!\wedge\!\wedge\!\wedge\!\wedge -\!\!\bigcirc\!\!- SO_3^- Na^+$$

a biodegradable synthetic detergent

$$\wedge\!\wedge\!\wedge\!\wedge\!\wedge\!\wedge\!\wedge\!\wedge -\!\!\bigcirc\!\!- SO_3^- Na^+$$

Figure 9.5b
Structure of detergent molecules (schematic).

Another use for fats and oils is to provide carboxylic acids as starting materials for the synthesis of organic compounds with eight or more carbon atoms (see table 9.5). The synthesis of C_{even} compounds is straightforward but C_{odd} compounds require the introduction of an extra carbon atom (or removal of one atom). The extra problems in obtaining C_{odd} compounds are reflected in their cost.

Number of carbon atoms	Naturally occurring compound	Systematic name	Trivial name	Source
8	$CH_3(CH_2)_6CO_2H$	octanoic acid	caprylic acid	coconut oil
10	$CH_3(CH_2)_8CO_2H$	decanoic acid	capric acid	coconut oil
12	$CH_3(CH_2)_{10}CO_2H$	dodecanoic acid	lauric acid	coconut oil
14	$CH_3(CH_2)_{12}CO_2H$	tetradecanoic acid	myristic acid	nutmeg seed fat
16	$CH_3(CH_2)_{14}CO_2H$	hexadecanoic acid	palmitic acid	palm oil
	$CH_3(CH_2)_{14}CH_2OH$	hexadecanol	cetyl alcohol	sperm whale oil
18	$CH_3(CH_2)_{16}CO_2H$	octadecanoic acid	stearic acid	animal fats
	$CH_3(CH_2)_7CH=CH(CH_2)_7CO_2H$	octadec-9-enoic acid	oleic acid	olive oil
	$CH_3(CH_2)_7CH=CH(CH_2)_7CH_2OH$	octadec-9-en-1-ol	oleyl alcohol	sperm whale oil

Table 9.5
Some naturally occurring carboxylic acids and alcohols.

Summary of reactions of carboxylic acids

A number of long-chain carboxylic acids are available from naturally occurring esters and therefore serve as a useful source of long-chain carbon compounds e.g. coconut oil contains esters of dodecanoic (lauric) acid, $C_{11}H_{23}CO_2H$, and spermaceti (sperm-whale oil) contains esters of hexadecanoic (palmitic) acid, $C_{15}H_{31}CO_2H$.

1   Reduction

Carboxylic acids are reduced to alcohols by lithium aluminium hydride, $LiAlH_4$. (Aldehydes can also be reduced to alcohols.) Esters can be reduced to alcohols by the action of sodium in the presence of ethanol, e.g. with an ethyl ester

$$RCO_2C_2H_5 + 4Na + 2C_2H_5OH \longrightarrow RCH_2O^-Na^+ + 3C_2H_5O^-Na^+$$

$$\downarrow H_2O$$

$$RCH_2OH + 3C_2H_5OH$$

2   Formation of halogenoalkanes

If silver salts of carboxylic acids are suspended in an inert solvent and treated with bromine at room temperature, bromoalkanes can be obtained in yields of 60 to 90 per cent:

$$RCO_2^-Ag^+ + Br_2 \longrightarrow RBr + Ag^+Br^- + CO_2$$

3   Replacement of the —OH group by halogen

Treatment of carboxylic acids with phosphorus pentachloride gives acyl chlorides in yields of 60 per cent or better:

$$RCO_2H + PCl_5 \longrightarrow RCOCl + POCl_3 + HCl$$

4   Formation of amides

The thermal dehydration of ammonium salts gives good yields of amides provided conditions are chosen to displace the possible equilibria in favour of the amide

$$RCO_2H + NH_3 \rightleftharpoons RCO_2^-NH_4^+ \rightleftharpoons RCONH_2 + H_2O$$

5   Formation of esters

Reaction of a carboxylic acid with an alcohol eliminates water and forms an ester

$$RCO_2H + R'OH \rightleftharpoons RCO_2R' + H_2O$$

An acid catalyst is usually added. By using appropriate conditions the reaction can be reversed and good yields of carboxylic acid and alcohol can be obtained by the hydrolysis of naturally occurring esters.

Revision question

You should now be able to deduce what reagents, organic and inorganic, are required for the synthesis in one step of the following products:

1   1-bromohexane
2   hexanamide
3   butyl hexanoate
4   pentan-1-ol

9.6   The behaviour of halogen atoms in carbon compounds

In the previous three sections you have examined some compounds in which the hydroxyl group is attached to three different arrangements of atoms. These arrangements, or *groups* were:

1   The ethyl group, $C_2H_5—$, an example of an *alkyl* group
2   The phenyl group, $C_6H_5—$, an example of an *aryl* group
3   The acetyl group, $CH_3CO—$, an example of an *acyl* group.

You will have noticed that these groups have a marked effect on the properties of the hydroxyl group.

In this section you are going to examine the behaviour of halogen atoms attached to these same three groups, alkyl, aryl, and acyl. You will be comparing the properties of some halogenoalkanes with those of chlorobenzene, C_6H_5Cl, and acetyl chloride, CH_3COCl (see figure 9.6a). Derivatives of butane are suitable halogenoalkanes for study; members of this series of lower molecular weight are volatile and this makes them inconvenient for our experiments.

Figure 9.6a
Space-filling models of acetyl chloride, chlorobutane, and chlorobenzene.

Try to use the experience gained in comparing the properties of the hydroxyl group in different molecular situations when making this comparison.

Experiment 9.6a
A comparison of the rates of hydrolysis of various halogenoalkanes

Procedure

   1   Can you recall the action of silver nitrate solution on solutions containing chloride, bromide, and iodide ions (experiment 5.1b)? If you cannot, turn up your results and repeat the experiment if necessary.

   2   Examine 1-bromobutane by the same technique as (1). Place a few *drops* of 1-bromobutane into 2–3 cm³ of approximately 0.1 M silver nitrate solution, shake them together and leave to stand for a few minutes.

Do you think that 1-bromobutane contains bromide ions? How else might the bromine atom be combined in 1-bromobutane? Have the substances mixed together well? If not, can you think of a better way in which they might be brought together?

Are bromide ions formed after a few minutes? If so, how do they arise? Try to write an equation for the change that might have taken place, and have it verified by your teacher before you continue.

   3   You may have concluded from the preliminary experiments of (2) that 1-bromobutane reacts slowly with water.

You can now carry out an experiment to compare the behaviour of 1-chloro-butane, 1-bromobutane, and 1-iodobutane. In each of three test-tubes place 1 cm³ of ethanol. Using separate teat pipettes, in one test-tube place 2 drops of 1-chlorobutane, in the next 2 drops of 1-bromobutane, and in the third 2 drops of 1-iodobutane. Stand the test-tubes in a beaker of water at about 60–65 °C and place another test-tube containing about 5 cm³ of approximately 0.1 M silver nitrate solution in the warm water. Wait until the contents of the tubes have reached approximately 60° and then place 1 cm³ of silver nitrate solution in each of the other test-tubes, and quickly shake the tubes to mix the contents. Note and time carefully what you observe throughout the next five minutes.

Does changing the halogen atom have an effect on the rate of hydrolysis? If so, which is fastest and which slowest? Does the higher temperature have an effect on the performance of 1-bromobutane?

Record exactly how you do the experiment, and the results that you obtain. Draw what conclusions you can, and state them clearly. Consider in particular the possible polarization of the C—Hal bonds and the relative ease of formation of the three halide anions.

Experiment 9.6b
The hydrolysis of chlorobenzene

Do you think the hydrolysis of chlorobenzene will be quicker or slower than that of chlorobutane? Test your prediction by attempting the hydrolysis of chlorobenzene as described for bromobutane, experiment 9.6a, part 2. If no hydrolysis appears to take place, raise the temperature, if necessary until the contents of the tube are boiling.

Was your prediction correct?

Experiment 9.6c
The hydrolysis of acetyl chloride

Place 5 cm^3 of aqueous silver nitrate solution in a 100 cm^3 beaker and from a teat pipette add *with great caution* 3 drops of acetyl chloride.

How readily is acetyl chloride hydrolysed?

You will be shown some of the other reactions of acetyl chloride.

Draw up a table comparing the behaviour towards water of the 1-halogenobutanes, chlorobenzene and acetyl chloride.

Background reading
Manufacture and uses of halogen compounds

Halogen compounds are almost entirely absent from the field of naturally-occurring materials and so almost all of them must be made synthetically. Those few which do occur in nature are found in rather obscure situations. Examples include the iodine compound *thyroxine*, a hormone produced by the thyroid gland, a lack of which causes goitre and cretinism; and the bromine compound *Tyrian purple*, present in the viscera (gut) of a type of sea-snail called *Purpura aperta*, which was extracted and used by the Romans for dyeing their leading statesmen's robes.

Owing to the considerable reactivity of the halogen atoms in halogenoalkanes many of these compounds are manufacturered as 'intermediates', that is, for conversion into other substances. An example is chloroethane. This is made by the action of chlorine on a mixture of ethane and ethylene obtained by cracking hydrocarbons in an oil refinery.

$$
\boxed{\begin{array}{l} CH_3CH_3 \\ CH_2{=}CH_2 \end{array}} \xrightarrow[\text{at 400°}]{Cl_2} \boxed{\begin{array}{l} CH_3CH_2Cl \\ CH_2{=}CH_2 \end{array}} + HCl \xrightarrow[\text{at 200°}]{AlCl_3 \text{ catalyst}} \boxed{\begin{array}{l} CH_3CH_2Cl \\ CH_3CH_2Cl \end{array}}
$$

    refinery gas   first stage product   final product

The reaction is carried out at 400 °C. Under these conditions ethane undergoes a substitution reaction which does not proceed beyond the formation of chloro-ethane, but at this high temperature ethylene does not react with chlorine. If the mixed gases are then cooled to about 200 °C and passed over a catalyst of aluminium chloride, ethylene reacts with the hydrogen chloride formed in the first reaction to make more chloroethane.

Chloroethane is made principally for conversion to tetraethyl lead. It is heated with a lead-sodium alloy under pressure.

$$4C_2H_5Cl + 4Na/Pb \longrightarrow (C_2H_5)_4Pb + 4Na^+Cl^- + 3Pb$$

Tetraethyl lead, as previously mentioned on page 35, is an additive for motor spirit (gasoline), in which it is used as an 'antiknock'. This is a substance which promotes even burning of the fuel-air mixture in the engine, and prevents explosive combustion or *knocking*. Its inclusion in gasoline accounts for the manufacture of another halogen compound, 1,2-dibromoethane.

When tetraethyl lead burns, lead oxide is formed. This is a non-volatile solid, which would slowly coat the cylinders, unless steps were taken to prevent this happening. To avoid this, 1,2-dibromethane is also added to gasoline. On com-bustion this ensures that the lead in the tetraethyl lead is converted to lead(IV) bromide, which is volatile, and is swept away with the exhaust gases. 1,2-dibromo-ethane is made by addition of bromine to ethylene. Almost all of the European consumption of bromine for this compound is made at Amlwch, Anglesey, from bromine extracted from sea water. This works is shown in the Nuffield film loop 'Bromine Manufacture'.

A number of halogen compounds are manufactured for use as insecticides. Well-known examples include:

BHC ('Benzene HexaChloride')
This is made by passing chlorine through liquid benzene irradiated by ultraviolet light. It is particularly valuable in the fight against the locust. The devastation caused by locust swarms can be judged from figure 9.6b.

Figure 9.6b
Devastation caused by locusts. *Top:* a typical productive orange grove near Tripoli, Libya. *Bottom:* a similar orange grove after a locust attack in the Souss Valley, Libya. *Photo, Shell.*

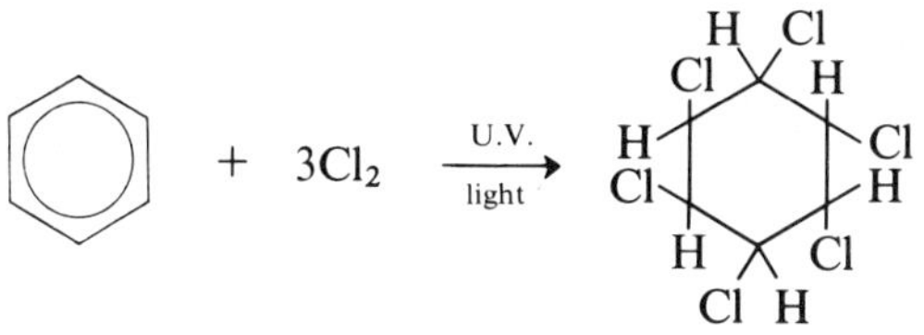

DDT ('Dichloro-Diphenyl-Trichloroethane')

This is made by reaction between chlorobenzene and trichloroacetaldehyde (chloral). It has been used extensively against mosquitoes in an effort to eliminate malaria from various regions—in particular Sicily and Southern Italy.

The equation for the manufacture of DDT is as follows:

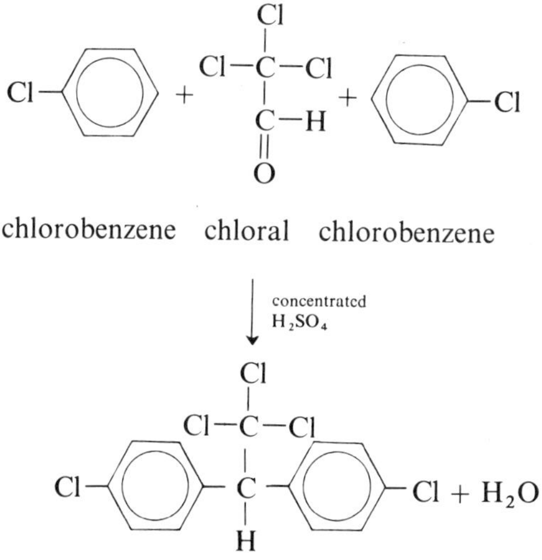

chlorobenzene   chloral   chlorobenzene

concentrated H$_2$SO$_4$

DDT

The uses of insecticides

Insecticides are used to kill insects for two main reasons: because insects transmit serious diseases, or because they eat crops, thus competing with man for the World's available food resources. Suspicion has been cast on a number of insecticides as they are not selective in their action, killing both harmful and beneficial insects, as well as birds and small animals, and their use is controlled in some parts of the World. This controlled usage is very necessary because misapplication can have serious secondary effects. In the Canadian province of New Brunswick, for example, the application of only sixty milligrammes of DDT per square metre of forest to control the spruce budworm has twice wiped out almost an entire year's production of young salmon in the Miramichi River. Rain washed DDT into the river where it was absorbed by plankton; and when fish feed on the DDT-tainted plankton a multifold concentration of DDT results, sufficient to kill young salmon.

One must set against the risks of misapplication the very great benefit to man arising from the use of insecticides. The reduction of pain and suffering, and the increase in the expectation of life, brought about by the reduction of diseases caused by insect carriers, and the vast increases in crop production brought about by chemical means of protection, are striking examples of the influence of chemistry upon everyday life.

In tropical climates especially, insect damage to stored crops can be considerable. For example in Kano, Nigeria, it was found in one experiment that when two water traps each of about 250 cm^2 were placed in and near the doorway of a groundnut store a catch of eight hundred groundnut beetles, *Tribolium castaneum*, was made in twenty-four hours.

Another study carried out in Rhodesia evaluated the damage to beans. Beans are an important crop for Rhodesian Africans, but in storage the beans are often much damaged by beetles belonging to the family *Bruchidae*. G. F. Cockbill, an entomologist, was asked therefore to find out if the application of a BHC dust would be an economical method of reducing damage.

Two stacks of beans, each of a hundred and five bags in layers of 5 × 3 bags, were prepared. One stack was used as a control and left untreated. The other stack was treated with a dust containing 0.65 per cent gamma BHC, each layer of bags receiving dust at the rate of one gramme per square metre.

Before building the stacks twelve bags were selected at random, the beans were weighed, and a sample was removed to assess the content of moisture and the damage done by insects. Six of the sampled bags were used in the building of each stack, three bags occupied outside positions in different layers and three central positions.

The stacks were left undisturbed for five months in a covered store after which time the sampled bags were recovered and re-examined for weight, moisture content, and beetle damage.

The moisture content of the beans had increased during storage so in assessing the losses due to insect attack a net loss in weight was determined which allowed for variation in moisture content.

	% loss in weight in control stack	% loss in weight in treated stack
outside bags	13.0	1.5
central bags	13.5	5.5
mean	13.3	3.5

Table 9.6
Loss in weight in beans affected by Brucid beetles

The results indicated that treatment saved approximately 10 per cent of the total quantity of beans from damage by insects. Position in the stack gave no protection in the case of the control stack although in the treated stack the central bag from the middle layer (one of the few bags not partially exposed) was one of the least attacked, suffering only a 0.7 per cent net loss in weight.

The number of beans damaged was also determined. This affects the eating quality because in heavy attack little is left of a bean but its tough seed coat.

	% increase in damaged beans in control stack	% increase in damaged beans in treated stack
outside bags	41.7	7.0
central bags	17.3	16.7
mean	29.5	11.8

Table 9.7
Increase in damage in beans attacked by Brucid beetles

The mean results indicate that during the five months of the experiment the number of damaged beans in the control stack increased by about 30 per cent but treatment constrained the increase in damage to about 12 per cent.

The effectiveness of the BHC treatment in killing Brucid beetles was found by recovering and counting beetles from the sample sacks. A mean of fifteen dead beetles was found in sacks from the control stack and twice as many in the treated sacks.

As a final test samples of beans were prepared and cooked. No unpleasant odour or flavour was detected in the treated beans which were preferred 'because there were fewer skins'.

Since the cost of treatment was 3 p per hundred bags and this effected a saving of about 10 per cent in a crop valued at £300 per hundred bags (1952 prices) the treatment was concluded to be effective, economical, and practicable.

Summary of reactions of halogenoalkanes

1 *Replacement of halogen by hydroxyl group*

To obtain alcohols from the hydrolysis of halogenoalkanes it is necessary to reflux with an aqueous suspension of a base, for example calcium carbonate, so as to displace the equilibrium

$$RCH(Hal)R + H_2O \rightleftharpoons RCH(OH)R + H^+Hal^-$$

The reaction is of very limited preparative value and does not give worthwhile yields from primary halogenoalkanes.

2 *Elimination of hydrogen halide*

Alkenes are obtained by the action of boiling alcoholic potassium hydroxide solution on halogenoalkanes

$$RCH_2CH_2Hal + K^+OH^- \longrightarrow RCH{=}CH_2 + K^+Br^- + H_2O$$

Yields are poor but the reaction is useful for introducing double bonds into complex molecules.

3 *Replacement of halogen by cyanide group*

Primary halogenoalkanes are converted to alkyl cyanides in yields of better than 60 per cent by refluxing with an alcoholic solution of potassium cyanide.

$$RHal + K^+CN^- \longrightarrow RCN + K^+Hal^-$$

This reaction is useful in synthesis because the cyanide group is reactive (see later) and an additional carbon atom has been introduced into the molecule.

4 *Replacement of halogen by amino group*

If halogenoalkanes are heated under pressure with an excess of alcoholic ammonia, amines are obtained

$$RHal + 2NH_3 \longrightarrow RNH_2 + NH_4^+Hal^-$$

The yield is 40 per cent or better; further reaction can occur to produce R_2NH and similar products.

5 *Reduction to alkanes*

Sodium dissolved in mercury in the presence of an alcohol produces hydrogen which will reduce halogenoalkanes to alkanes.

$$RHal + Na + C_2H_5OH \longrightarrow RH + Na^+Hal^- + C_2H_5O^-Na^+$$

Iodides and bromides react best to give good yields of alkane.

Revision question

You should now be able to determine what reagents, organic and inorganic, are required for the synthesis in one step of the following products.

1 propylamine
2 octyl cyanide
3 hexan-1-ol (by two routes)
4 hex-1-ene (by two routes)

Summary of reactions of acyl chlorides

1 *Replacement of halogen by the* $-NH_2$ *group*

When acyl chlorides are added with caution to cold concentrated aqueous ammonia, amides are obtained in yields of 50 per cent or better.

$$RCOCl + 2NH_3 \longrightarrow RCONH_2 + NH_4^+ Cl^-$$

2 *Reduction by hydrogen*

By the use of a palladium catalyst 'poisoned' to reduce its activity, acyl chlorides can be reduced by hydrogen at atmospheric pressure to give aldehydes in good yield. If the catalyst were not 'poisoned' reduction to alcohols would take place.

$$RCOCl + H_2 \longrightarrow RCHO + HCl$$

3 *Formation of esters*

Acyl chlorides will react with alcohols to give esters in excellent yield. Unlike the reaction of carboxylic acids with alcohol this is not an equilibrium reaction.

$$RCOCl + R'OH \longrightarrow RCO_2R' + HCl$$

The reaction is conveniently carried out in a basic solvent which will absorb the hydrogen chloride and prevent it reacting with the alcohol.

Revision question

You should now be able to deduce what reagents, organic and inorganic, are required for the synthesis in one step of the following products.

1 butanamide
2 pentyl butanoate
3 propanal (propionaldehyde) (by two routes)

9.7 Simple synthetic routes

A common problem of organic chemists is the synthesis of compounds of known structure. This can be important, for example, in the synthesis of drugs. When an active compound is discovered it is necessary to synthesize as many similar compounds as possible as it is found that minor changes in structure can profoundly alter the effectiveness of a drug. By the synthesis of all possible varieties of an active compound the most effective compound with the minimum of side effects can be made available to doctors. For example a range of similar compounds has been found all having useful properties as local anaesthetics:

H_2N—⟨◯⟩—$CO_2CH_2CH_3$ benzocaine

H_2N—⟨◯⟩—$CO_2CH_2CH_2N(C_2H_5)_2$ procaine

C_4H_9HN—⟨◯⟩—$CO_2CH_2CH_2N(CH_3)_2$ amethocaine

These are now used by doctors where previously the dangerous addictive drug cocaine might have been used.

Before you can tackle effectively simple synthetic problems involving changing the functional group in compounds, it is necessary to organize your knowledge into a pattern that reveals the relationships between the various functional groups, and also, if your syntheses are to be plausibly based, you should be aware of the starting materials available.

To see the pattern that interconnects organic compounds it is worth attempting to construct a 'flow-chart' based on the reactions you have met in the laboratory, together with the other important reactions described in the summaries given in each section. A number of reactions will be dead-ends not leading on to further interesting compounds while others will give useful products in good yields. For example, in the reactions of alcohols

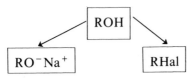

which product is going to be more useful for the synthesis of further compounds?

Before assembling your flow chart you should note the reactions of the cyanide group because alkyl cyanides can be prepared in good yield and being unsaturated, $-C\equiv N$, the group takes part in addition reactions. An important use of the cyanide group is to introduce a further carbon atom into molecules.

Summary of the chemistry of the cyanide group

1 Dehydration of amides, for example with phosphorus pentoxide, yields alkyl cyanides.

$$RCONH_2 + P_2O_5 \longrightarrow RCN + 2HPO_3$$

2 Replacement of halogen by refluxing halogenoalkanes with an alcoholic solution of potassium cyanide yields alkyl cyanides.

$$RHal + K^+CN^- \longrightarrow RCN + K^+Hal^-$$

3 Cyanides can be hydrolysed, in what amounts to a reversal of (1) above, by heating with strong acid or base. Carboxylic acids are obtained because the intermediate amides are also hydrolysed in the experimental conditions.

$$RCN + 2H_2O + H^+Cl^- \longrightarrow [RCONH_2 + H_2O + H^+Cl^-]$$
$$\longrightarrow RCO_2H + NH_4^+Cl^-$$

4 If cyanides are refluxed with moist alcohols (rather than water as in (3) above), in the presence of acids, they are 'alcoholysed' to esters

$$RCN + R'OH + H_2O + H^+ \longrightarrow RCO_2R' + NH_4^+$$

5 Addition of hydrogen to the unsaturated cyanide group produces amines in yields of 50 per cent or more.

$$RCN + 2H_2 \longrightarrow RCH_2NH_2$$

The reaction is carried out using hydrogen gas with a nickel catalyst.

Now consider the availability of starting materials. The petrochemical industry, by the cracking of the alkanes in crude oil, makes available the short-chain alkenes in large quantities. By appropriate reactions (see Topic 13 later) these alkenes can be converted to primary alcohols. Such processes make available the primary alcohols from C_2, ethanol, to C_8, octanol.

Naturally-occurring fats and oils are a source of carboxylic acids in the range C_8 to C_{18} (and beyond) by hydrolysis, or of alcohols by reduction. The main restriction as noted in section 9.5 is that only C_{even} compounds are available. If C_{odd} compounds are required it is necessary to introduce an extra carbon atom via the cyanide group.

With the information assembled let us consider possible routes for the synthesis of tridecylamine, $C_{13}H_{27}NH_2$.

Amino groups can be produced by two alternative reactions:

$$C_{13}Hal + NH_3 \longrightarrow C_{13}NH_2 + NH_4^+ Hal^- \tag{1}$$

or

$$C_{12}CN + 2H_2 \longrightarrow C_{12}CH_2NH_2 \tag{2}$$

(using C_{13} and C_{12} to represent $C_{13}H_{27}-$ and $C_{12}H_{25}-$)

Considering the first reaction, how can halogenoalkanes be produced? Either by addition of halogen hydrides to alkenes, or by substitution for hydroxyl groups. But C_{13} alkenes are not available from petrochemicals and C_{13} alcohols (or carboxylic acids) are not available from natural sources, being C_{odd} compounds.

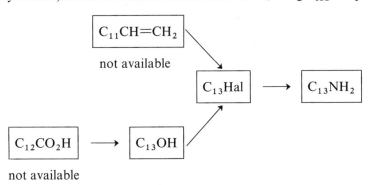

A further method of producing halogenoalkanes would be from the silver salt of a carboxylic acid treated with bromine.

$$\boxed{C_{13}CO_2H} \longrightarrow \boxed{C_{13}Hal} \longrightarrow \boxed{C_{13}NH_2}$$
available

The required carboxylic acid, being C_{even}, will be available so this is a possible route to tridecylamine. On a large scale the use of this route would be restricted by the capital cost of the silver required and the need to recover it from the reaction mixture.

If the production of a cyanide group for the second reaction is now considered it is known to be available from an amide group by dehydration. But the amide required would have to be produced from a C_{13} carboxylic acid and this is not available.

$$\boxed{C_{12}CO_2H} \longrightarrow \boxed{C_{12}CO.NH_2} \longrightarrow \boxed{C_{12}CN}$$

not available

$$\longrightarrow \boxed{C_{13}NH_2}$$

Alternatively cyanide groups can be produced by substitution for halogen and the halogenoalkane required would be obtained from an alcohol

$$\boxed{C_{11}CO_2H} \longrightarrow \boxed{C_{12}OH} \longrightarrow \boxed{C_{12}Hal} \longrightarrow \boxed{C_{12}CN}$$

available

$$\longrightarrow \boxed{C_{13}NH_2}$$

and a C_{12} carboxylic acid being a C_{even} compound should be available from natural sources.

Hence two possible routes to tridecylamine have been found.

When you have produced your own flow charts attempt the synthesis of the following compounds:

1   $C_8H_{17}NH_2$, octylamine
2   $C_{12}H_{25}CHO$, tridecanal
3   $C_3H_7CONH_2$, butanamide
4   $C_4H_9CO_2C_2H_5$, ethyl pentanoate

Appendix
Some rules for naming carbon compounds

Isomerism, i.e. the existence of a number of compounds having the same molecular formula, is very common amongst carbon compounds. This makes it important to have a systematic method for naming compounds so that different isomers can be indicated clearly, and also so that structural formulae can be deduced from names. Apart from a few very simple compounds, such as methane, CH_4, and carbon dioxide, CO_2, in order to name a given compound systematically its structural formula must be known. Molecular formulae are generally useless for this purpose; there are, for example, 802 possible compounds of molecular formula $C_{13}H_{28}$ and more than 350 000 compounds of formula $C_{20}H_{42}$.

The rules for naming compounds have been settled by international agreement through the International Union of Pure and Applied Chemistry. They are usually known as IUPAC rules. The complete set of rules is a very lengthy document and occupies quite a large book. We shall need only a few rules to start with so this short account deals mainly with simple hydrocarbons and hydroxy-compounds. Once you are accustomed to naming these you should not find it difficult to extend your knowledge of the rules to cover other classes of compound as you go along.

Hydrocarbons

Saturated hydrocarbons

These are of two main types.

1   Compounds in which the molecules are made up of chains of carbon atoms, either unbranched or branched. The general name for these is *alkanes*. Names for individual compounds all have the ending *-ane*.

2   Compounds whose molecules contain one or more rings of carbon atoms (to which side chains may be attached). The general name for these is *cycloalkanes*. Names for individual compounds contain the syllable *cyclo-* and have the ending *-ane*.

1   Alkanes

a   Those containing unbranched carbon atom chains

The names of the first four hydrocarbons in the series, containing 1, 2, 3, and 4 carbon atoms respectively, are methane, ethane, propane, and butane. These do not follow any logical system and must be learned. The rest of the hydrocarbons in the series are named by using a Greek numeral root and the ending -ane, e.g. pentane (5 carbon atoms in an unbranched chain), hexane (6 carbon atoms), and heptane (7 carbon atoms). The roots are the same as those used in naming geometric figures (pentagon, hexagon, heptagon, etc.) A list of examples is given below.

Number of carbon atoms in chain	Molecular formula	Name	Number of carbon atoms in chain	Molecular formula	Name
1	CH_4	methane	8	C_8H_{18}	octane
2	C_2H_6	ethane	9	C_9H_{20}	nonane*
3	C_3H_8	propane	10	$C_{10}H_{22}$	decane
4	C_4H_{10}	butane	11	$C_{11}H_{24}$	undecane*
5	C_5H_{12}	pentane	12	$C_{12}H_{26}$	dodecane
6	C_6H_{14}	hexane	13	$C_{13}H_{28}$	tridecane
7	C_7H_{16}	heptane	14	$C_{14}H_{30}$	tetradecane
			15	$C_{15}H_{32}$	pentadec-ane
			16	$C_{16}H_{34}$	hexadecane
			18	$C_{18}H_{38}$	octadecane
			etc.		

* Latin roots used here instead of the Greek 'ennea' for 9 and 'hendeca' for 11.

b Those containing branched chains
In order to name these compounds, *alkyl* groups must be used. These are derived from hydrocarbons with unbranched carbon chains by removing one hydrogen atom from the end carbon atom of the chain, e.g. $CH_3CH_2CH_3$ becomes $CH_3CH_2CH_2-$ (one bond unoccupied). Alkyl groups are named from the parent hydrocarbon by substituting the ending -yl for the ending -ane, thus $CH_3CH_2CH_2-$ is the propyl group. A list of alkyl groups is given below.

Hydro-carbon	Alkyl group	Formula for alkyl group	
methane	methyl	CH_3-	
ethane	ethyl	CH_3CH_2-	(C_2H_5-)
propane	propyl	$CH_3CH_2CH_2-$	(C_3H_7-)
butane	butyl	$CH_3CH_2CH_2CH_2-$	(C_4H_9-)
pentane	pentyl	$CH_3CH_2CH_2CH_2CH_2-$	$(C_5H_{11}-)$
hexane	hexyl	$CH_3CH_2CH_2CH_2CH_2CH_2-$	$(C_6H_{13}-)$
and so on			

Branched chain hydrocarbons are named by combining names of alkyl groups with the name of an unbranched chain hydrocarbon. A branched chain hydro-carbon cannot have less than 4 carbon atoms per molecule. The simplest is

CH_3CHCH_3 which is called 2-methylpropane.
    |
    CH_3

The hydrocarbon name is always derived from the *longest* continuous chain of carbon atoms in the molecule. The '2-' is obtained by numbering the carbon atoms in the propane chain

$$3 \quad 2 \quad 1$$
$$C-C-C$$

A similar procedure is applied to all branched chains, the numbering being done so that the lowest numbers possible are used to indicate the side chain (or chains). Thus

$$CH_3CH_2CH_2\underset{\underset{CH_3}{|}}{C}HCH_3$$

is named 2-methylpentane, *not* 4-methylpentane which would be obtained by numbering from the other end of the chain.

When more than one substituent alkyl group of the same kind is present the figures indicating their positions are separated by commas, e.g.

$$CH_3\underset{\underset{CH_3}{|}}{C}H-\underset{\underset{CH_3}{|}}{C}HCH_3 \quad \text{is 2,3-dimethylbutane}$$

Different alkyl groups are placed in alphabetical order in the name for a branched chain hydrocarbon, e.g.

$$CH_3CH_2\underset{\underset{\underset{\underset{CH_3}{|}}{CH_2}}{|}}{C}HCH_2\underset{\underset{CH_3}{|}}{C}HCH_3 \quad \text{is named 4-ethyl-2-methylhexane}$$

2   Cycloalkanes

These are named from the corresponding unbranched chain hydrocarbon by adding the prefix, cyclo-. The simplest is

$$\begin{array}{c} H \\ | \\ H-C \\ | \quad \diagdown C \diagup H \\ H-C \diagup \quad \diagdown H \\ | \\ H \end{array} \quad \text{or} \quad \begin{array}{c} H_2C \\ | \quad \diagdown CH_2 \\ H_2C \diagup \end{array} \quad \text{or} \quad \triangleright$$

which is named cyclopropane.

All the carbon atoms in an unsubstituted cycloalkane ring are equivalent, so far as substitution is concerned, so that if only one radical is added as a substituent there is no need to number the carbon atoms. Thus

$$H_2C-C{\overset{CH_3}{\underset{H}{}}} \quad \text{or} \quad \square CH_3 \quad \text{is methylcyclobutane}$$
$$H_2C-CH_2$$

Unsaturated hydrocarbons

These contain double or triple bonds between some of the carbon atoms. We shall consider double bonds only here. Those compounds whose molecules contain chains of carbon atoms, whether branched or unbranched, are called *alkenes*. Compounds whose molecules contain rings of carbon atoms are called *cycloalkenes*. Individual names for both groups of compounds end in *-ene*. The rules for naming unsaturated hydrocarbons are similar to those used for the corresponding saturated compounds. The only additional problem which arises is that of locating the double bond. This is done by using the lowest numbered carbon atom in the longest unbranched chain of carbon atoms in the hydrocarbons. There is no need to do this for the first two members of the series, i.e. $CH_2=CH_2$ is ethene, and $CH_3-CH=CH_2$ is propene. (These are often called ethylene and propylene, which are permissible alternative names.) With four carbon atoms and one double bond two compounds are possible

$CH_2=CHCH_2CH_3$ which is but-1-ene

and $CH_3CH=CHCH_3$ which is but-2-ene

The same rule is used for compounds containing more than one double bond,

e.g. $CH_2=CHCH=CHCH_3$ is penta-1,3-diene

('di' indicates two double bonds; note also that 'a' is added to the hydrocarbon root when more than one double bond is present.)

Branched chain alkenes are dealt with as for alkanes, e.g.

$CH_2=CCH_2CH_3$ is 2-methylbut-1-ene.
$\quad\;\; |$
$\quad\; CH_3$

Cycloalkenes follow similar rules to cycloalkanes.

$$\begin{array}{c} HC \\ \| \\ HC \end{array}\!\!\!> CH_2 \quad \text{is cyclopropene}$$

$$\begin{array}{c} H_2 \\ C \\ H_2C \diagup \diagdown CH \\ | \qquad \| \\ H_2C \diagdown \diagup CH \\ C \\ H_2 \end{array} \quad \text{is cyclohexene}$$

Arenes

These were originally called the aromatic hydrocarbons. The only ones with which we shall be concerned here have non-systematic names.

benzene (C_6H_6) naphthalene $(C_{10}H_8)$

The phenyl group, $C_6H_5—$, is derived from benzene but most of the substitution products, when one substituent only is involved, have non-systematic names, e.g.

toluene $(C_6H_5CH_3)$

When drawing such structures the convention is that where a group is attached to the benzene ring a hydrogen atom has been removed.

Hydroxy compounds

We shall deal here mainly with the alcohols, which are formed when one or more —OH groups are substituted for hydrogen atoms in hydrocarbons.

The alcohols are named by using the name of the hydrocarbon from which they are derived, omitting the terminal -e, and adding -ol in its place, e.g.

CH_3OH methanol
CH_3CH_2OH ethanol

Two alcohols can be derived from propane. They are distinguished by numbering the carbon atoms as for the hydrocarbons.

$CH_3CH_2CH_2OH$ propan-1-ol
$CH_3CH(OH)CH_3$ propan-2-ol

This method is used for all the higher alcohols.

$CH_3CH_2CH_2CH_2OH$ butan-1-ol
$CH_3CH_2CH(OH)CH_3$ butan-2-ol

Alcohols with side chains are named in the same way as the corresponding hydrocarbons.

CH_3CHCH_2OH 2-methylpropan-1-ol
    |
    CH_3

$CH_3C(OH)CH_3$ 2-methylpropan-2-ol
    |
    CH_3

For alcohols with more than one hydroxyl group (polyhydric alcohols) the full name of the hydrocarbon is used, followed by the number of the carbon atom to which the —OH groups are attached, followed by di, tri, tetra, etc. to indicate the total number of —OH groups, with -ol as the ending, e.g.

$HOCH_2CH_2OH$ ethane-1,2-diol (but this is usually called glycol)

$HOCH_2CH_2CH_2OH$ propane-1,3-diol

$HOCH_2CH(OH)CH_2OH$ propane-1,2,3-triol (usually called glycerol).

Hydroxy-arenes
The simpler members of this class of compounds are not named systematically. The only one that we need to consider at this stage of the course is

OH
|
phenol (C_6H_5OH)

Other substitution products of hydrocarbons
A few simple examples will show how compounds with other functional groups can be named in a similar way to those dealt with above.

CH_3Cl	chloromethane
CH_3CH_2Cl	chloroethane
CH_3CHBr_2	1,1-dibromoethane
CH_2BrCH_2Br	1,2-dibromoethane
CH_3CHCHI_2	1,1-di-iodo-2-methylpropane
|	
CH_3	

Writing structural formulae from systematic names

For compounds whose molecules contain unbranched or branched carbon chains, the hydrocarbon root tells you the longest unbranched carbon chain. Two examples will help to explain this.

1 What is the structural formula for *hexan-3-ol*?

'hexan' tells you that there is an unbranched chain of 6 carbon atoms

$$C-C-C-C-C-C$$

with an $-OH$ group on the third carbon atom from the end of the chain.

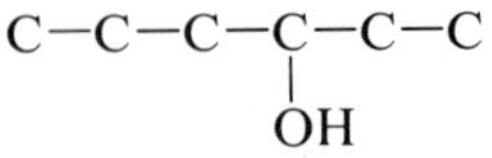

$$C-C-C-\underset{\underset{OH}{|}}{C}-C-C$$

The formula is completed by adding the appropriate number of hydrogen atoms

$$CH_3CH_2CH_2CH(OH)CH_2CH_3$$

2 What is the structural formula for *2-bromo-3-methylbutan-1-ol*? 'butan' indicates the longest chain, of 4 carbon atoms.

$$C-C-C-C$$

Put an $-OH$ group on the first carbon atom, a bromine atom on the second, and a methyl radical (CH_3) on the third from one end, to give

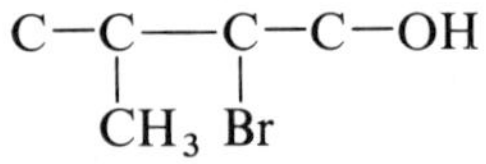

$$C-\underset{\underset{CH_3}{|}}{C}-\underset{\underset{Br}{|}}{C}-C-OH$$

add hydrogen atoms

$$CH_3\underset{\underset{CH_3}{|}}{C}HCHBrCH_2OH$$

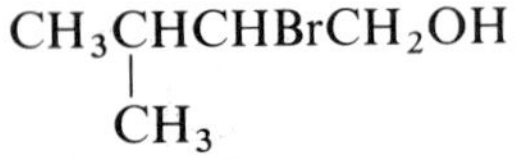

This can also be written

$$CH_3CH(CH_3)CHBrCH_2OH$$

the bracket indicating a side chain.

For cyclic compounds the number of carbon atoms in the ring is indicated by the name, e.g. cyclopentane – 5 carbon atoms in a ring.

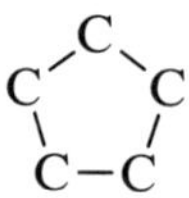

Add hydrogen atoms

$$\begin{array}{c} CH_2 \\ CH_2 \quad CH_2 \\ CH_2 - CH_2 \end{array} \quad \text{or} \quad \pentagon$$

For some carbon compounds you will have to rely largely on memory for the formula – benzene, phenol, naphthalene, etc.

Names and structures of some functional groups

In this table the structures of the functional groups are printed out in the second column so as to show the atomic linkages. When these structures are repeated in the examples given in the fourth column they are printed on one line only, so as to show this abbreviated method of writing them.

Class of compound	Structure of functional group	Example of compound Name	Formula
Alkene	$\underset{/}{\overset{\backslash}{C}}=\underset{\backslash}{\overset{/}{C}}$	but-1-ene	$CH_2{=}CHCH_2CH_3$
Alkyne	$-C{\equiv}C-$	acetylene (ethyne)	$CH{\equiv}CH$
Alcohol	$-OH$	ethanol	CH_3CH_2OH
Ether	$-O-$	diethyl ether	$CH_3CH_2OCH_2CH_3$
Aldehyde	$\underset{/}{\overset{H\backslash}{C}}=O$	butanal	$CH_3CH_2CH_2CHO$
Ketone	$\underset{/}{\overset{\backslash}{C}}=O$	butanone	$CH_3CH_2COCH_3$
Carboxylic acid	$-C\underset{\backslash\backslash O}{\overset{/OH}{}}$	butanoic acid	$CH_3CH_2CH_2CO_2H$
Carboxylate	$-C\underset{\backslash\backslash O}{\overset{/O^-}{}}$	sodium butanoate	$CH_3CH_2CH_2CO_2^-\,Na^+$
Acyl chloride	$-C\underset{\backslash\backslash O}{\overset{/Cl}{}}$	butanoyl chloride	$CH_3CH_2CH_2COCl$

Acid anhydride	(structure)	butanoic anhydride	$(CH_3CH_2CH_2CO)_2O$
Amide	(structure)	butanamide	$CH_3CH_2CH_2CONH_2$
Primary amine	$-NH_2$	ethylamine	$CH_3CH_2NH_2$
Cyanide	$-C\equiv N$	methyl cyanide	CH_3CN
Chloride	$-Cl$	chlorobenzene	C_6H_5Cl
Bromide	$-Br$	1-bromobutane	$CH_3(CH_2)_2CH_2Br$
Iodide	$-I$	iodoethane	CH_3CH_2I
Nitro compound	$-N$	nitrobenzene	$C_6H_5NO_2$
Sulphonic acid	$-S-OH$	benzenesulphonic acid	$C_6H_5SO_3H$
Sulphonyl chloride	$-S-Cl$	benzenesulphonyl chloride	$C_6H_5SO_2Cl$
Sulphonamide	$-S-NH_2$	benzenesulphonamide	$C_6H_5SO_2NH_2$

Trivial names

When an organic compound is first obtained, for example, from a plant, its structure may be quite unknown and it is usually given a name indicative of its source or its properties. Thus from varieties of the *Cinnamomum* tree is obtained oil of cinnamon which gives us the name for cinnamic acid.

$$\langle\!\bigcirc\!\rangle\!-\!CH\!=\!CHCO_2H$$

Another example is $H_2N(CH_2)_4NH_2$ which occurs in rotten fish and is known as putresine.

The compounds quoted, being of known structure, can now be given systematic names and the 'trivial' names, although interesting and evocative, will be decreasingly used in the future. Other trivial names, and special names such as trademarks, are likely to continue in use: often the systematic name for a complex natural product would be far too clumsy for use in conversation.

At the present time therefore in different books the same organic compounds are likely to go under different names. Some of the names will be trivial names, some trademarks, and yet others will be systematic names derived from a non-IUPAC system.

To help you in your reading of other books a table of alternative names is now given. The main differences are compounds with branched-chain alkyl groups, and acids and their derivatives.

Alternative name	Systematic name (mainly IUPAC)
paraffin	alkane
olefin	alkene
aromatic hydrocarbon	arene
alkyl halide	halogenoalkane
formic acid (*Latin*: ant)	[methanoic acid]
acetic acid (*L.* vinegar)	[ethanoic acid]
propionic acid (*Greek*: first fat)	propanoic acid
butyric acid (*L.* butter)	butanoic acid
valeric acid (valerian root)	pentanoic acid
caproic acid (*L.* goat)	hexanoic acid
enanthic acid (*Greek*: vine blossom)	heptanoic acid
caprylic acid (*L.* goat)	octanoic acid
pelargonic acid (pelargonium)	nonanoic acid
capric acid (*L.* goat)	decanoic acid
etc. see table 9.5	
butyraldehyde	butanal
sodium butyrate	sodium butanoate
butyryl chloride	butanoyl chloride
butyric anhydride	butanoic anhydride

butyramide	butanamide
normal or *n*-propyl alcohol	propan-1-ol
isopropyl alcohol	propan-2-ol
secondary or *s*-butyl alcohol	butan-2-ol
tertiary or *t*-butyl alcohol	2-methylpropan-2-ol
n-amyl alcohol	pentan-1-ol
acetonitrile	methyl cyanide
ethyl alcohol, alcohol	ethanol

Problems

1   Here are the empirical formulae of some compounds.

A CH_2O

B CH_2

C C_3H_7Cl

D NH_3

E $COCl_2$

i   Which one of the compounds is a hydrocarbon?

ii   Which one of the compounds would not form carbon dioxide when heated with copper(II) oxide?

iii   Which one of the compounds would not form water when heated with copper(II) oxide?

iv   Two of the compounds might form a white precipitate when boiled with nitric acid and silver nitrate solution. Which two?

2   Refer to question 1 and its answers.
Suppose that 1 *mole* each of compounds A, B, and C were completely oxidized so that all their carbon is converted to carbon dioxide. From the information given, which of the following would be the most reasonable conclusion?

A   Compound A would form the greatest weight of carbon dioxide.

B   Compound B would form the greatest weight of carbon dioxide.

C   Compound C would form the greatest weight of carbon dioxide.

D   Compound C would form three times as much carbon dioxide as either compound A or compound B.

E   It is not possible to decide which of the compounds would give the greatest weight of carbon dioxide.

State briefly the reasons for your answer.

3   The following table shows the results obtained from experiments to find the number of moles of each element present in 100 g each of four different compounds, A, B, C, and D. Work out the molecular formula of each compound.

	C	H	O	N	Molecular weight
A	4.35	12.9	2.17	—	46
B	7.13	16.2	—	—	86
C	3.12	12.4	3.12	—	32
D	3.39	8.42	1.69	1.69	59

4 One of the compounds specified in question 3 has no isomers. Which one?

5 For each of the three compounds in question 3 for which more than one structure can be devised, draw structural formulae for two of the possible isomers.

6 Two hydrocarbons have a molecular weight of 58. Deduce their structural formulae.

7 1.00 g of a gaseous compound gave, on oxidation, 1.46 g of carbon dioxide and 0.600 g of water. The compound was composed of carbon, hydrogen, and oxygen only. 1.00 g of the compound had a volume of $747 \, cm^3$ at s.t.p. Calculate the molecular formula of the compound.

8 $10 \, cm^3$ of a gaseous hydrocarbon reacted with $50 \, cm^3$ of oxygen to form $30 \, cm^3$ of carbon dioxide and some water. No oxygen or hydrocarbon remained. Determine the molecular formula of the hydrocarbon, by the following steps (all volumes measured at the same temperature and pressure).
  i Express a simple ratio:
Volume of hydrocarbon : volume of oxygen : volume of carbon dioxide.
  ii Express a simple ratio:
Moles of hydrocarbon : moles of oxygen : moles of carbon dioxide.
  iii How many atoms of carbon must be contained in 1 molecule of hydrocarbon?
  iv How many molecules of oxygen would be required to convert your answer to (*iii*) into carbon dioxide?
  v How many molecules of oxygen must have been used to convert the hydrogen in 1 molecule of the hydrocarbon into water?
  vi How many atoms of hydrogen must be contained in 1 molecule of the hydrocarbon?
  vii From your answers to (*iii*) and (*vi*) deduce the molecular formula of the hydrocarbon.

9   A compound was found to be composed of carbon, hydrogen, and chlorine in the ratio of 1 g-atom C : 2 g-atoms H : 1 g-atom Cl. The molecular weight of the compound was 99. Devise two structural formulae for the compound.

10   Which of the following would you expect to have the *highest* boiling point? Give a reason for your answer.

    A CH_3NH_2

    B $CH_3CH_2NH_2$

    C $CH_3CH_2CH_2NH_2$

    D $CH_3CH_2CH_2CH_2NH_2$

    E $CH_3CH_2CH_2CH_2CH_2NH_2$

11   Name the following compounds.

  i $CH_3CH_2CH_2CH_2CH_3$

  ii $CH_3CHCH_2CH_2CH_3$
           |
           CH_3

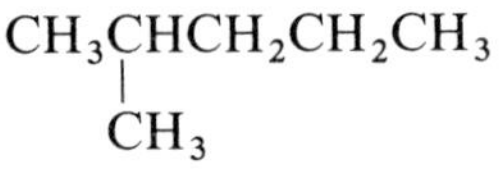

  iii $CH_3CHCH_2CHCH_3$
           | |
           CH_3 CH_3

  iv $CH_3CH-CH-CHCH_3$
             | | |
             CH_3 CH_3 CH_3

                  CH_3
                  |
  v $CH_3CH_2CCH_2CH_3$
                 |
                 CH_3

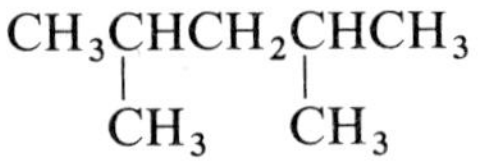

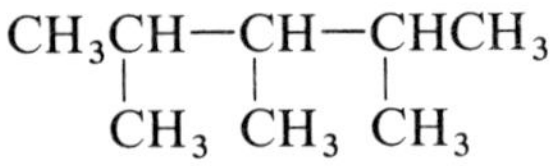

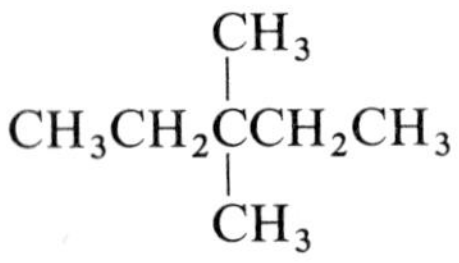

12   Draw the structural formula of each of the following.

  i   2-methylbutane

  ii 2-methylpropane

  iii butan-2-ol

  iv 2-chloro-3-methylpentane

  v 2-chloro-2-methylpentane

13   When a substance of the formula

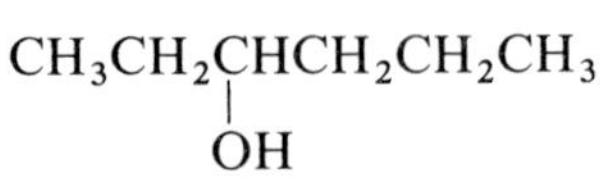

    $CH_3CH_2CHCH_2CH_2CH_3$
           |
           OH

is vaporized and passed over heated pumice stone, one mole of water is eliminated from each mole of the original substance.

   i   Name the original substance.

   ii   Deduce the formula of each of the two possible isomers which result from the reaction given above.

   iii   One mole of each of the isomers in (*ii*) reacts with one mole of bromine; give the formula of each of the products.

Questions 14 and 15

Each of the following two groups of chloro-compounds were intended to contain substances of similar properties and structures. The classification could be improved by transferring one substance from each group to the other group:

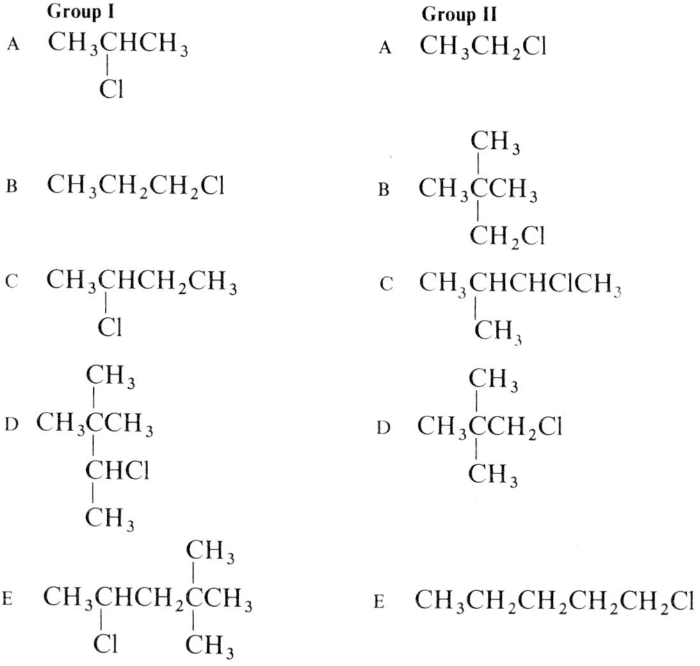

Group I

A CH_3CHCH_3
         $|$
         Cl

B $CH_3CH_2CH_2Cl$

C $CH_3CHCH_2CH_3$
         $|$
         Cl

          CH_3
          $|$
D CH_3CCH_3
          $|$
          $CHCl$
          $|$
          CH_3

                      CH_3
                      $|$
E $CH_3CHCH_2CCH_3$
         $|$ $|$
         Cl CH_3

Group II

A CH_3CH_2Cl

          CH_3
          $|$
B CH_3CCH_3
          $|$
          CH_2Cl

C $CH_3CHCHClCH_3$
         $|$
         CH_3

          CH_3
          $|$
D CH_3CCH_2Cl
          $|$
          CH_3

E $CH_3CH_2CH_2CH_2CH_2Cl$

14   Which substance would you transfer from group I to group II? On what basis did you make your decision?

15   Which substance would you transfer from group II to group I? On what basis did you make your decision?

16 Classify the following alcohols into two groups putting those of which you would expect similar properties into the same group.

A CH_3CH_2OH

B CH_3CHCH_3
        |
        OH

        CH_3
        |
C CH_3CCH_3
        |
        CH_2OH

        CH_3
        |
D CH_3CCH_3
        |
        $CHOH$
        |
        CH_3

E $CH_3CH_2CH_2OH$

F CH_3CHCH_2OH
        |
        CH_3

G $CH_3CH-CHCH_3$
        | |
        CH_3 OH

On what basis did you make your decision?

17 Draw the structural formula of each product which would result from the following treatment of a compound of the formula

        CH_3CHCH_2OH
        |
        CH_3

i   Action of sodium.
ii   Action of *prolonged* boiling with sodium dichromate and dilute sulphuric acid.
iii   Distilling with potassium bromide and concentrated sulphuric acid.
iv   Passing its vapour over heated pumice stone.

18 How would you expect the two substances of the formulae

A

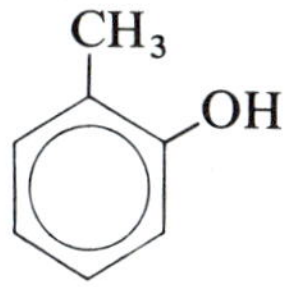

and   B CH_3CHCH_2OH
                 |
                 CH_3

to differ under the following conditions?
  i Their action with sodium hydroxide.
 ii Their action with acetic acid.

19
  i Arrange the following in order of their strength as acids, putting the weakest acid first.

A

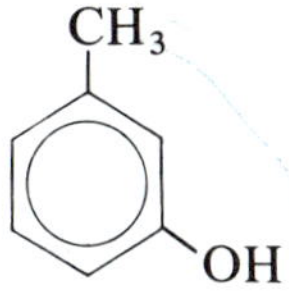

B CH_3CHCH_3
            |
            OH

C $CH_3CH_2\overset{\displaystyle O}{\underset{\displaystyle ||}{C}}{-}OH$

D $CH_3\overset{}{\underset{\displaystyle Cl}{C}}H\overset{\displaystyle O}{\underset{\displaystyle ||}{C}}{-}OH$

 ii Which group is the acidic group in these compounds?
iii Show the change which takes place when the group referred to in *(ii)* acts as an acid.
 iv Explain why the acidic strength of these compounds differs.

20
  i Arrange the following halides in order of the rate at which they would be expected to undergo hydrolysis, putting the one with the *slowest* rate first.

A $CH_3CH_2\overset{\displaystyle }{\underset{\displaystyle O}{C}}{-}Cl$

B

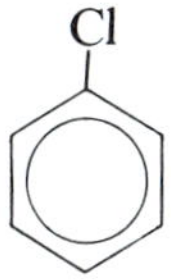

C $-CH_2Cl$

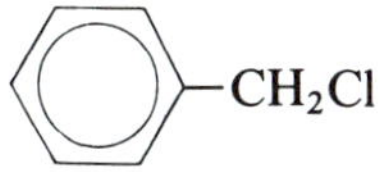

D $-CH_2I$

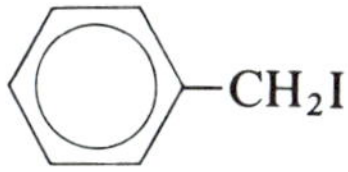

ii Give the formulae of the products of hydrolysis.

iii Describe simple experiments which would demonstrate this difference in rate of hydrolysis.

There now follows a group of questions concerned with the conversion of one carbon compound into another. A good deal of guidance is given to help you solve the problems. Work through these carefully and try to remember the way in which the problems are solved. You will then be better able to solve problems in which little or no guidance is given. All reactions involved in this group of questions have been met in Topic 9, or are given in the questions.

21   This problem is to find a way of converting 1-bromobutane into butanoic acid.

i   What are the structural formulae of these compounds?

ii   Are there the same number of carbon atoms in the product as in the starting material?

(*Note :* if the number of carbon atoms had to be changed, special reactions would be necessary.)

iii   To what series of compound does the product belong?

iv   This series of compounds can be prepared by an oxidation reaction. What would be a suitable compound to treat in this manner?

v   Can this compound be made from 1-bromobutane, and if so, how?

vi   Write down the equations, using full structural formulae, for the changes you have suggested.

22   This problem is to make butanoic acid from propan-1-ol.

i   What are the structural formulae of these compounds?

ii   Are there the same number of carbon atoms in the product as in the starting material?

(*Note :* if more carbon atoms have to be introduced, a reaction involving the cyanide ion would increase the number by one.)

iii   What type of reaction must the desired conversion therefore involve.

iv   Carboxylic acid functional groups can be made by oxidation of $-CH_2OH$ groups or by hydrolysis of $-CN$ groups. Does this suggest a suitable compound from which butanoic acid might be made, bearing in mind your answer to (*iii*), and if so, what is its structural formula?

v   Think of a way by which the compound can be made from something easily obtained from propan-1-ol. Write down your suggested route using structural formulae.

23   Some 1,2-dibromobutane has to be made from butan-1-ol.

i   What class or series of compounds does butan-1-ol belong to?

ii Write down any reactions of this class of compounds that you know. Use words to describe these, rather than equations.

iii Write down the structural formula of the desired product.

iv What sort of reaction can place two bromine atoms on neighbouring carbon atoms?

v What compound must you have in order to make 1,2-dibromobutane by this sort of reaction? Write its name and its structural formula.

vi You should be able to make the compound you have named in (*v*) from butan-1-ol by one of the reactions you mentioned in (*ii*). Write the equations, using structural formulae, both for making this intermediate compound from butan-1-ol, and for converting it into the desired product.

Topic 10

Intermolecular forces

The strong and sometimes violent interactions between molecules which are classed as chemical reactions produce significant changes in properties. But there are other weak interactions between molecules which may easily be overlooked because they leave chemical properties largely unaffected. These weak interactions can nevertheless greatly affect physical properties, and they can be vital in promoting essential biochemical reactions.

The study of weak interactions, which is the subject of this topic, leads to suggestions as to why chlorine is a gas and iodine a solid, or why water is a liquid but hydrogen sulphide a gas, to quote two examples. The study of weak interactions also contributes to our understanding of the structure of proteins.

10.1 The behaviour of mixtures of liquids: ideal and non-ideal solutions

If two liquids of similar structure are mixed in equimolar amounts, the boiling point of the mixture is often found to be close to the average of the boiling points of the two components.

The boiling points of hexane, C_6H_{14}, and heptane, C_7H_{16}, are 69 and 98 °C respectively, while an equimolar mixture of these hydrocarbons has a boiling point of 83 °C.

If the boiling point of a mixture of hexane and heptane is plotted against its composition, as in figure 10.1a, it is seen that the relationship is very nearly a linear one.

Another pair of liquids which have an approximately linear relationship between boiling point and composition is propan-1-ol and propan-2-ol. These compounds have molecular formulae $CH_3CH_2CH_2OH$ and $CH_3CH(OH)CH_3$ respectively.

When mixtures of liquids having dissimilar structures are boiled, a rather different result is usually obtained. Examples of such mixtures are trichloromethane and acetone ($CHCl_3$ and CH_3COCH_3), and ethanol and cyclohexane (C_2H_5OH and C_6H_{12}). The first experiment in this topic is to find out what sort of relationship holds in these cases.

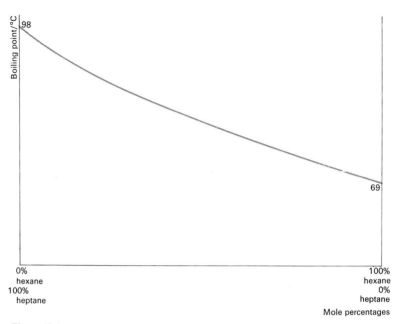

Figure 10.1a

Experiment 10.1a

How does the boiling point of a mixture of liquids vary with its composition?

In this experiment, you should investigate one of the two systems:

   A   trichloromethane and acetone, or

   B   ethanol and cyclohexane.

Set up the apparatus illustrated in figure 10.1b. The thermometer should dip into the liquid mixture, but must not touch the walls of the flask.

The purpose of the reflux condenser is to prevent loss of vapour and *reduce the risk of fire*. Remember when handling inflammable liquids and especially when adding samples to the apparatus to *move the Bunsen burner well away*.

Measure 10.0 cm³ of one liquid directly into the flask and heat until a slow steady reflux is obtained. Record the temperature as the boiling point of the pure liquid.

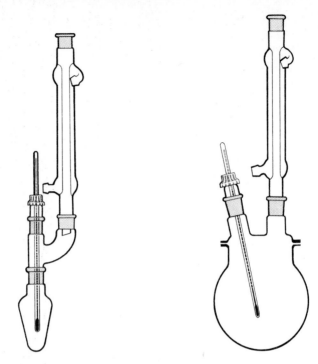

Figure 10.1b
Suitable forms of reflux apparatus for experiment 10.1a.

Measure 2.0 cm^3 of the second liquid into a test-tube and, after moving the Bunsen burner away and allowing the apparatus to cool briefly, add the second liquid by pouring it down the condenser. Reheat until a slow steady reflux is obtained and record the boiling point of the mixture.

Take further readings after additions of 2.0 cm^3 portions until a total of 10.0 cm^3 of the second liquid has been added.

Repeat the experiment starting with 10.0 cm^3 of the second liquid in the flask and make 2.0 cm^3 additions of the first liquid.

Plot a graph of boiling point against percentage composition by volume. Is the result a linear relationship between boiling point and composition? If deviations occur can you suggest any reasons?

Note that for a proper comparison of chemical behaviour the graph of boiling point should be plotted against mole fraction. If this is done, the change in shape of the graph will not be sufficient to invalidate the simple deductions required at this stage.

If time is available the near-ideal solution pair, propan-1-ol and propan-2-ol, can be studied in the same way.

Vapour pressures and Raoult's law

In order to understand these boiling point-composition relationships we must first study the vapour pressures of the liquids. When two liquids are mixed and the partial vapour pressures of each are measured, it is found that the experimental results can be expressed in a generalization known as *Raoult's law*.

> For a mixture of liquids, the partial vapour pressure of each component is proportional to its mole fraction in the mixture.

Few mixtures follow this generalization *exactly*; those that do are known as *ideal solutions*. For an ideal solution of A and B, then:

$$\frac{\text{partial vapour pressure of A}}{\text{vapour pressure of mixture}} = \frac{\text{number of particles of A}}{\text{number of particles of (A + B)}}$$

$$= \frac{\text{moles of A}}{\text{moles of (A + B)}}$$

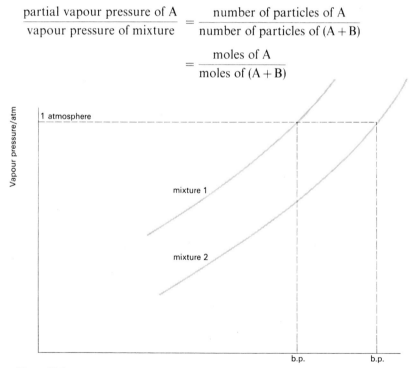

Figure 10.1c
Relationship between vapour pressure and temperature for mixtures.

The boiling point of a mixture is the special case when its vapour pressure becomes equal to the atmospheric pressure. From the graph (figure 10.1c) it can be appreciated that when the composition of a mixture is changed a decrease in vapour pressure will result in a proportional increase in boiling point.

For an ideal solution the relationship between vapour pressure and composition can be expressed in the form of a graph (figure 10.1d). This graph should be compared with the boiling point against composition graph (figure 10.1a).

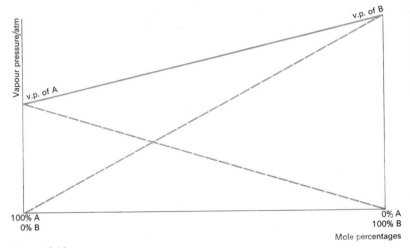

Figure 10.1d

The relationship at constant temperature between vapour pressure and mole fraction for an ideal solution. The dashed lines represent the partial pressures of the components.

The results from experiment 10.1a will have indicated that many liquids form *non-ideal solutions*, which deviate from the behaviour expressed in Raoult's law. Such deviations can be positive or negative.

If the vapour pressure of a mixture is greater than that predicted for an ideal solution, this is known as a *positive deviation* from Raoult's law. Because of the inverse relationship between vapour pressure change and boiling point change, a vapour pressure *greater* than ideal results in a boiling point *less* than ideal.

Negative deviations from Raoult's law are opposite in sense to those described above. Both types of deviation are illustrated in figure 10.1e.

Classify the trichloromethane-acetone system and the ethanol-cyclohexane system into their appropriate category.

Positive deviations from Raoult's Law

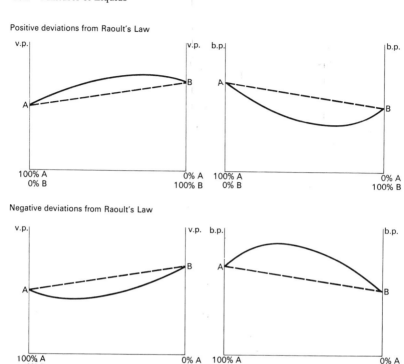

Figure 10.1e
Vapour pressure and boiling point curves for non-ideal solutions.

Another type of system is obtained by adding a non-volatile solute to a volatile solvent, for example by adding sodium chloride to water. It is found that at low concentrations the lowering of the vapour pressure of the water is roughly proportional to the concentration of the sodium chloride. What then is the influence of added salt on (a) the boiling point and (b) the freezing point of water?

The interpretation of non-ideal behaviour
If two liquids mix to form an ideal solution, it seems likely that there has been little or no net change in the molecular interactions occurring in the liquid state.

In contrast, if there are large deviations from ideal behaviour, it is possible that there are considerable interactions of new types between the different molecules.

To investigate the possibility of bond making and bond breaking, experiments can be conducted to find out whether enthalpy changes take place when liquids are mixed.

Experiment 10.1b

Do enthalpy changes occur when non-ideal solutions are formed? Investigate whichever system you studied before, either

A   trichloromethane and acetone, or
B   ethanol and cyclohexane.

Remember that three of these liquids are inflammable, and all of them are volatile. Make sure that no Bunsen burners are alight, and see that the laboratory is well ventilated.

Place each pure liquid in a burette. Place a boiling-tube in a beaker and surround the tube with a heat-insulating material, such as cotton wool. Keep a second boiling-tube available in a test-tube rack.

Run $18.0 \, cm^3$ of one liquid into the insulated boiling-tube, and then run $2.0 \, cm^3$ of the other liquid into the second boiling-tube. Record the temperature of each liquid.

Tip the liquid in the free tube into the insulated tube; stir gently with the thermometer, and record the temperature.

Repeat for $15.0 \, cm^3$ of the first liquid and $5.0 \, cm^3$ of the second, and continue the process until a complete range of values has been obtained. Use a total of $20.0 \, cm^3$ of mixture on each occasion.

Volume/cm³					
Liquid A 18	15	12	8	5	2
Liquid B 2	5	8	12	15	18

Plot a graph of temperature change against volume composition.

How does this graph compare with the graph of boiling point against volume composition?

How many moles of A and how many moles of B are there at any point of particular interest in the graph?

The densities of the liquids are:

acetone	$0.79 \, g \, cm^{-3}$
cyclohexane	$0.78 \, g \, cm^{-3}$
ethanol (industrial methylated spirits)	$0.81 \, g \, cm^{-3}$
trichloromethane	$1.5 \, g \, cm^{-3}$

What interpretation in terms of bonding would you put on
    A   an endothermic change
    B   an exothermic change?

It is apparent from the experimental work on non-ideal mixtures and their heat of mixing that intermolecular bonding occurs when substances such as trichloromethane and acetone are allowed to mix. Further qualitative tests can now be carried out to find out which atoms are involved in this bonding.

Experiment 10.1c

    To discover which atoms participate in the intermolecular bonding in the trichloromethane-acetone system

Measure the temperature rise on mixing $10 \, cm^3$ portions of the following liquids:
1   dichloromethane and acetone
2   trichloromethane and acetone
3   tetrachloromethane and acetone

As a result of these experiments would you suggest that the substituted methanes bond to acetone by means of a hydrogen or a chlorine atom?

Repeat the measurement of temperature rise using:
    4   trichloromethane and hexane, $CH_3(CH_2)_4CH_3$
    5   trichloromethane and triethylamine, $(CH_3CH_2)_3N$

Would you suggest that trichloromethane bonds to a methyl group or an oxygen atom of acetone? Are other atoms such as nitrogen also able to take part in this bonding?

Warning. Triethylamine has a strong odour. It should be handled in a fume cupboard only, and the residues from the experiment placed in a bottle, not poured into a sink.

10.2 Hydrogen bonding

Experiments such as 10.1c show us that for very many examples of intermolecular bonding a hydrogen atom is necessary, and that the bonding takes place by means of this atom. Such bonds are known as *hydrogen bonds.*

An approximate value for the heat of formation of the hydrogen bond can be found by a simple calorimeter experiment, using, for example, the reaction between trichloromethane and acetone.

$$\begin{array}{c}\text{Cl} \\ | \\ \text{Cl}-\text{C}-\text{H} \\ | \\ \text{Cl}\end{array} + \text{O}=\text{C}\begin{array}{c}\nearrow \text{CH}_3 \\ \searrow \text{CH}_3\end{array} \longrightarrow \begin{array}{c}\text{Cl} \\ | \\ \text{Cl}-\text{C}-\text{H}\cdots\cdots\text{O}=\text{C} \\ | \\ \text{Cl}\end{array}\begin{array}{c}\nearrow \text{CH}_3 \\ \searrow \text{CH}_3\end{array}$$

The strength of this hydrogen bond can then be compared with the strengths of comparable covalent bonds, using the values of the appropriate bond energy terms (see Topic 7).

Bond	$\bar{E}/\text{kJ mol}^{-1}$
O—H	+463
N—H	+391

Experiment 10.2a
The approximate strength of the hydrogen bond

Devise a method for determining a possible value for the enthalpy change on the formation of one mole of hydrogen bonds, that is, the strength of the hydrogen bond for either

    A   trichloromethane and acetone, or

    B   trichloromethane and triethylamine.

You will need an insulated calorimeter, a thermometer graduated in 0.1 °C, and a measuring cylinder.

You will also require some of the data in table 10.2a.

Compound	Molecular weight	Density /g cm^{-3}	Specific heat capacity/J g^{-1} K^{-1}
Trichloromethane	119.4	1.49	0.96
Acetone	58.1	0.79	2.22
Triethylamine	101.2	0.73	2.18

Table 10.2a
Data for experiment 10.2a

Do not use large quantities of liquid; a *total* volume of about 40 cm^3 for the mixture is suitable.

The interpretation of properties in terms of hydrogen-bond formation

1 Bonding interpretations of non-ideal behaviour

The vapour pressure of a liquid is a measure of the tendency of the molecules within it to escape. Thus if one liquid is added to another and a solution is formed whose boiling point is higher than expected on the basis of ideal behaviour some phenomenon must be reducing the escaping tendency. There must be a greater measure of intermolecular attraction than before; in instances where marked deviation from the ideal occurs the attraction is strong enough to be classed as a weak bond. In the instance of trichloromethane and acetone this weak bond is a hydrogen bond.

$$
\begin{array}{c}
Cl \\
| \\
Cl-C-H\cdots\cdots O=C \\
| \\
Cl
\end{array}
\begin{array}{c}
CH_3 \\
\diagup \\
\diagdown \\
CH_3
\end{array}
$$

If a solution is formed whose boiling point is lower than expected on the basis of ideal behaviour, then some phenomenon is occurring which is increasing the escaping tendency. It must be reducing the intermolecular attractions existing in the pure liquids. Thus in a mixture of ethanol and cyclohexane the intermolecular attraction must be less than in pure ethanol or pure cyclohexane. This may be interpreted if one visualizes molecules of one liquid becoming so numerous in the mixture that they interfere with intermolecular attraction between the molecules of the other liquid.

In pure ethanol a very high proportion of the molecules are hydrogen bonded, for instance

$$
\begin{array}{c}
C_2H_5-O \\
\diagdown \\
H \\
\vdots \\
O-H\cdots\cdots O \\
\diagup \\
C_2H_5
\end{array}
\qquad
\begin{array}{c}
H \\
\diagdown \\
O-C_2H_5 \\
\vdots \\
H \\
\diagup \\
O \\
\diagdown \\
C_2H_5
\end{array}
$$

On adding cyclohexane, the cyclohexane molecules get in between the molecules of ethanol, breaking up the weak hydrogen bonds, and markedly reducing the previous intermolecular attraction. The vapour pressure thus rises and the boiling point falls more than expected in an ideal system.

Further evidence for the presence of hydrogen bonds in alcohols, and the effects on such bonds of the addition of non-polar molecules, is obtained from a study of infra-red spectra.

In ideal mixtures such as hexane and heptane, the hexane-hexane attraction, the heptane-heptane attraction, and the hexane-heptane attraction are so nearly the same that interference of hexane with heptane makes no appreciable difference.

Many negative deviations from Raoult's law may be attributed to hydrogen bond formation in the mixture, and a range of positive deviations may be attributed to the breaking of hydrogen bonds in one component by the interfering action of the other. A further example of a pair of compounds showing a negative deviation is acetone and aniline; a further example of a positive deviation is given by a mixture of ethanol and benzene.

It must be remembered, however, that other weak intermolecular forces such as those discussed later in this topic may also be responsible in part for observed deviations from Raoult's law.

Anomalous properties of the second short period hydrides

The boiling points of the group IV hydrides (figure 10.2a) decrease with decreasing molecular weight from tin to carbon. But in groups V, VI, and VII this is not so, for extrapolation of the general trend in each group would give a much lower boiling point for the first member than actually occurs. Graphs of melting points, and of latent heats of vaporization, show similar patterns.

In liquid hydrogen fluoride, water, and liquid ammonia, there must be appreciable intermolecular attraction.

Consideration of models suggests that hydrogen fluoride and ammonia can form chains or rings of molecules linked by hydrogen bonds.

For a hydrogen bond to form, a hydrogen atom and a non-bonded pair of electrons are required. In the hydrogen fluoride molecule there are *one* hydrogen atom and *three* non-bonded pairs, so, in principle, four hydrogen bonds might form. However, in one hundred hydrogen fluoride molecules although there are three hundred non-bonded electron pairs there are only one hundred hydrogen atoms and therefore only one hundred hydrogen bonds can be formed – an average of one bond per molecule.

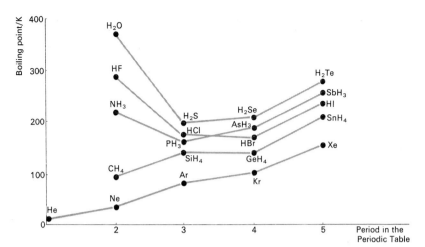

Figure 10.2a
Boiling points of some hydrides and the inert gases.

Ammonia has *three* hydrogen atoms and *one* non-bonded electron pair, and by a similar argument will only form on average *one* hydrogen bond per molecule (figure 10.2b).

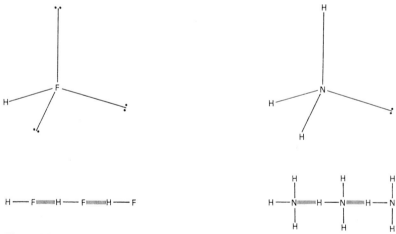

Figure 10.2b
Hydrogen bonding in hydrogen fluoride and ammonia.

Water molecules however have *two* hydrogen atoms and *two* non-bonded electron pairs each and so can form an average of *two* hydrogen bonds each.

There is therefore the possibility of water molecules being bound by hydrogen bonds into a three-dimensional lattice. This is so in ice, which commonly has the wurtzite structure (figure 10.2c); but under certain conditions of formation a diamond structure is found.

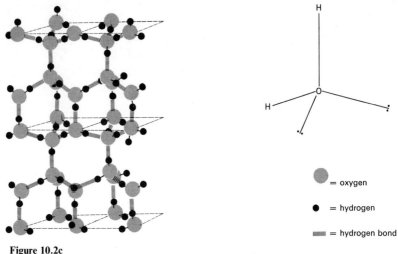

Figure 10.2c
The structure of ice.

Although these structures account for very many of the properties of ice they do not account for all of them, and the structure of ice is not fully understood. The structure of water is even less certain and is still under discussion. In water the strong hydrogen bonding still succeeds in retaining some four coordination of oxygen atoms, and there is a short-range order but no long-range ordered structure. As hydrogen bonds are constantly being broken and made in water the regions of short-range order are continually changing, in a flickering manner.

What explanation can you offer in terms of hydrogen bonding for:
   1   The temperature dependence of the density of water (figure 10.2d)?
   2   The high surface tension of water which enables objects such as needles to be 'floated' on an undisturbed water surface?

Examine the data given in table 10.2b. What do you suppose is the effect of hydrogen bonding on the heats of fusion, $\Delta H_{fus}^{\ominus}$, heats of vaporization, $\Delta H_{vap}^{\ominus}$, and melting points of the hydrides of the p-block elements?

Compound	Melting point /°C	Boiling point /°C	$\Delta H_{fus}^{\ominus}/kJ$ mol^{-1}	$\Delta H_{vap}^{\ominus}/kJ$ mol^{-1}
CH_4	−183	−162	0.92	8.20
SiH_4	−185	−112	0.67	12.1
GeH_4	−165	−90	0.84	14.1
SnH_4	−150	−52	—	18.4
PbH_4	—	−13	—	—
NH_3	−78	−33	5.65	23.4
PH_3	−134	−88	1.13	14.6
AsH_3	−116	−55	2.34	17.5
SbH_3	−88	−17	—	—
BiH_3	—	+22	—	—
OH_2	0	+100	6.02	40.7
SH_2	−85	−60	2.39	18.7
SeH_2	−66	−41	2.51	19.3
TeH_2	−48	−2	—	23.2
HF	−83	+19	4.56	7.74
HCl	−114	−85	2.01	16.2
HBr	−86	−66	2.43	17.6
HI	−51	−35	2.89	19.8
Ne	−249	−246	0.33	1.76
Ar	−189	−186	1.18	6.53
Kr	−157	−153	1.64	9.04
Xe	−112	−108	2.29	12.6

Table 10.2b
Data on the hydrides of the p-block elements and the inert gases

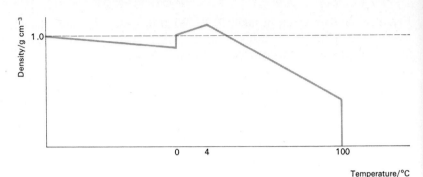

Figure 10.2d
The temperature dependence of the density of water (schematic).

2   The dimerization of organic acids

When acetic acid, CH_3CO_2H (m.w. 60), and benzoic acid, $C_6H_5CO_2H$ (m.w. 122) are dissolved in benzene, it is found that the solute particles have molecular weights of nearly 120 and 244 respectively. To account for these observations it is suggested that the molecules must exist in solution as dimers, held together by hydrogen bonds:

$$CH_3-C \overset{\displaystyle O\cdots\cdots H-O}{\underset{\displaystyle O-H\cdots\cdots O}{}} C-CH_3$$

When acetic acid is dissolved in water or alcohols the solute particles have the expected molecular weight of 60. Why do dimers not form in these cases?

Further evidence for the existence of dimers of organic acids has been obtained from X-ray diffraction studies of crystals of sorbic acid, $CH_3(CH{=}CH)_2CO_2H$ (see figure 10.2e).

3   Liquid flow in some alcohols

Suppose we have a long sealed tube containing a liquid, and a small air bubble trapped at the top. If the tube is inverted, the time which it takes for the air bubble to travel through the length of the tube can be taken as a measure of the intermolecular forces in the liquid.

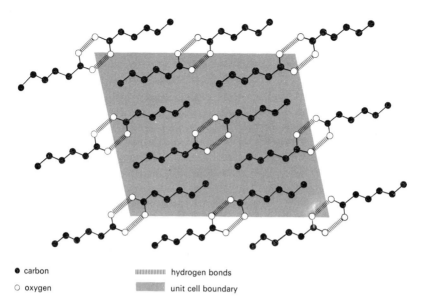

● carbon   ⅢⅢⅢⅢⅢⅢⅢ hydrogen bonds
○ oxygen   ▓▓▓▓▓▓▓ unit cell boundary

Figure 10.2e
Sorbic acid structures (hydrogen atoms omitted).

Set up three such tubes of similar cross-section and length, and compare the
rate of bubble movement in propan-1-ol, propane-1,2-diol, and propane-
1,2,3-triol. Can you relate the results to the number of hydroxyl groups in the
molecule? Draw the structures of the compounds and suggest a possible
interpretation of your observations.

What further compounds would give additional experimental evidence to help
confirm your interpretation?

4   Crystal cleavage
Experiment 10.2b
   The cleavage of some crystals
Examine some crystals of gypsum, $CaSO_4$, $2H_2O$, for hardness by scratching
them with your fingernail, and for cleavage by attempting to split the crystals
using a penknife or a spatula.

The structure of gypsum is illustrated in figure 10.2f. It consists of layers of
calcium and sulphate ions, the layers being linked together through the hydrogen
bonds of the water molecules attached to the sulphate ions.

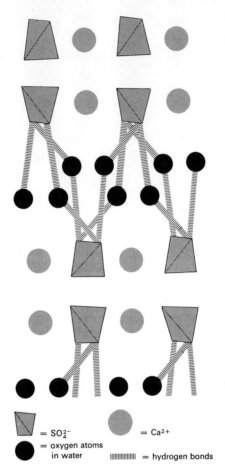

Figure 10.2f
The structure of gypsum, $CaSO_4, 2H_2O$.

What relative strengths would you expect the crystal to have in its ionic layers and hydrogen-bonded layers? Does this explain satisfactorily the properties of gypsum?

Anhydrite, $CaSO_4$, has a purely ionic structure (figure 10.2g). How would you expect its properties to compare with gypsum? Examine some anhydrite crystals for hardness and cleavage.

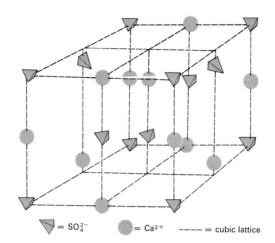

Figure 10.2g
The structure of anhydrite. CaSO₄.

Crystals of orthoboric acid, B(OH)₃, may also be examined. What shape would you expect for a B(OH)₃ molecule? What bond angles would you expect? What structure might be expected for the crystals? Would they be isotropic or anisotropic?

5   Structures in living organisms

It is probably no exaggeration to say that in living processes the hydrogen bond is as important as the carbon–carbon bond. It is responsible for a very wide range of structural features and chemical processes, and the full extent is certainly greater than has so far been discovered.

To select a few examples, the structures of proteins are maintained by means of hydrogen bonds. Thus the various varieties of silk are maintained in puckered sheet arrangements that depend on hydrogen bonds to hold the amino acid chains together (see figure 10.2h, and figure 18.4k in Topic 18).

Since enzymes are proteins, the whole range of enzyme-catalysed reactions upon which life depends are also determined by hydrogen bonding.

The spiral structure of DNA (page 217), and its ability to replicate are dominated by hydrogen bonds (figure 10.2j).

The structure of proteins is considered further in Topic 18.

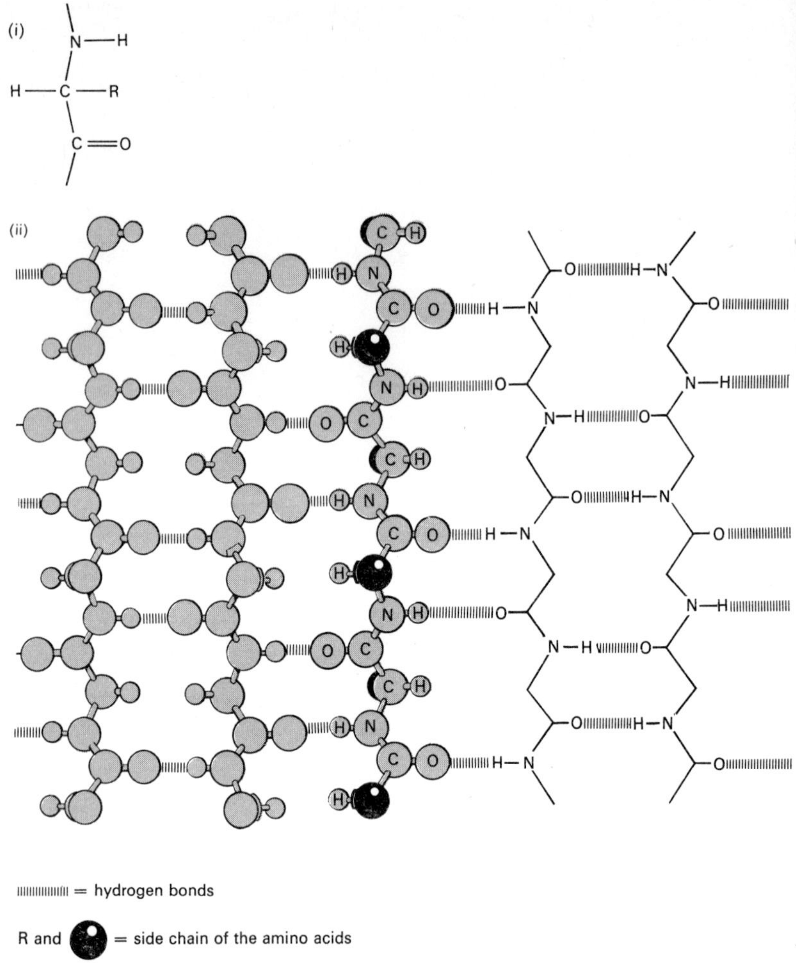

IIIIIIIIIIIIII = hydrogen bonds

R and ⬤ = side chain of the amino acids

Figure 10.2h
(*i*) The amino acid sub-unit in a protein chain. (*ii*) The linking by hydrogen bonds of the protein chains in silk.

The solubility in water of organic compounds is due to hydrogen-bonding involving electronegative atoms in the organic molecules and the oxygen and hydrogen atoms in water. An example is sugar, the large number of —OH groups per molecule giving rise to the possibility of extensive hydrogen bonding with water molecules, and hence solvation; an animal's ability to transport energy supplies rapidly to those parts of the body where they are needed depends upon water-solubility. Fats, although relatively richer in energy than sugars, are not suitable for this purpose because they are not soluble in blood.

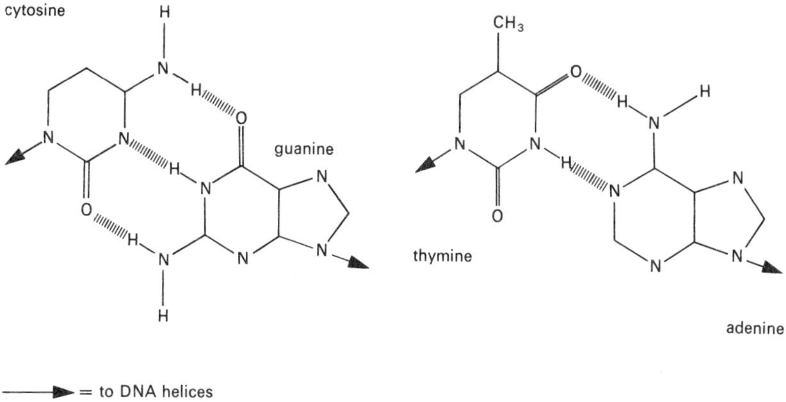

— = to DNA helices

Figure 10.2j
The hydrogen bonded base pairs in DNA.

Also of interest is the way in which cotton, hair, and wool absorb moisture while some synthetic materials such as nylon do not. Why should cotton, a sugar polymer (figure 10.2k), be able to absorb moisture? Why should hair and wool which are protein fibres absorb moisture?

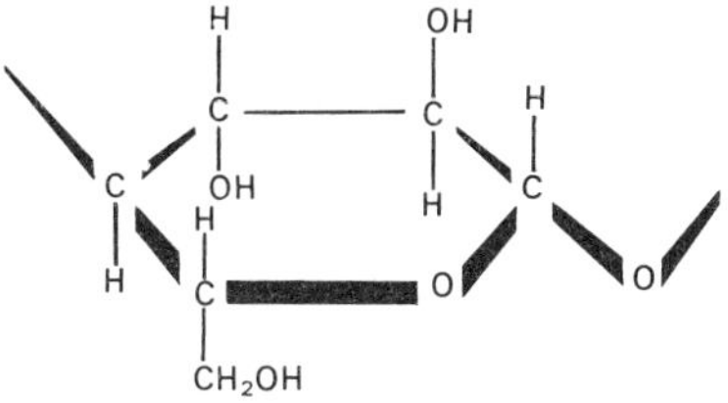

A six-membered ring;
part of a cellulose polymeric
molecule (a three–dimensional
representation)

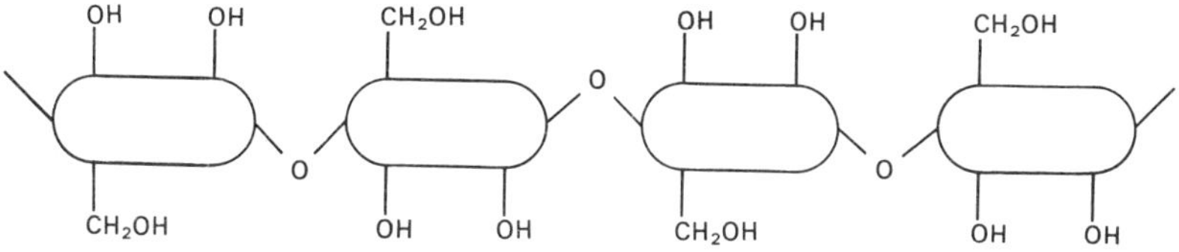

Figure 10.2k
Several of the ring members of a cellulose molecule such as cotton (a two-dimensional representation).

10.3 Dipole–dipole attractions

Earlier in the topic the behaviour of tetrachloromethane and acetone on mixing was studied, and a slight temperature drop was observed. Does this suggest an increase or decrease in intermolecular bonding when the pure liquids are mixed? What properties of the mixture might be affected by changes in intermolecular bonding?

The tetrachloromethane molecule contains no hydrogen atom to form a hydrogen bond to an acetone molecule and it is also a symmetrical molecule. Is the acetone molecule symmetrical in regard to distribution of charge due to its electrons? What previous experimental tests have indicated the charge distribution in acetone?
Can the distribution of charge in the acetone molecule be used to explain the heat of mixing with tetrachloromethane?

10.4 Van der Waals' forces

The inert gases provide evidence for the existence of cohesive forces between molecules. Helium, which does not form normal bonds and has symmetrical atoms, condenses to a liquid and ultimately freezes to a solid at very low temperatures; energy is evolved in this process showing that cohesive forces are operating.

The energy of sublimation of solid helium is only $0.105 \text{ kJ mol}^{-1}$ ($\Delta H_{fus} + \Delta H_{vap}$). This should be compared with the energy of sublimation of ice, 46.9 kJ mol^{-1}, which is used to overcome hydrogen bonding, and the dissociation energy of the oxygen molecule, 494 kJ mol^{-1}, required to break the two covalent bonds.

The weak forces of attraction, independent of normal bonding forces, which are found to exist between atoms and molecules in the solid, liquid, and gas states are known as *van der Waals' forces*.

Van der Waals' forces are considered to be due to continually changing dipole-induced dipole interactions between atoms.

These dipoles are thought to arise because the electron charge-cloud in an atom is in continual motion. In the turmoil it frequently happens that rather more of the charge-cloud is on one side of the atom than on the other. This means that the centres of positive and negative charge do not coincide, and a fluctuating dipole is set up. This dipole induces a dipole in neighbouring atoms, and a force of attraction results. These flickering atomic dipoles and induced dipoles produce a cohesive force between neighbouring atoms and molecules.

The greater the number of electrons in an atom, the greater will be the fluctuation in the asymmetry of the electron charge-cloud and the greater will be the van der Waals' attraction set up. The rise in boiling point down group VII, fluorine, chlorine, bromine, and iodine, is due more to the increase in electrons present in the atoms and the consequent increased van der Waals' attractions, than to the increase in the mass of the atoms.

The increase in boiling point up the homologous series of alkanes is due more to the increased number of electrons in the molecules and the increased total van der Waals' attractions than to the increase in mass of the molecules. Similarly the difference in boiling point between isomers can be explained in terms of van der Waals' attractions.

The boiling points of two of the C_5H_{12} isomers (figure 10.4a) are 36 and 9 °C. Think what a scale model (e.g. of polystyrene spheres) of each molecule would look like, and then put forward a suggestion for the difference in boiling points.

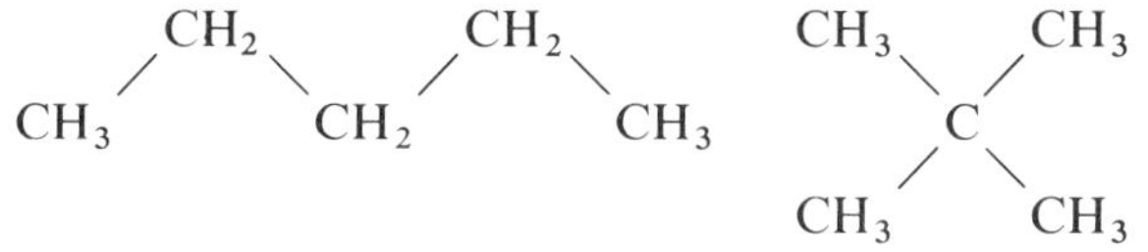

pentane	2,2-dimethylpropane
m.w. = 72	m.w. = 72
b.p. = 36 °C	b.p. = 9 °C

Figure 10.4a
Two isomers of formula, C_5H_{12}.

Van der Waals' radii

The normal bonding forces in molecules are concentrated within the molecules themselves; they are intramolecular. In crystals individual molecules are held to each other by van der Waals' forces. Examples are iodine, solid carbon dioxide, and naphthalene. (As the forces are weak, the melting points of molecular crystals tend to be low. If permanent dipoles, or hydrogen bonding are present as well, then the melting point will be higher.)

In molecular crystals the van der Waals' forces draw molecules together until their electron charge-clouds repel each other to the extent of balancing the attraction. Thus for argon the atoms are drawn together until the atomic nuclei have a separation of about 0.4 nm (figure 10.4b).

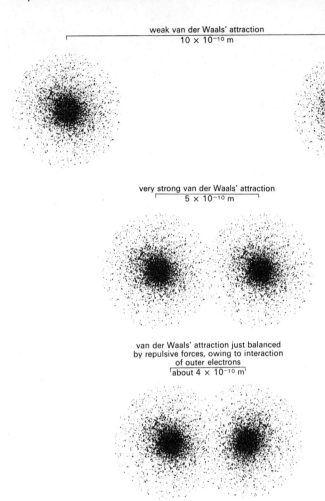

weak van der Waals' attraction
10×10^{-10} m

very strong van der Waals' attraction
5×10^{-10} m

van der Waals' attraction just balanced
by repulsive forces, owing to interaction
of outer electrons
about 4×10^{-10} m

Figure 10.4b
Van der Waals' attraction between argon atoms.

The atomic distances within simple molecules and between simple molecules are not the same. The covalent radius is one half of the distance between two atoms in the same molecule. The van der Waals' radius is one half of the distance between the nuclei of two atoms in adjacent molecules (see figure 10.4c). The relative values of these radii can be seen in table 10.4a.

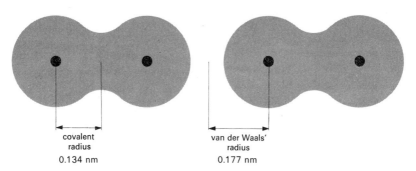

Figure 10.4c
Covalent radius and van der Waals' radius for molecules of I_2 in an iodine crystal.

Atom	Covalent radius/nm	Van der Waals' radius/nm
H	0.037	0.12
N	0.070	0.15
O	0.066	0.14
P	0.110	0.19
S	0.104	0.185

Table 10.4a
Comparative values for covalent and van der Waals' radii

The importance of van der Waals' forces

Although the van der Waals' forces between individual atoms as in argon give only a small bonding energy, the total van der Waals' bonding energy can be significant in large molecules with many contacts between atoms. For example, a well-crystallized sample of polythene (the high density form) has a tensile strength of $30 \times 10^6 \, N \, m^{-2}$ while polythene of poorer crystallinity, that is, with less orderly packing (the low density form), has reduced van der Waals' binding energy and a tensile strength of $10 \times 10^6 \, N \, m^{-2}$.

Van der Waals' forces can also make an important contribution to the structure of globular proteins. Molecules such as proteins tend to form aggregates in aqueous solution because of their hydrocarbon side chains. These aggregates can be dispersed by the addition of urea, which is known to break weak bonds. Thus on the addition of urea, haemoglobin will break into two different sub-units. If the urea is then removed, the sub-units will associate to reform haemoglobin molecules. The process is known to be exact because the reformed haemoglobin is physiologically active.

Problems

1   Figure 10.5 shows the vapour pressure at constant temperature of mixtures, of various composition, of two liquids X and Y. State which of the following pairs of liquids are most likely to behave like X and Y, and give reasons for your choice.

   A   propan-1-ol and propan-2-ol

   B   propan-1-ol and toluene

   C   benzene and toluene

   D   benzene and heptane

   E   propan-2-ol and ethanol

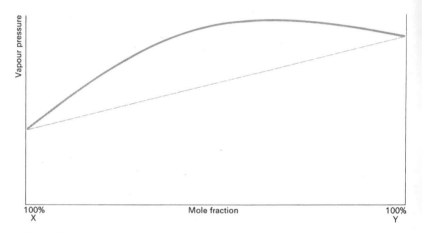

Figure 10.5

2   Arrange the following in the order you should expect for their boiling points, putting the one with the highest boiling point first.

   A $CH_3CH_2CH_2CH_3$

   C $CH_3CH_2CH_2CH_2Cl$

$$\text{B} \quad CH_3-\underset{\underset{\textstyle CH_3}{|}}{\overset{\overset{\textstyle CH_3}{|}}{C}}-H$$

$$\text{D} \quad CH_3-\underset{\underset{\textstyle CH_3}{|}}{\overset{\overset{\textstyle CH_3}{|}}{C}}-Cl$$

Give reasons for your answer.

3 The vapour pressures of octane and 2-methylheptane at 30 °C are 19.00 mmHg and 27.4 mmHg respectively.

i Calculate the vapour pressure of a mixture of the two liquids containing the mole fraction 0.4 of octane at 30 °C.

ii What assumption about the mixture did you need to make before you could calculate its vapour pressure? On what evidence did you decide that the assumption was reasonable?

4 Classify the following mixtures of liquids into:
    A   those likely to obey Raoult's law
    B   those likely to show a positive deviation from Raoult's law
    C   those likely to show a negative deviation from Raoult's law
  i   acetone and butanone
 ii 1,1,2,2-tetrachloroethane and acetaldehyde
iii toluene and xylene
 iv trichloromethane and ethylpropyl ether
  v benzyl alcohol and benzene.
State briefly the reasons for your answers.

5 Arrange each of the following groups of liquids in the order you would expect for their boiling points, putting the liquid with the highest boiling point first.
  i   helium, neon, argon
 ii propane, butane, pentane
iii hydrogen fluoride, hydrogen chloride, sodium chloride
 iv hydrazine, silicoethane, diborane
  v benzoic acid, *p*-hydroxybenzoic acid, benzene
Give reasons for your answers to (*iv*) and (*v*).

6 The graph shown below (figure 10.6) represents the partial vapour pressures of pyridine and acetic acid above various mixtures of the two liquids at constant temperature.

i What would be the vapour pressure of the following over a mixture containing a mole fraction 0.4 of acetic acid?
    A   acetic acid
    B   pyridine
    C   the whole mixture
ii Show the following in the form of graphs:
    A   The vapour pressures of acetic acid, pyridine, and the whole mixture, if the mixtures were *ideal*.
    B   The *actual* total vapour pressures of mixtures of pyridine and acetic acid.

iii   Your answer to (*ii*, B) should show a very large negative deviation from that of an ideal mixture, the greatest deviation being at a mole fraction of 0.5. Suggest reasons for this deviation.

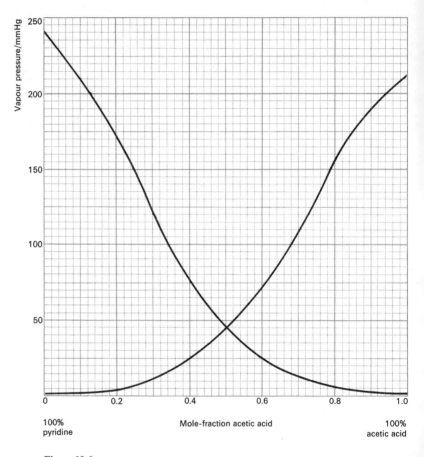

Figure 10.6

Topic 11

Solvation

Most of the reactions conducted by chemists in the laboratory and in industry are done in solution, and most of the reactions which occur in nature do so in solution. An understanding of solubility is therefore important. Only general trends can be looked for at this stage because many factors are involved: the nature of the solute, the nature of the solvent, and the possible interactions of the particles, will all influence the possibility of solution occurring.

So far, chemists have only a very partial understanding of solubility. A quantitative understanding is rarely possible, and there are many unexplained observations; but several broad principles can be seen, and these are developed in this topic.

11.1 The influence of the solvent on the nature and rate of a reaction

When substances react together in solution, the principal function of the solvent is to enable the substances to mix intimately with each other thus producing conditions leading to reaction. The effect is well illustrated by barium bromide and silver nitrate. The reaction

$$BaBr_2 + 2AgNO_3 \longrightarrow Ba(NO_3)_2 + 2AgBr(s)$$

is exothermic, $\Delta H = -188 \, kJ \, mol^{-1}$. When the solids are mixed, however, even in powdered form, there is almost no detectable reaction. If the reagents are dissolved in water and then mixed the reaction is almost instantaneous, and is evident by the precipitation of silver bromide and the evolution of a large quantity of heat.

The choice of a solvent for a reaction can have an important influence on the outcome of a reaction; for example, in a different solvent the above reaction can be made to go in the opposite direction. So that with liquid ammonia as solvent barium bromide proves to be insoluble and the reaction is

$$2AgBr + Ba(NO_3)_2 \longrightarrow BaBr_2(s) + 2AgNO_3$$

There is also the possibility of chemical reaction between the solvent and the compounds whose reactions are being studied. For example, it is not possible to study the solution chemistry of phosphorus trichloride, or of disulphur dichloride, using water as a solvent, because each of these substances reacts with water. But by using another solvent the study may be made.

The nature of the solvent can also exert a profound influence on the rate of a reaction.

The reaction between ethyl iodide and triethylamine to give a substituted ammonium salt may be conducted in solution in hexane.

$$C_2H_5I + \underset{\underset{C_2H_5}{|}}{\overset{\overset{C_2H_5}{|}}{N}}-C_2H_5 \longrightarrow \left[C_2H_5-\underset{\underset{C_2H_5}{|}}{\overset{\overset{C_2H_5}{|}}{N}}-C_2H_5 \right]^+ + I^-$$

ethyl
iodide   triethylamine   tetraethylammonium iodide

The rate of this reaction in hexane can be measured ; when it is conducted using benzene as the solvent the rate is found to be 80 times faster than in hexane. In nitrobenzene it is 2800 times faster than in hexane.

Since the solvent plays so influential a part in solution chemistry it is important to understand something of the relationship between solute and solvent particles, and in particular the nature of the forces between them, and the structural relations. The remainder of this topic is devoted to considering these aspects.

11.2 The pattern of solubility

Experiment 11.2a
Is there any general relationship between solubility and the natures of solutes and their solvents?
You will be supplied with the following solvents :
hexane or cyclohexane, tetrachloromethane, methanol, propan-1-ol, acetone, and water.

The solubility of one of the compounds, anhydrous sodium iodide, anhydrous calcium chloride, and anhydrous iron(III) chloride, in each of these solvents, should be tested. All the members of the class will test the solubility of iodine in hexane, methanol, and water and the solubility of the three solvents in one another.

Three criteria may be used to judge whether the ionic compounds have dissolved or not : they are :

Does the solid disappear?
Is there any temperature change on dissolving?
Does the solution become electrically conducting?

Proceed as follows:

1   Test the methanol, propan-1-ol, and acetone to ensure that they are
anhydrous. Add a small quantity of anhydrous copper sulphate powder to
one or two cm³ of the liquid in a test-tube.

2   If you are studying sodium iodide or calcium chloride, and if it is not in
finely powdered form, place a few grammes in a mortar, grind them quickly,
and transfer them without delay to a corked test-tube or a weighing bottle with
a cap. Each of the solids rapidly absorbs moisture from the air.

3   Place a very small quantity of solid in a test-tube (using only enough of it
to be clearly visible in the bottom of the tube).
The solids being tested absorb moisture from the atmosphere, and the stoppers
of the stock bottles should be replaced securely immediately after use.
Add one liquid to a depth of about 4 cm. Shake the test-tube and place it in a
test-tube rack. Repeat for the other liquids. Shake the tubes periodically.
If a solid dissolves, add a second quantity to confirm the observation, and
finally add a rather larger quantity on the end of a spatula.
Record your observations.

4   For each liquid which dissolved the solid, place liquid in a test-tube to a
depth of about 4 cm, record the temperature, add solid heaped on the end of a
spatula, replace the thermometer and stir, and note the temperature of the
solution. Record any temperature change.

5   Test the electrical conductivity of the solution. First, place a few cm³
of pure solvent in a small beaker and test its conductivity. Then place the
solution in the beaker and test its conductivity. Record the results.

6   Test the solubility of iodine in hexane, in methanol, and in water.
Test the solubility of hexane in water, hexane in methanol, and methanol in
water.
Temperature and conductivity measurements need not be made in this section.

7   Compare results with groups which have investigated the other ionic
solids and draw up a table of combined results.

Table 11.2a shows the values of two important physical properties of solvents.

Liquid	Dipole moment $/10^{-30}$ C m	Relative permittivity*
Hexane	0	1.9
Cyclohexane	0	2
Tetrachloromethane	0	2.2
Benzene	0	2.3
Ammonia	3.0	22
Trichloromethane	3.3	4.8
Formic acid	5.0	58
Sulphur dioxide	5.3	16
Propan-1-ol	5.7	20
Ethanol	5.7	24
Methanol	5.7	33
Water	6.3	79
Hydrogen fluoride	6.7	84
Acetone	10	21
Hydrogen cyanide	10	114

* Relative permittivity (also known as dielectric constant) is $\varepsilon_{liquid}/\varepsilon_{vacuum}$, where ε_{liquid} is the permittivity of the liquid, and ε_{vacuum} the permittivity of a vacuum.

Table 11.2a
Dipole moment and relative permittivity of some solvents

The dipole moment is a measure of the polarity of the molecules of the compound.

What type of solvent dissolves ionic compounds? What type of solvent does not dissolve ionic compounds?

Solubility of iodine at 10 °C in various solvents/g per 100 g of solution	
Hexane	12 (25 °C)
Benzene	10
Tetrachloromethane	3
Methanol	8
Acetone	3
Water	0.02

Table 11.2b
Solubility of iodine in various solvents

In general, what type of liquids are good solvents for iodine, and what type are poor solvents? Does this correspond to the solubility of hexane in different liquids?

The behaviour of non-polar solvents, and solvents of low and high polarity
The general pattern emerging from these experiments, and substantiated by other experiments, is that non-polar liquids are good solvents for non-polar substances, and polar liquids are good solvents for polar substances.

Iodine molecules are held to each other in the solid by weak van der Waals' forces, and benzene molecules are held to each other by the same type of forces (see figure 11.2a). These are similar in magnitude to the iodine–benzene van der Waals' forces. Thus it is easy for benzene molecules to penetrate into the iodine crystal and to separate the iodine molecules; similarly it is easy for iodine molecules to penetrate between the benzene molecules, and so distribute themselves throughout the liquid. In this way the iodine dissolves.

Similarly, hexane–hexane, benzene–benzene, and hexane–benzene attractions are weak and comparable to each other, and hexane dissolves in benzene.

In liquids of moderate polarity such as methanol the situation is more complicated and solubility effects cannot be predicted with confidence.

For liquids of high polarity such as water a general pattern of behaviour can be seen. In the instance of water, the water–water attraction is markedly increased by the existence of hydrogen bonding (see figure 11.2a); these water–water attractions are very much stronger than either the iodine–iodine attractions or the iodine–water attractions, and the iodine molecules cannot force their way into the water structure. Iodine is thus almost insoluble in water.

A similar situation exists for water and benzene (or hexane), and these two do not dissolve.

It must be borne in mind that the pattern of solubility which has emerged from the investigation, while true for many common substances, is untrue for an important number of special cases. The important criterion is that the new solute–solvent interactions must outweigh the existing solute–solute and solvent–solvent interactions if solution is to occur.

What processes occur in the dissolving of an ionic solid?
As an aid to discussing this question some experiments may be conducted to find the magnitude of the energy changes taking place, and whether there is any overall volume change.

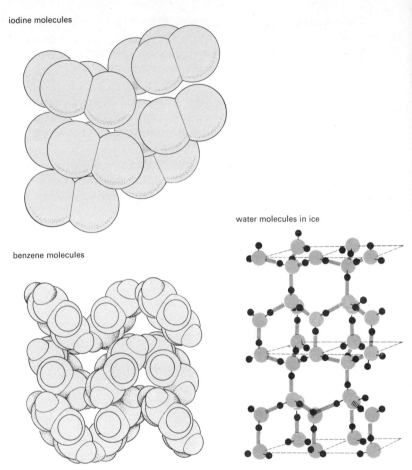

iodine molecules

benzene molecules

water molecules in ice

Figure 11.2a
The crystal structures of iodine, benzene, and ice (not to scale).

Experiment 11.2b
An investigation of temperature changes and volume changes accompanying the dissolving of some ionic compounds

A suitable series is formed by the anhydrous compounds: lithium chloride, sodium chloride, potassium chloride, calcium chloride, iron(III) chloride. A convenient concentration to prepare is 2M.

1   Temperature changes

Place 50 cm^3 of water in an insulated calorimeter. Record its temperature. Add the correct weight of powdered solid to produce a 2M solution, stir to dissolve and record the temperature.

Calculate the temperature change. Calculate also the heat transferred on dissolving 1 mole to give a 2M solution.

Since the solids absorb moisture from the atmosphere it is necessary to weigh them in stoppered containers. It is also essential to replace the stopper of the stock reagent bottle immediately after use.

2   Volume changes

Devise a method for finding the overall volume change which accompanies the solution process, that is:
(volume of solution) − (volume of pure solvent + volume of solid) = final volume − original volume.

Two 50 cm^3 burettes will be needed.

Anhydrous compound	Density/g cm^{-3}	Formula weight
Lithium chloride	2.1	42.4
Sodium chloride	2.2	58.4
Potassium chloride	2.0	74.6
Calcium chloride	2.5	111
Iron(III) chloride	2.8	162

Table 11.2c
Densities of some anhydrous compounds

Tabulate the temperature and volume changes recorded for the series.

Is there any trend to be seen on descending the group I chlorides?

Are the temperature and volume changes you have recorded related to the charge on the ions?

What happens to the ions in the crystal lattice when a solid is dissolved in a liquid? Assuming that a very dilute solution is formed, approximately how much energy would you expect to be involved in the process? On this basis are the actual energy changes (as revealed by the recorded temperature changes) surprising? How may they be explained?

The dissolving of ionic solids

For an ionic solid to dissolve the crystal lattice must be broken down, and the positive and negative ions separated from each other. This requires a substantial amount of energy. If the ions were separated to an infinite distance apart the energy would be numerically equal to the lattice energy; thus in the formation of a fairly dilute solution the separation of the ions from the lattice must involve the absorption of a quantity of energy almost equal to the lattice energy.

For the alkali metal halides the lattice energies are in the region of 400–800 kilojoules per mole; the absorption of so much energy from the solvent and its immediate surroundings would produce a freezing of most solvents. But the dissolving of many ionic solids in a solvent produces a rise in temperature, and if a drop in temperature is produced it is usually small. Some additional phenomenon which is exothermic must therefore be occurring, and the energy released by it must be about the same in value as the lattice energy.

The most likely source of such a large quantity of energy is the occurrence of some form of bonding. It has been seen that polar molecules are best for dissolving ionic compounds; the negative end of a polar solvent molecule becomes attracted to a positive ion, and the positive end of a solvent molecule becomes attracted to a negative ion, and thus ion-solvent molecule bonds are established with a consequent release of energy. This release of energy enables the ions to become detached from the lattice.

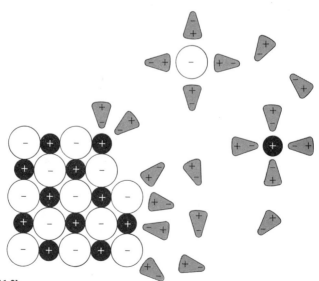

Figure 11.2b
Polar solvent molecules attaching themselves to ions in a crystal lattice, and solvating the ions.

Figure 11.2b represents the process. Copy the diagram into a notebook where positive ions may be coloured in red, the negative blue or green, and the polar solvent molecules some other suitable colour.

The attachment of polar solvent molecules to ions is known as *solvation*, and the ions are said to be solvated. If the solvent is water then the process is known as *hydration*, and the ions are said to be hydrated. Figure 11.2c represents two hydrated ions, one positive and the other negative.

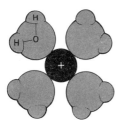
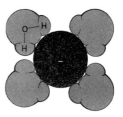

Figure 11.2c
A hydrated positive ion, and a hydrated negative ion (a two-dimensional representation).

The electrostatic attraction of an ion for polar solvent molecules extends beyond one layer of attached molecules, and a second, and perhaps third, layer or sheath of solvent molecules is loosely associated with a given ion. The situation is represented in figure 11.2d.

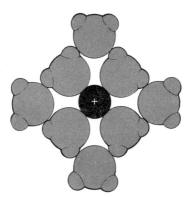

Figure 11.2d
A positive hydrated ion showing an inner layer of solvating molecules and an outer, looser layer (a two-dimensional representation).

These diagrams may also be copied and their clarity improved by colouring; the positive ions might be coloured red, the negative ions green or blue, and the oxygen atoms of the water molecules some other suitable colour.

If the arrangement of solvent molecules around an ion of single positive charge is approximately as shown in figure 11.2d, what do you suppose might be the arrangement around an ion of the same diameter but with $2+$ charges? And what might be the situation around an ion of the same diameter but $3+$ charges?

The solvent number of an ion is the number of solvent molecules which can at any given instant be regarded as firmly or loosely associated with that ion.

Table 11.2d shows approximate values for the solvation numbers of some ions in a range of solvents.

| Solvent | Ion and approximate solvation number (± 1 unless stated) | | | | | | | | |
	Li^+	Na^+	K^+	Cl^-	Br^-	I^-	Mg^{2+}	Ca^{2+}	Al^{3+}
Methanol	7	5	4	4	2–4	1			
Ethanol	6	4	3	4	4	2			
Acetone	4	4	4	2	1	1			
Water	5	5	4	1	1	1	15 ± 2	13 ± 2	26 ± 5

Table 11.2d
Some approximate solvation numbers

From the foregoing discussion it is clear that most ions in solution do not exist as isolated entities; they are surrounded by a sheath of oriented solvent molecules which may be several molecules thick. As the outer molecules are only loosely held there is a continual exchange between them and unattached molecules nearby; the situation is never static.

Because of the surrounding sheath of attached solvent molecules most ions diffuse more slowly in solution than would be expected for unencumbered ions. For the same reason most ions move more slowly in electrolysis cells than would be expected if they were existing on their own in solution. Measurement of the speed of migration of ions in electrolysis provides one method for estimating solvation numbers. Other methods are available, and the values obtained vary with the method used. Solvation numbers are therefore only approximate. This is partly due to the impossibility of defining precisely where a sheath of loosely attached molecules ends.

When an ionic solid dissolves, the removal of the ions from the surface of the lattice is begun by the attaching of polar solvent molecules to the ions; it is aided by an effect due to the relative permittivity of the solvent. The force F between two charges Q_1 and Q_2 is given by:

$$F = \frac{Q_1 Q_2}{4 \pi \varepsilon r^2}$$

where r is the distance apart of the centres of the charges, and ε is the permittivity of the medium separating them.

The value of ε for a vacuum is 8.85×10^{-12} F m^{-1}, and for all other media it is greater than this. Thus the force between two charges is lessened when they are in anything other than a vacuum. An examination of table 11.2a shows that the polar liquids listed all have relative permittivities very much larger than that of a vacuum. Thus as a polar solvent begins to surround an ion on a crystal lattice it reduces to some extent the force of attraction holding the ion to the lattice, and this assists the detachment of the ion.

Thus two physical processes occur; there is the solvation of ions by polar molecules, whose polarity is measured by their dipole moment; and there is the weakening of the inter-ionic attraction by the effect of the relative permittivity. The more important of the two is solvation. A large dipole moment is more important than a large relative permittivity.

Energy changes and solution

The energy released when a mole of a substance in the form of gaseous ions is solvated is the solvation energy; if the solvent is water then the term used is hydration energy. The energy relationships principally involved in the dissolving of a solid may be represented by an energy cycle as in figure 11.2e.

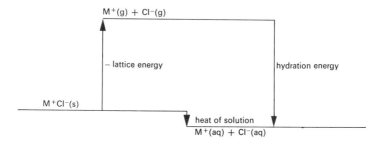

Figure 11.2e
Principal energy relationships in the dissolving of an ionic solid. (aq) represents the full hydration of the ions.

The heat of solution is usually small compared with the lattice energy, and it may be positive or negative.

Use the table of lattice energies given in Topic 7 (page 195) to produce values for the hydration energies of the alkali metal halides, on the basis of your results for the heat transferred on making 2M solutions, in experiment 11.2b.

The structure of some hydrated ions

Copper sulphate crystals have the formula $CuSO_4$, $5H_2O$. Structure determinations show that four of the water molecules are distributed in a plane around the copper ion, and that the sulphate ion is joined by hydrogen bonds to the fifth water molecule and to the hydrated copper ion (see figure 11.2f). The formula could therefore be written as $Cu(H_2O)_4^{2+}SO_4^{2-}(H_2O)$.

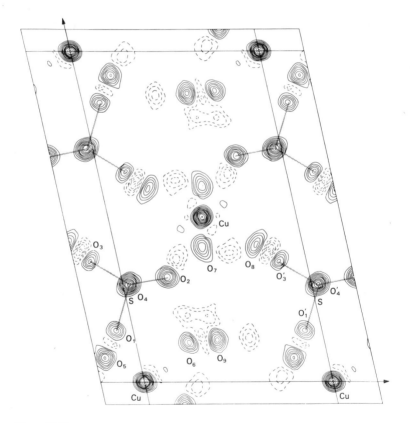

Figure 11.2f
A contour map indicating the layout of the atoms in the unit cell of a crystal of $CuSO_4$, $5H_2O$ projected onto a principal plane.

The contour lines in figure 11.2f indicate the position of the nuclei, the broken lines being the contours of the hydrogen atoms. The S–O bonds within the sulphate group, which includes the oxygen atoms O_1, O_2, O_3, and O_4, have been drawn in. The sulphate group is a tetrahedron but is distorted by the planar projection. The remaining oxygen atoms O_5, O_6, O_7, O_8, and O_9 belong to the five water molecules.

Four oxygen atoms O_5, O_6, O_7, and O_8 are coordinated to the copper atom and their hydrogen atoms either form hydrogen bonds to sulphate groups, e.g. O_7—H⬛⬛⬛⬛O_4 and O_7—H⬛⬛⬛⬛O_3', or to a sulphate group and the fifth water molecule e.g. O_6—H⬛⬛⬛⬛O_2 and O_6—H⬛⬛⬛⬛O_9. The fifth water molecule forms hydrogen bonds to sulphate groups O_9—H⬛⬛⬛⬛O_2 and O_9—H⬛⬛⬛⬛O_1'.

The hydrogen bonds vary in length and angle but are about 0.28 nm long and the bond angle is about 170°. The water molecules are somewhat distorted having a mean bond angle of 110° compared with 109.5° for an exact tetrahedron and 105° for the value found in steam.

Hydrated iron(III) chloride crystals have the formula $FeCl_3$, $6H_2O$. Structure determinations have shown this to be $Fe(H_2O)_6^{3+}(Cl^-)_3$, with the six water molecules distributed on three axes mutually at right-angles (octahedral coordination). The representations of the ions are shown in figure 11.2g.

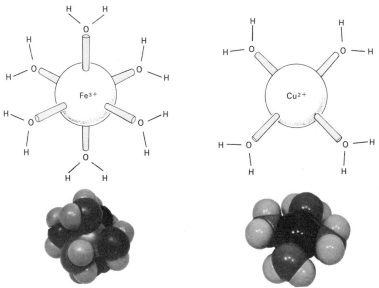

Figure 11.2g
Representation of hydrated ions in the solid state. The water molecules would, in fact, be in contact with the ions.

Copy figure 11.2g and join the oxygen atoms by means of coloured lines to show the octahedral faces and the square plane. Use thick lines for edges close to the observer and thinner ones for edges further away from the observer.

6 coordination of water molecules around a positive ion is common in crystals, and 4 coordination is fairly common. Other examples are $CrCl_3, 6H_2O$ which is $Cr(H_2O)_6^{3+}(Cl^-)_3$, and $BeCl_2, 4H_2O$ which is $Be(H_2O)_4^{2+}(Cl^-)_2$. When these ions dissolve in water the inner sheath of solvent molecules probably retains the 6 and 4 coordinated structure.

Solvent molecules coordinated around an ion can often be replaced by another type of polar molecule, or by another ion. For instance if concentrated ammonia solution is added to copper sulphate solution a very deep blue coloured solution results; the solid obtained from this solution contains $Cu(NH_3)_4^{2+}$ ions. The four ammonia molecules have replaced four water molecules, retaining a planar distribution.

What shape is an ammonia molecule? Draw a diagram to represent the structure and orientation of the $Cu(NH_3)_4^{2+}$ ion.

The dissolving of organic compounds in water

Small ionic organic compounds are likely to be soluble in water because of the effects which have just been discussed. Thus sodium acetate, $CH_3CO_2^- Na^+$, and sodium butanoate, $CH_3CH_2CH_2CO_2^- Na^+$, are soluble.

A very wide range of non-ionic organic compounds is also soluble; but examination of the formulae shows that they contain at least one polar group, and it is the polar group that confers water-solubility.

Thus propylamine, $CH_3CH_2CH_2NH_2$, is very soluble in water; as in ammonia, the nitrogen end of the molecule is negative with respect to the other end, and a dipole exists.

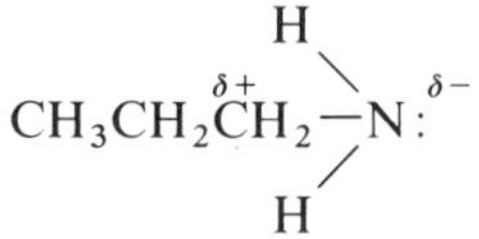

$$CH_3CH_2\overset{\delta+}{C}H_2 \overset{H}{\underset{H}{\diagdown}} N \overset{\delta-}{:}$$

Not only can dipole–dipole interaction with the water result, but hydrogen bonding can also take place:

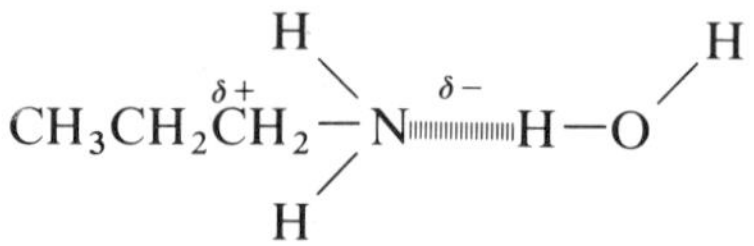

$$CH_3CH_2\overset{\delta+}{C}H_2 \overset{H}{\underset{H}{\diagdown}} N\overset{\delta-}{\cdots\cdots}H-O\overset{H}{\diagup}$$

A further possibility of hydrogen bonding exists, since the two H atoms attached to the nitrogen atom may take part. The situation is shown for one H atom.

$$CH_3CH_2\overset{\delta+}{C}H_2-\overset{\delta-}{N}\cdots\cdots H-O$$

The molecule has thus become solvated by hydrogen bonding, and the amine dissolves.

Molecules containing a nitrogen atom, N, or an oxygen atom, O, will be polar, and these electronegative atoms with their one or two lone pairs of electrons give opportunities for hydrogen bonding.

Thus reasonably small molecules containing the following groups will normally be water-soluble:

$$-O\diagup^{H} \qquad \text{alcohols}$$

$$-O\diagup \qquad \text{ethers}$$

$$\diagup\diagdown C=O \qquad \text{aldehydes and ketones}$$

$$\diagup\diagdown -N \qquad \text{primary, secondary, and tertiary amines}$$

$$\diagdown_{C}\diagup^{N-}_{\overset{\|}{O}} \overset{H}{\underset{}{|}} \qquad \text{peptides (of low molecular weight)}$$

Are you able to think of any examples of these classes of compounds which you have encountered that do not follow this general guiding rule?

Cyclohexane, a six-membered ring compound containing six carbon atoms, is insoluble in water. Yet glucose, a six-membered ring compound containing six carbon atoms, is very soluble in water. Why is this? The structural formulae are given in figure 11.2h.

Figure 11.2h
The structural formulae of glucose and cyclohexane.

On this planet the solvent upon which life is based is water. Plant fluids, blood, and body fluids are aqueous solutions. Sugars, amino acids, peptides of low molecular weight, and amines can be transported rapidly in a living organism because they are water-soluble. Without the polarity conferred upon carbon compounds by a nitrogen or an oxygen atom and the opportunity which this provides for hydrogen bonding with water, life could not have evolved as we know it.

In what other context is hydrogen bonding of central importance in living organisms?

Solvent–solute interaction, and solubility
In every example in this topic where a substance has been dissolved in a liquid there has been some interaction between the substance and the solvent.

Cyclohexane and iodine interact through van der Waals' forces. Ions and polar solvents interact through ion–dipole attractions. Molecules made polar by the presence of a nitrogen or oxygen atom can form hydrogen bonds with water.

Without a solvent–solute interaction a substance cannot dissolve, but for solution this interaction must outweigh the previously existing forces of attraction within the solvent and also those within the solute.

11.3   **Some other solvents**

In this section some other solvents are discussed.

Liquid ammonia

Liquid ammonia boils at $-33\,°C$, and therefore all reactions with it have to be conducted in a vacuum flask or similar vessel.

The molecules of liquid ammonia are highly associated through hydrogen bonding.

Its other important physical constants, from the point of view of solvent qualities, are:

dipole moment$/10^{-30}\,C\,m = 3.0$
relative permittivity $= 22$ (at $-34\,°C$)
dissociation constant, $K = 10^{-30}$ ($2NH_3 \rightleftharpoons NH_4^+ + NH_2^-$)

Anhydrous copper sulphate dissolves readily in liquid ammonia to give a solution containing $Cu(NH_3)_4^{2+}SO_4^{2-}$ in which the structure of the complex ion is the same as that of the hydrated copper(II) ion. Chromium(III) chloride also dissolves to give a complex ion $Cr(NH_3)_6^{3+}Cl_3^-$. The structure of this is the same as hydrated iron(III) chloride.

Silver nitrate dissolves to the extent of 86 g per 100 g of liquid ammonia, and sodium iodide to the extent of 162 g per 100 g of ammonia.

The alkali metals, lithium, sodium, and potassium, and the alkaline earth metals, beryllium, magnesium, calcium, strontium, and barium dissolve in liquid ammonia to form blue solutions. Unlike their behaviour in water they do not decompose the solvent; evaporation yields the metal again. Careful evaporation of the solutions of calcium, strontium, and barium yields solids with the composition $Ca(NH_3)_6$, $Sr(NH_3)_6$, and $Ba(NH_3)_6$. Can you suggest structures for these?

Alkanes are insoluble in liquid ammonia, as is to be expected.

Liquid ammonia is a particularly good solvent in which to conduct reactions with strong reducing agents.

The following are some typical reactions which can be conducted in liquid ammonia. Attempt to classify them, for example as reactions involving a change in oxidation number or as acid-base reactions. In the case of acid-base reactions identify the proton donor and write equations for comparable reactions in water as solvent.

1 $NH_4Br + KNH_2 \longrightarrow KBr + 2NH_3$
2 $Fe + 2NH_4Br \longrightarrow FeBr_2 + H_2 + 2NH_3$
3 $SiCl_4 + 8NH_3 \longrightarrow Si(NH_2)_4 + 4NH_4Cl$
4 $ZnI_2 + 2KNH_2 \longrightarrow Zn(NH_2)_2 + 2KI$

then $Zn(NH_2)_2 + 2KNH_2 \longrightarrow K_2[Zn(NH_2)_4]$
5

6 $+ NH_3 \rightleftharpoons CH_3C\overset{O}{\overset{\|}{-}}NH^-NH_4^+$

(Look at this again after studying Topic 13.4)

Liquid sulphur dioxide
Liquid sulphur dioxide boils at $-10\,°C$, and must be used in a vacuum flask or similar vessel, or under pressure.

Dipole moment$/10^{-30}\,C\,m = 5.3$
Relative permittivity $= 16$ (at $0\,°C$)

Liquid sulphur dioxide dissolves a range of ionic compounds, and a few organic compounds are soluble in it.

Sodium iodide dissolves, and on careful evaporation yields a solvate $NaI, 2SO_2$; aluminium chloride behaves similarly, yielding $AlCl_3, 2SO_2$.

Large quantities of liquid sulphur dioxide are used in the petroleum refining industry, advantage being taken of the fact that some hydrocarbons are soluble in the liquid.

High temperature solvents
All of the solvents which have been discussed so far boil below or at $100\,°C$. This is not a very high temperature. Few molecular solvents are of use much above $100\,°C$.

In the temperature range $100\,°C$ to $2000\,°C$ fused salts or fused metals are used if solvents are required.

Many metals, non-metals, and compounds are soluble in suitably chosen fused salts, such as molten sodium chloride or molten potassium thiocyanate.

A range of metals and non-metals is soluble in several fused metals. Liquid sodium is a solvent that is frequently used. When a substance is dissolved in a molten metal it is surrounded by a sea of free electrons, and this is a very different environment from being dissolved in a molecular solvent and being surrounded by molecules. Consequently chemistry conducted in liquid metals is very different from that in low temperature solvents. This branch of the subject is in a relatively early stage of development, and many aspects of it are not yet understood.

Metallurgy and the production of ceramics are particularly concerned with reactions in high-temperature solvents.

Problems

1 In which of the following liquids would you expect sodium chloride to be most soluble?

  A hexane
  B cyclohexane
  C benzene
  D acetic acid (non-aqueous)
  E bromine

Give reasons for your choice.

2 The lattice energy of potassium iodide is $-640\,\text{kJ mol}^{-1}$ (by direct measurement). The heat of solution of potassium iodide in water is $+21\,\text{kJ mol}^{-1}$. Calculate the hydration energy of potassium iodide and explain how you arrive at your answer.

3 Give the formulae for the following compounds. Arrange them in the order you would expect for their solubility in water, putting the most soluble first.

  A ethane-1,2-diol
  B cyclohexanol
  C cyclopentane

Give reasons for your answer.

4 The lattice energy of rubidium chloride is $-674\,\text{kJ mol}^{-1}$ and its heat of solution is $+19\,\text{kJ mol}^{-1}$. Calculate the hydration energy of rubidium chloride, showing how you arrive at your answer.

5   The lattice energy of calcium fluoride is $-2611 \text{ kJ mol}^{-1}$; its heat of solution is -11 kJ mol^{-1}. Calculate the hydration energy of calcium fluoride, showing how you arrive at your answer.

6   Use the following data to deduce the trend in solvation energies of the sodium, potassium, and rubidium ions. Explain how you arrived at your answer and comment on the trend.

	Lattice Energy	Heat of solution
Sodium chloride	-770	$+\ 5.0$
Potassium chloride	-703	$+17.6$
Rubidium chloride	-674	$+18.8$

Values are in kJ mol^{-1}.

Topic 12
Equilibria: gaseous and ionic

12.1 General introduction. The equilibrium law

You will know of a number of changes which, when allowed to start, do not proceed to completion, that is, all the starting materials (called the reactants) are not changed into new forms or into new substances (called the products). Some examples are:

1 If some liquid bromine is poured into a bottle which is then stoppered, some of the liquid evaporates to form a brown gas, but, unless the volume of liquid used is small, some liquid remains. We say that a *state of equilibrium* has been reached between bromine liquid and bromine gas. This state of equilibrium is represented by

$$Br_2(l) \rightleftharpoons Br_2(g)$$

The symbol $\rightleftharpoons$ indicates that we are considering a system in which equilibrium has been reached and also that the equilibrium state can be reached from either direction; we could start with bromine gas only, at a higher temperature, and cool it, when some would liquefy and give an equilibrium between liquid and gas. A similar equilibrium state is reached by any system of liquid and gas in a *closed* container.

2 When a mixture of water and sodium nitrate is shaken, if there is more sodium nitrate than is required to saturate the water at the temperature concerned, some solid will remain undissolved and an equilibrium state represented by

$$NaNO_3(s) \rightleftharpoons NaNO_3(aq)$$

will be reached. We can approach this equilibrium state from either direction, by dissolving sodium nitrate in water at a given temperature, or by cooling a hot saturated solution of sodium nitrate to the same temperature. If the system of solid sodium nitrate and saturated solution is kept in a closed container at a constant temperature after equilibrium has been attained, no further change in bulk properties can be detected. By 'bulk properties' in this case we mean the concentration of the solution and the amount of undissolved solid. Since, however, the equilibrium state can be approached from either direction it is not unreasonable to suppose that some solid sodium nitrate may dissolve whilst at the same time some dissolved sodium nitrate crystallizes. If this is so, the two processes must go on at the same rate, so that the bulk, or *macroscopic* properties do not alter. This possibility can be investigated by using radioactive isotopes, which can be detected by a Geiger tube. If some solid sodium nitrate in which some of the sodium ions are of a radioactive isotope of sodium (for example $^{24}_{11}$Na) is added to an equilibrium mixture of solid sodium nitrate and its solution, radioactivity can be detected in the solution after some time has elapsed. Thus there must be an interchange of solute between solution and solid. This means that equilibrium must be a dynamic state (and not a static state) in chemical systems.

3 $CaCO_3(s) \rightleftharpoons CaO(s) + CO_2(g)$

When calcium carbonate is heated in a sealed vessel at a fixed temperature an equilibrium state is reached in which calcium carbonate, calcium oxide, and carbon dioxide are all present. (In an open vessel carbon dioxide will escape and the reaction can then go to completion.)

4   An equilibrium state in which hydrogen iodide, hydrogen, and iodine are all present is reached by heating hydrogen iodide gas in a sealed vessel at a fixed temperature.

$2HI(g) \rightleftharpoons H_2(g) + I_2(g)$

5   The formation of esters from carboxylic acids and alcohols, for example the formation of ethyl acetate from ethanol and acetic acid, involves an equilibrium state.

$CH_3CO_2H(l) + C_2H_5OH(l) \rightleftharpoons CH_3CO_2C_2H_5(l) + H_2O(l)$
acetic acid   ethanol   ethyl acetate   water

Equilibrium is attained very slowly indeed at ordinary temperatures in this system. It is attained more rapidly by heating the reagents or by using a catalyst (hydrogen ions).

Characteristics of the equilibrium state

1   A stable state of equilibrium can only be attained in a *closed* system – one that cannot exchange matter with its surroundings. In an open system, which permits matter to enter or leave, stable equilibrium is not possible. (*Note*. A state of equilibrium can also be achieved in an *isolated* system – one that cannot exchange either matter *or energy* with its surroundings. An example of such a system is the contents of a sealed and insulated vessel. We shall not be concerned with such systems in this topic.)

2   The equilibrium state can be approached from either direction, that is, reactants and products can be interchanged. Changes of this kind are said to be *reversible* and the reactions which take place as equilibrium is approached are called *reversible reactions*.

3   Under given conditions of temperature, pressure, and initial concentrations of reactants, properties such as density and concentration of the various parts of the system do not change. These are *intensive properties* (not dependent on the total quantity of matter present) as measured on the macroscopic scale. Hence we can say that the equilibrium state is characterized by constancy of intensive properties. (Properties which do depend on the total quantity of matter in a system are called *extensive properties*. These include mass, volume, and energy content.)

4   Equilibrium is a dynamic state in that opposing changes on the molecular level are continually taking place. The net result of these is that the macroscopic properties of a system in equilibrium do not change. This dynamic aspect of equilibrium means that it is stable under fixed conditions but sensitive to alteration in these conditions. The existence of an equilibrium state can be recognized by taking advantage of its sensitivity to changes in conditions. For example, if a change in temperature, pressure, or concentration leads to an obvious change in a system it is likely to have been at equilibrium before the change took place.

Relative concentrations under equilibrium conditions

It has been stated above that the equilibrium state is characterized by constancy of intensive properties, which include concentration. For some equilibria the concentrations of substances in the various parts of the equilibrium mixture do not depend on the relative proportions of substances used initially, that is before equilibrium is approached, provided that other conditions (temperature and pressure) are constant.

Examples of this type are the equilibria

$$Br_2(l) \rightleftharpoons Br_2(g)$$

and $$NaNO_3(s) \rightleftharpoons NaNO_3(aq)$$

discussed earlier.

For other types of equilibria we would not expect this to happen. For example, in the equilibrium

$$CH_3CO_2H(l) + C_2H_5OH(l) \rightleftharpoons CH_3CO_2C_2H_5(l) + H_2O(l)$$

it might be expected that the concentrations of acetic acid, ethanol, ethyl acetate, and water would depend in some way on the relative proportions of acetic acid and ethanol taken originally. It thus becomes of interest to explore if there is any relationship between the equilibrium concentrations in systems of this kind when different proportions of starting materials are used. Three investigations for this purpose are described briefly below.

1   The distribution of a solute between two immiscible solvents

There have been many investigations of the ratio in which a dissolved substance distributes itself between two liquids in contact with each other, which do not mix, when equilibrium has been established. A typical example, the results of which are given in table 12.1a, used ammonia as the solute, and trichloromethane ($CHCl_3$) and water as the immiscible solvents. Separate solutions of ammonia in water of different concentrations were shaken with trichloromethane until equilibrium was established. How could this be ensured? After the layers had settled on standing, known volumes of each layer were titrated with acid of known molarity.

$$NH_3(aq) + H^+(aq) \longrightarrow NH_4^+(aq)$$

This is quite possible with the trichloromethane layer if some water is added before titration and the mixture shaken well during titration. What happens during this process? From the titration results the concentration of ammonia in each layer at equilibrium, in moles per cubic decimetre ($mol\,dm^{-3}$) was calculated. The results are shown in table 12.1a.

Temperature: 292 K (19°C)

Concentration in CHCl₃ layer, $[NH_3(CHCl_3)]_{eqm}$/mol dm^{-3}	Concentration in water layer, $[NH_3(H_2O)]_{eqm}$/mol dm^{-3}	$\dfrac{[NH_3(H_2O)]_{eqm}}{[NH_3(CHCl_3)]_{eqm}}$
0.0574	1.35	23.5
0.0429	1.05	24.5
0.0348	0.825	23.7
0.0261	0.648	24.8
0.0208	0.513	24.7
0.0163	0.405	24.8

Table 12.1a

Note. Square brackets, [ ], are used to denote concentrations in moles per cubic decimetre. The subscript eqm indicates that concentrations at equilibrium are being considered.

Questions

   i   Why are units given for the figures in the first and second columns but not for those in column three?

   ii   What can you say about the ratio

$$\frac{[NH_3(H_2O)]_{eqm}}{[NH_3(CHCl_3)]_{eqm}}?$$

This ratio is often called the *distribution ratio* (or partition coefficient) for ammonia between trichloromethane and water.

2   The equilibrium between hydrogen, iodine, and hydrogen iodide, all in the gas phase; $H_2(g) + I_2(g) \rightleftharpoons 2HI(g)$

This equilibrium has been extensively investigated. It is established very slowly indeed at room temperature but is attained in a day or so at higher temperatures. The method of investigation is to seal mixtures of hydrogen and iodine (or pure hydrogen iodide, so that equilibrium can be established from both directions) in glass or silica vessels. The vessels are then heated at a known temperature until equilibrium is attained. (How could this be checked?) The vessels are then rapidly cooled to 'freeze' the equilibrium, that is, to preserve the concentrations of reactants and products as they were at high temperature by taking advantage of the fact that any changes at low temperatures will be very slow. The contents of the vessels are then analysed.

One method of doing this is to open the tubes which initially contained hydrogen and iodine, under potassium iodide solution when the iodine and hydrogen iodide dissolve. The solution is then titrated with sodium thiosulphate solution to estimate the iodine. The solution resulting from this titration (from which the iodine has now been removed) is titrated with alkali to find the amount of hydrogen iodide present. (Why is it necessary to remove the iodine first?) The hydrogen can be estimated from the fact that conversion of 1 mole of iodine to hydrogen iodide is accompanied by similar conversion of 1 mole of hydrogen. The tubes which contained hydrogen iodide initially, are opened similarly under potassium iodide solution and the iodine estimated by titration with sodium thiosulphate solution. If the initial weight of hydrogen iodide in the tube is known, the composition of the equilibrium mixture can be calculated. From the known volume of the sealed vessel, equilibrium concentrations of hydrogen, iodine, and hydrogen iodide can be calculated.

Concentrations at equilibrium for the system: $H_2(g) + I_2(g) \rightleftharpoons 2HI(g)$*

Temperature : 698 K (425 °C)

a   *Results obtained by heating hydrogen and iodine in sealed vessels*

$[H_2(g)]_{eqm}$ /mol dm^{-3}	$[I_2(g)]_{eqm}$ /mol dm^{-3}	$[HI(g)]_{eqm}$ /mol dm^{-3}
4.56×10^{-3}	0.74×10^{-3}	13.54×10^{-3}
3.56×10^{-3}	1.25×10^{-3}	15.59×10^{-3}
2.25×10^{-3}	2.34×10^{-3}	16.85×10^{-3}

Table 12.1b

Note. $[H_2(g)]_{eqm}$ means 'the concentration of hydrogen gas at equilibrium, in moles of hydrogen molecules H_2 per cubic decimetre', and the other symbols have corresponding meanings.

* (Adapted from Taylor and Crist (1941), *Journal of the American Chemical Society* **63**, 1381.)

Can you find a constant numerical relationship between the equilibrium concentrations of the reactants and products? Try

$$\frac{[HI(g)]_{eqm}}{[H_2(g)]_{eqm}[I_2(g)]_{eqm}} \quad \text{and} \quad \frac{[HI(g)]^2_{eqm}}{[H_2(g)]_{eqm}[I_2(g)]_{eqm}}$$

Which gives the better constant? Check your conclusion by using the results given in table 12.1c. These were obtained by approaching the equilibrium from the reverse direction, that is starting from hydrogen iodide.

b   *Results obtained by heating hydrogen iodide in sealed vessels*

$[H_2(g)]_{eqm}$ /mol dm^{-3}	$[I_2(g)]_{eqm}$ /mol dm^{-3}	$[HI(g)]_{eqm}$ /mol dm^{-3}
0.48×10^{-3}	0.48×10^{-3}	3.53×10^{-3}
0.50×10^{-3}	0.50×10^{-3}	3.66×10^{-3}
1.14×10^{-3}	1.14×10^{-3}	8.41×10^{-3}

Table 12.1c

Questions
  i   The symbol K_c is used to represent the constant obtained.
Hence

$$K_c = \frac{[HI(g)]^2_{eqm}}{[H_2(g)]_{eqm}[I_2(g)]_{eqm}}$$

when this equilibrium is represented by

$$H_2(g) + I_2(g) \rightleftharpoons 2HI(g)$$

K_c is called the *equilibrium constant*. The subscript c indicates that it is expressed in concentrations. By convention, the concentrations of the substances on the *righthand side* of the equation are always put in the *numerator* (top line of fraction) of the equilibrium constant and those of the substances on the *lefthand side* in the *denominator*.
The equilibrium can also be represented by

$$\tfrac{1}{2}H_2(g) + \tfrac{1}{2}I_2(g) \rightleftharpoons HI(g)$$

What expression must be used to calculate K_c from this equation? Will the values be the same as those obtained from tables 12.1b and 12.1c? If not, how will the two sets of values be related?

ii Find the average value of K_c from the calculations that you have done on the results in tables 12.1b and 12.1c, correct to 2 significant figures. If an equilibrium mixture of hydrogen, iodine, and hydrogen iodide at 698 K contains 0.5 mole hydrogen and 5.43 mole hydrogen iodide, how much iodine is present? Why do you not need to know the volume of the system in order to obtain an answer?

iii Why are the concentrations of hydrogen and iodine the same in each of the first two columns of table 12.1c?

3   The equilibrium between ethyl acetate, water, acetic acid and ethanol,
$$CH_3CO_2C_2H_5(l) + H_2O(l) \rightleftharpoons CH_3CO_2H(l) + C_2H_5OH(l)$$
What is the expression for the equilibrium constant, K_c for this reaction? Is your answer confirmed by the experimental results shown in table 12.1d? Why is the total volume of the equilibrium mixture not needed to calculate K_c?

The equilibrium constant can be determined by allowing mixtures of ethyl acetate, water, and catalyst (2 cm^3 hydrochloric acid) to stand at room temperature and then analysing the contents when equilibrium has been reached by titrating the contents of each flask with alkali. 2 cm^3 of concentrated hydrochloric acid are titrated separately and the volume of alkali needed is subtracted from all the other results to allow for acid added as catalyst. The amount of acetic acid (*d* mole) in the equilibrium mixture is calculated. As the original mixture contained *b* mole ethyl acetate and *c* mole water there must be *d* mole acetic acid, *d* mole ethanol, (*b–d*) mole ethyl acetate, and (*c–d*) mole water at equilibrium.

The results in table 12.1d are adapted from the work of sixth form students of Dauntsey's School, Devizes, Wilts., in 1966. The experiment was carried out at about 293 K (20 °C) and results were obtained by allowing mixtures of ethyl acetate and water, of differing proportions (between 29 cm^3 ester + 1 cm^3 water and 15 cm^3 ester + 15 cm^3 water) plus a fixed volume (2 cm^3) of concentrated hydrochloric acid, to stand for one week. Hydrogen ions from the hydrochloric acid act catalytically in enabling equilibrium to be attained more rapidly.

The quite remarkable effect of the catalyst in this set of experiments can be appreciated when it is known that a mixture of ethyl acetate and water alone would take several years to reach equilibrium at room temperature.

Amount of ethyl acetate at equilibrium/mole	Amount of water at equilibrium /mole	Amount of acetic acid at equilibrium/mole	Amount of ethanol at equilibrium /mole
0.231	0.079	0.065	0.065
0.204	0.118	0.082	0.082
0.150	0.261	0.105	0.105
0.090	0.531	0.114	0.114

Table 12.1d

Taking the value of K_c as 0.27, how many moles of ethyl acetate would you expect to obtain at equilibrium from a starting mixture of 3 mole ethanol and 2 mole acetic acid? What weight of ethyl acetate (in grammes) is this?

The equilibrium law
From many investigations such as those described above, the following general statement emerges.

'For any system in equilibrium, there is a simple relationship between the concentrations of the substances present. If the reaction for the equilibrium is represented by the equation

$$mA + nB \rightleftharpoons pC + qD$$

the expression

$$\frac{[C]_{eqm}^{p}[D]_{eqm}^{q}}{[A]_{eqm}^{m}[B]_{eqm}^{n}} = \text{a constant at a given temperature} = K_c,$$

K_c is called the *equilibrium constant*'. This is known as *the equilibrium law*.

Two important conclusions arise from this law:
1   If K_c is large the equilibrium mixture will contain a high proportion of products, that is, the reaction has gone nearly to completion. If K_c is small, the reaction does not proceed very far at the temperature concerned and the concentration of products is low.
2   The addition of more reactants to a system in equilibrium will result in the formation of more products, the system adjusting itself so that the concentrations again satisfy the value for K_c. Similarly, on addition of more products the equilibrium will move in the opposite direction and the concentrations of reactants will increase. Once K_c for a reaction is known, the relative proportions of reactants and products at equilibrium for any mixture of reactants used initially can be calculated.

When stating the value of K_c for a particular reaction, it is important to indicate the equation on which the constant is based. For example, in the reaction between hydrogen and bromine to form hydrogen bromide, if we write the equation in the form

$$H_2(g) + Br_2(g) \rightleftharpoons 2HBr(g)$$

$$K_c = \frac{[HBr(g)]_{eqm}^2}{[H_2(g)]_{eqm}[Br_2(g)]_{eqm}}$$

At 500 K, the value of this constant is about 10^{12} (reaction very nearly complete).

The same equilibrium can also be represented by

$$\tfrac{1}{2}H_2(g) + \tfrac{1}{2}Br_2(g) \rightleftharpoons HBr(g)$$

but then there is a different equilibrium constant

$$K_c' = \frac{[HBr(g)]_{eqm}}{[H_2(g)]_{eqm}^{1/2}[Br_2(g)]_{eqm}^{1/2}}$$

It will be seen that $K_c' = \sqrt{K_c}$
thus K_c' at 500 K $\approx \sqrt{10^{12}} \approx 10^6$

For reactions in which the number of particles on each side of the equation is the same, as in the example above, the concentration units cancel, and K_c has no units. For all other reactions this is not the case and units for K_c must be stated. Thus in the equilibrium

$$2SO_2(g) + O_2(g) \rightleftharpoons 2SO_3(g)$$

which is the basis of an important process for manufacturing sulphuric acid

$$K_c = \frac{[SO_3(g)]_{eqm}^2}{[SO_2(g)]_{eqm}^2[O_2(g)]_{eqm}}$$

If a particular equilibrium mixture contains x mol dm^{-3} SO_3, y mol dm^{-3} SO_2, and z mol dm^{-3} O_2

$$K_c = \frac{(x \text{ mol dm}^{-3})^2}{(y \text{ mol dm}^{-3})^2(z \text{ mol dm}^{-3})}$$

$$= \frac{x^2}{y^2 z \text{ mol dm}^{-3}} = \frac{x^2}{y^2 z} \text{ dm}^3 \text{ mol}^{-1}$$

One final point: it is often convenient to assume that, in principle, all reactions can proceed to an equilibrium state, even if for practical purposes many of them appear either to go completely to products, or not to start at all. The usefulness of this assumption will become evident later in this topic, and in Topic 15.

12.2   The effect of pressure and temperature on equilibrium
The equilibrium constant expressed in terms of partial pressures

When dealing with reactions involving gases it is often found more convenient to use an equilibrium constant expressed in terms of pressure (K_p) rather than. in terms of concentration (K_c).

The possibility of using a pressure relationship for an equilibrium constant when gases are involved in a reaction arises from the fact that the total pressure of a gas mixture is the sum of separate pressures (the *partial pressures*) exerted by each of the gases in the mixture. The partial pressure of each gas in the mixture is the pressure that it would exert if it alone occupied the volume occupied by the mixture. This was first pointed out by John Dalton, in 1801, and is known as the law of partial pressures. Strictly speaking, it applies to 'ideal' gases only, but under conditions of not too high pressures it is a sufficiently accurate statement to serve for most purposes.

From the general gas equation (discussed in Topic 3),

$$pV = nRT \quad (n = \text{number of moles of gas})$$

$$\therefore \quad \frac{p}{RT} = \frac{n}{V}$$

but $\dfrac{n}{V}$ = concentration of gas = [gas]

$$\therefore \quad p = RT\,[\text{gas}]$$

This means that, for a fixed temperature, the concentration of a gas is proportional to its pressure.

$$[\text{gas}] \propto p$$

Also, since

$$p = \frac{n}{V}RT$$

the partial pressure of each gas in a mixture of gases is proportional to the number of moles of each gas in the mixture. For example, if a mixture of ammonia, hydrogen, and nitrogen contains 1 mole ammonia, 3.6 mole hydrogen and 13.5 mole nitrogen (total 18.1 mole), and the total pressure is 2 atmospheres

partial pressure of ammonia, $p_{NH_3} = \dfrac{1}{18.1} \times 2 = 0.11$ atm

partial pressure of hydrogen, $p_{H_2} = \dfrac{3.6}{18.1} \times 2 = 0.40$ atm

partial pressure of nitrogen, $p_{N_2} = \dfrac{13.5}{18.1} \times 2 = 1.49$ atm

Since [gas] $\propto p$ (the partial pressure of a gas), if we represent an equilibrium situation involving gases only as

$$aX(g) + bY(g) \rightleftharpoons cZ(g)$$

we can write

$$\frac{p^c_{Zeqm}}{p^a_{Xeqm} p^b_{Yeqm}} = K_p$$

Thus, if the proportions of ammonia, hydrogen, and nitrogen given above are in equilibrium at a temperature T kelvin, from the equation

$$N_2(g) + 3H_2(g) \rightleftharpoons 2NH_3(g)$$

$$K_p = \frac{p^2_{NH_3 eqm}}{p_{N_2 eqm} p^3_{H_2 eqm}}$$

$\therefore$ at T K, $K_p = \dfrac{(0.12 \text{ atm})^2}{(1.49 \text{ atm})(0.40 \text{ atm})^3}$

$$= \frac{1.5 \times 10^{-1}}{\text{atm}^2}$$

$$= 1.5 \times 10^{-1} \text{ atm}^{-2}$$

We can, of course, also use the expression

$$\frac{[Z(g)]^c_{eqm}}{[X(g)]^a_{eqm} [Y(g)]^b_{eqm}} = K_c$$

if this is more convenient, but the values obtained for the equilibrium constant will not necessarily be the same, and the units will always be different except, of course, for a gas reaction in which the number of gas molecules is the same on each side of the stoichiometric equation; in this case neither K_c nor K_p has units. K_c and K_p are, however, related to each other by the expression

$$K_p = K_c(RT)^n$$

where n = number of gas molecules on the righthand side of the equilibrium equation minus the number of gas molecules on the lefthand side of the equilibrium equation.

The effect of pressure on equilibrium

For an equilibrium involving gases, if the partial pressure of one of the gases is changed, the system will move in such a way that the partial pressures of reactants and products again give the same value for K_p. If the *total* pressure of the system is increased, the various partial pressures also increase.

Consider the equilibrium

$$2SO_2(g) + O_2(g) \rightleftharpoons 2SO_3(g)$$

for which

$$K_p = \frac{p^2_{SO_3eqm}}{p^2_{SO_2eqm}p_{O_2eqm}}$$

If at a given temperature,

$$p_{SO_3eqm} = a \text{ atm}$$

$$p_{SO_2eqm} = b \text{ atm}$$

$$p_{O_2eqm} = c \text{ atm}$$

$$K_p = \frac{a^2}{b^2c} \text{ atm}^{-1}$$

Now let the pressure be doubled so that the three partial pressures are doubled also. This will mean that momentarily,

$$p_{SO_3} \text{ becomes } 2a \text{ atm}$$

$$p_{SO_2} \text{ becomes } 2b \text{ atm}$$

$$p_{O_2} \text{ becomes } 2c \text{ atm}$$

so that

$$\frac{p^2_{SO_3}}{p^2_{SO_2}p_{O_2}} = \frac{(2a)^2}{(2b)^2 2c} = \frac{a^2}{b^2c} \times \frac{1}{2} \text{ atm}^{-1}$$

The system then adjusts itself so that the partial pressures again satisfy the value for K_p, that is more SO_2 and O_2 must react to form SO_3.

This change results in a decrease in volume, 3 moles of gas changing to 2 moles.

In general, *increasing the pressure on a system in equilibrium produces a change which tends to a decrease in volume.* Conversely, reducing the pressure favours the change which results in an increase in volume. Thus for the sulphur trioxide equilibrium

pressure increase
$\longrightarrow$

$$2SO_2(g) + O_2(g) \rightleftharpoons 2SO_3(g)$$

$\longleftarrow$
pressure decrease

Questions
1 What would be the result of an increase of pressure on the following equilibria?

$$N_2O_4(g) \rightleftharpoons 2NO_2(g)$$

$$CO(g) + 2H_2(g) \rightleftharpoons CH_3OH(g)$$

$$H_2(g) + Br_2(g) \rightleftharpoons 2HBr(g)$$

$$CO_2(g) + NO(g) \rightleftharpoons CO(g) + NO_2(g)$$

2 How would you alter the pressure in order to increase the yield of ethylene from the equilibrium given below?

$$C_2H_6(g) \rightleftharpoons C_2H_4(g) + H_2(g)$$

The effect of temperature change on the value of the equilibrium constant for a reaction

Changes in K_p value for an equilibrium state with temperature
Table 12.2 shows some values of K_p for a number of reactions at different temperatures, together with the heat changes involved when the reactions go to completion.

$$N_2(g) + 3H_2(g) \rightleftharpoons 2NH_3(g)$$

$$\Delta H^{\ominus}_{298} = -92 \text{ kJ}$$

T/K	K_p/atm^{-2}
500	1×10^{-1}
700	8×10^{-5}
1100	5×10^{-8}

$$N_2(g) + O_2(g) \rightleftharpoons 2NO(g)$$

$$\Delta H^{\ominus}_{298} = +180 \text{ kJ}$$

T/K	K_p
700	5×10^{-13}
1100	4×10^{-8}
1500	1×10^{-5}

$$2SO_2(g) + O_2(g) \rightleftharpoons 2SO_3(g)$$

$$\Delta H^{\ominus}_{298} = -197 \text{ kJ}$$

T/K	K_p/atm^{-1}
500	2.5×10^{10}
700	3×10^4
1100	1.3×10^{-1}

$$H_2(g) + I_2(g) \rightleftharpoons 2HI(g)$$

$$\Delta H^{\ominus}_{298} = -9.6 \text{ kJ}$$

T/K	K_p
500	160
700	54
1100	25

Table 12.2
Variation of K_p values with temperature, and values of $\Delta H^{\ominus}_{298}$, for a series of reactions

The principle of Le Chatelier

From the information given in table 12.2 and from investigations that you may have done in class, it may be concluded that:

'An *increase in temperature* of an equilibrium system results in an *increase in the value of the equilibrium constant* if the reaction involved is *endothermic* (ΔH positive), and a *decrease in the value of the equilibrium constant* if the reaction is *exothermic* (ΔH negative).'

An endothermic reaction becomes exothermic when carried out in the reverse direction, for example

$$2SO_2(g) + O_2(g) \longrightarrow 2SO_3; \qquad \Delta H^{\ominus}_{298} = -197 \text{ kJ}$$

$$2SO_3(g) \longrightarrow 2SO_2(g) + O_2(g); \qquad \Delta H^{\ominus}_{298} = +197 \text{ kJ}$$

To increase the temperature of a system in equilibrium thermal energy must be added; the effect of this addition is to cause the equilibrium to change so that thermal energy is absorbed, that is in the direction of the endothermic reaction.

The effect of change in conditions on a system in equilibrium can be predicted qualitatively by a general statement put forward by H. L. Le Chatelier in 1885. This applies both to changes of *concentration* or *pressure* in which the system is adjusting itself so that the concentrations or partial pressures again satisfy the expressions for K_c or K_p; or to changes of *temperature* in which the system is adjusting itself *to a new value of K_c or K_p.*

Le Chatelier's principle of equilibrium
'If a system in equilibrium is subjected to a change, the processes which take place are such as to tend to counteract the effect of the change.'

For chemical systems the changes most frequently involved are those of concentration, pressure, and temperature. The results of such changes on systems in equilibrium are summarized below.

Concentration changes
If the concentration of one of the reactants or products in a system in equilibrium is increased, the effect of the change may be counteracted by the establishment of a new equilibrium so that the concentration of the added substance is reduced.

Example
If more gaseous iodine is added to the equilibrium

$$H_2(g) + I_2(g) \rightleftharpoons 2HI(g)$$

its effect is to cause more hydrogen iodide to be formed, thus reducing the effect of the change. Addition of hydrogen iodide would cause more hydrogen and iodine to be formed. (If the value of the equilibrium constant for the reaction is known, the actual concentration changes can be calculated.)

Pressure changes
If the pressure of the system is increased, the effect of this change may be counteracted by the establishment of a new equilibrium so that the volume is decreased. Conversely, decrease of pressure will cause a new equilibrium to be established such that the volume is increased. Obviously this is only applicable if there is a change in volume in the reaction.

Example
If the pressure of the equilibrium system

$$N_2(g) + 3H_2(g) \rightleftharpoons 2NH_3(g)$$

is increased, its effect is to cause more ammonia to be formed, thus reducing the effect of the change by decreasing the volume of the system and hence the pressure. Decrease of pressure would cause more nitrogen and hydrogen to be formed.

Temperature changes

Changing the temperature of a system in equilibrium results in the establishment of a new equilibrium constant. (This does not happen for changes in concentration or pressure.) Qualitatively, increase in temperature favours the change which takes place with absorption of thermal energy. This may be represented as follows:

temperature increase
→

$$A+B \rightleftharpoons C+D; \qquad \Delta H = +z \, kJ$$

←
temperature decrease

Example
For the reaction

$$H_2O(g)+C(s) \longrightarrow CO(g)+H_2(g); \qquad \Delta H^{\ominus}_{298} = 130 \, kJ,$$

high temperatures therefore favour the production of carbon monoxide and hydrogen, whilst at low temperatures the equilibrium lies almost completely on the lefthand side of the equation.

Summary

If the temperature, pressure, or concentration of one of the substances present in an equilibrium state is altered, there is a shift of equilibrium which can result in a change in the relative concentrations of the substances involved. Changes in pressure and concentration do not affect the value of the equilibrium constant for the reaction. A change in temperature always results in a new equilibrium being established for which the equilibrium constant is different from that for the initial conditions.

12.3 Heterogeneous equilibria

Most of the equilibrium systems discussed so far have involved one *phase* only. (A phase is any part of a system which is of the same composition throughout, that is, it is *homogeneous*, and is separated from the rest of the system by a distinct boundary.) These are called homogeneous equilibria, and they include equilibria in the gas phase and those in solution.

Equilibria which involve two or more phases are called *heterogeneous*.

A common example of a heterogeneous equilibrium is a solid salt in contact with its ions in solution. In experiment 12.3 an equilibrium of this type is investigated.

Experiment 12.3

To investigate the relationship between the concentrations of dissolved ions in equilibrium with a solid

The system to be studied is:

$$AgIO_3(s) \rightleftharpoons Ag^+(aq) + IO_3^-(aq)$$

Procedure

A series of mixtures is made up from different proportions of 0.005M solution of silver ions and 0.005M solution of iodate ions. The mixtures are shaken and allowed to stand at a constant temperature until equilibrium is reached. This occurs when no more precipitate of silver salt is formed.

$$Ag^+(aq) + IO_3^-(aq) \rightleftharpoons AgIO_3(s)$$

The precipitates are removed by filtration (without washing the precipitate) and the filtrates analysed to find the concentration of silver and of iodate ions in them. There will be different proportions for each mixture. The object of the experiment is to find out what the value of

$$[Ag^+(aq)]_{eqm}[IO_3^-(aq)]_{eqm}$$

is in each solution.

Analysis of the filtrates is carried out by titrating known volumes with a solution of iodide ions in the presence of acid to determine the silver ion concentration. The reaction

$$Ag^+(aq) + I^-(aq) \rightleftharpoons AgI(s)$$

is very nearly complete under these conditions. When all the silver has been precipitated thus, the iodide concentration rises and it reacts with the iodate to give iodine. This iodine gives a green colour with the starch indicator.

$$IO_3^-(aq) + 5I^-(aq) + 6H^+(aq) \longrightarrow 3I_2(aq) + 3H_2O(l)$$

This method involves a very small error since silver iodide is very slightly soluble, but this is not appreciable. If excess iodide is now added, the liberated iodine can be titrated with thiosulphate solution and the iodate ion concentration can be found from this.

Experimental directions

Suitable mixtures are given in table 12.3a.

	0.005M AgNO$_3$(aq)/cm^3	0.005M KIO$_3$(aq)/cm^3
A	60	40
B	55	45
C	50	50
D	45	55
E	40	60

Table 12.3a

You will be given one of these to investigate, so that the class covers the complete range. Measure the amounts of solution for your mixture, from burettes, into a clean dry flask. Cork the flask, shake the mixture well and allow it to stand overnight at a constant temperature. A water bath maintained at a fixed temperature is most suitable for this but quite satisfactory results can be obtained by keeping the mixture at room temperature. This temperature should be recorded.

Filter the mixture, using a dry funnel and a dry filter paper. (Why should these be dry?) Collect the filtrate in a second clean, dry flask. Using a measuring cylinder, transfer 75 cm^3 of the filtrate into another flask (250 cm^3) as accurately as possible. Add about 1 cm^3 of starch solution and about 20 cm^3 bench dilute (1M) sulphuric acid. Titrate with potassium iodide solution (for concentration, see table 12.3b). The end-point occurs with the formation of the first permanent blue-green colour. To this solution add about 1 g solid potassium iodide and a further 20 cm^3 bench dilute (1M) sulphuric acid, swirl the flask to dissolve the precipitate, and titrate the liberated iodine with sodium thiosulphate solution (for concentration, see table 12.3b). The end-point occurs when the blue colour is just discharged.

Mixture used	Concentration of iodide solution /mol dm^{-3}	Concentration of thiosulphate solution /mol dm^{-3}
A	0.004	0.0075
B	0.004	0.0075
C	0.0015	0.0075
D	0.0015	0.025
E	0.0015	0.025

Table 12.3b

From the titration results, calculate the concentration (in mol dm^{-3}) of silver and of iodate ions in the solution and hence find the value of

$$[Ag^+(aq)]_{eqm}[IO_3^-(aq)]_{eqm}$$

For sparingly soluble substances this product of concentration, called the *solubility product*, is reasonably constant with different ionic concentrations of cation and anion. The concentration of the solid phase is always included in the value of the constant. For fairly soluble substances, however, the Equilibrium Law does not apply. The crowding together of the ions and the forces of attraction and repulsion arising from their charges make the concentration terms inappropriate for this purpose. In the very dilute solutions given by sparingly soluble salts such interfering effects become negligible. Thus we do not use relationships such as

$$K_{sp} = [Na^+(aq)]_{eqm}[NO_3^-(aq)]_{eqm}$$

for the very soluble salt, sodium nitrate; but for silver chloride, which is only slightly soluble, the constant

$$K_{sp} = [Ag^+(aq)]_{eqm}[Cl^-(aq)]_{eqm}$$
$$= 1.7 \times 10^{-10} \, mol^2 \, dm^{-6} \, at \, 25 \, °C$$

applies with fair accuracy to all solutions containing silver ions and chloride ions in contact with solid silver chloride. Some solubility product values are given in the *Book of Data*.

Other heterogeneous equilibria

The fact that the concentration of solid silver iodate is constant in the equilibrium

$$AgIO_3(s) \rightleftharpoons Ag^+(aq) + IO_3^-(aq)$$

is an example of a general rule. *In heterogeneous equilibria the concentrations of all pure solid and pure liquid phases can be taken as constant.* Some examples will illustrate this.

For the equilibrium

$$CaCO_3(s) \rightleftharpoons CaO(s) + CO_2(g)$$

we could write

$$K = \frac{[CaO(s)]_{eqm}[CO_2(g)]_{eqm}}{[CaCO_3(s)]_{eqm}}$$

In fact, however, the concentrations of CaO and $CaCO_3$ are *constant*.

No change in the system can alter these concentrations. This means that, if we re-arrange the above expression to

$$\frac{K[CaCO_3(s)]_{eqm}}{[CaO(s)]_{eqm}} = [CO_2(g)]_{eqm}$$

the lefthand side will be a constant. By convention this is taken as the value of K_c for the system

$$K_c = [CO_2(g)]_{eqm} (= 3.4 \times 10^{-5} \text{ mol dm}^{-3} \text{ at 873 K})$$

Or, if the equilibrium constant is based on partial pressures,

$$K_p = p_{CO_2 eqm}$$

$[K_p = 2.4 \times 10^{-3} \text{ atm at 873 K (600 °C), and 1.4 atm at 1173 K (900 °C)}]$

Similarly in the equilibrium

$$Br_2(l) \rightleftharpoons Br_2(g)$$

The concentration of liquid bromine is constant, hence

$$K_p = p_{Br_2 eqm} (= 0.228 \text{ atm at 293 K})$$

Reactions between metals and metal ions are examples of ionic, heterogeneous equilibria, for example

$$Cu(s) + 2Ag^+(aq) \rightleftharpoons Cu^{2+}(aq) + 2Ag(s)$$

For this

$$K_c = \frac{[Cu^{2+}(aq)]_{eqm}}{[Ag^+(aq)]^2_{eqm}}$$

$$= 3 \times 10^{15} \text{ dm}^3 \text{ mol}^{-1} \text{ at 298 K}$$

Equilibria of this type are studied in section 15.1.

The equilibrium constant for the ionization of water

This is a very important equilibrium, and arises from the fact that even highly pure water is slightly dissociated into ions.

$$H_2O(l) \rightleftharpoons H^+(aq) + OH^-(aq)$$

The proportion of water molecules that ionize in this way is very small so that, although the ionization increases with temperature rise, the concentration of water, $[H_2O(l)]$, can be treated as constant. A special symbol, K_W, is used for the equilibrium constant of this reaction

$$K_W = [H^+(aq)]_{eqm}[OH^-(aq)]_{eqm}$$

$$K_W = 10^{-14} \text{ at 298 K, and } 9.5 \times 10^{-14} \text{ at 333 K}$$

Is the ionization of water an endothermic or an exothermic process?

This equilibrium will be studied in detail in section 15.4.

Problems

* Indicates that the *Book of Data* may be used.

1   11 g of ethyl acetate were mixed with 18 cm^3 of molar hydrochloric acid in a flask and allowed to stand at constant temperature until equilibrium had been reached.

$$CH_3CO_2C_2H_5(l) + H_2O(l) \rightleftharpoons CH_3CO_2H(l) + C_2H_5OH(l)$$

The contents of the flask were titrated with molar alkali solution and 106 cm^3 of the alkali were required. Calculate the equilibrium constant K_c. (Assume that 18 cm^3 of molar hydrochloric acid contains 18 g of water.)

2   Pentene (C_5H_{10}) reacts with acetic acid to produce pentyl acetate, the equilibrium

$$C_5H_{10} + CH_3CO_2H \rightleftharpoons CH_3CO_2C_5H_{11}$$

being established. When a solution of 0.02 mole pentene and 0.01 mole acetic acid in 600 cm^3 of an inert solvent was allowed to reach equilibrium at 15 °C, 0.009 mole pentyl acetate was formed.

 i   How many moles of (a) pentene and (b) acetic acid were present in the solution at equilibrium?

 ii   Write down the expression for the equilibrium constant, K_c, for the above reaction.

 iii   Complete the following expressions:

$$[C_5H_{10}]_{eqm} = \frac{\times}{} \text{ mol dm}^{-3}$$

$$[CH_3CO_2H]_{eqm} = \frac{\times}{} \text{ mol dm}^{-3}$$

$$[CH_3CO_2C_5H_{11}]_{eqm} = \frac{\times}{} \text{ mol dm}^{-3}$$

 iv   Use your answers to (*ii*) and (*iii*) to calculate the value of K_c at 15 °C.

3   The equilibrium

$$N_2O_4 \rightleftharpoons 2NO_2$$

can be established in chloroform solution at temperatures near 0 °C, and the composition of the equilibrium mixture can be calculated from the density of colour of the solution (N_2O_4 is colourless, NO_2 is brown). In a solution of this kind, at 10 °C, the concentration of NO_2 molecules was found to be 0.0014 mol dm^{-3} and the concentration of N_2O_4 molecules, 0.13 mol dm^{-3}. Calculate the value of K_c for the reaction at 10 °C.

4   For the equilibrium

$$PCl_5(g) \rightleftharpoons PCl_3(g) + Cl_2(g)$$

$$K_c = 0.19 \text{ mol } dm^{-3} \text{ at } 250 \text{ °C}.$$

2.085 g phosphorus pentachloride was heated to 250 °C in a sealed vessel of capacity 500 cm^3 and maintained at this temperature until equilibrium was established.

   i   What would be the concentration of PCl_5, in mol dm^{-3}, if no change took place? (Assume that the volume of the vessel remains constant.)

   ii   If the concentration of chlorine at equilibrium is x mol dm^{-3}, what are the equilibrium concentrations of (a) PCl_5, (b) PCl_3, in mol dm^{-3}?

   iii   What is the expression for the equilibrium constant, K_c, for the above reaction?

   iv   Insert the values obtained in (*ii*) and the value of K_c given above into the equilibrium constant expression, and thus calculate the value of x. (This involves solving a quadratic equation, one root of which is obviously absurd as an answer to this problem.)

   v   What are the concentrations of PCl_5, PCl_3, and Cl_2 present at equilibrium?

   (Atomic weights: P = 31 ; Cl = 35.5)

5   Acetone and hydrocyanic acid react in ethanol solution to form acetone cyanohydrin, according to the equilibrium equation

$$CH_3COCH_3 + HCN \rightleftharpoons CH_3-\underset{\underset{\displaystyle CN}{|}}{\overset{\overset{\displaystyle OH}{|}}{C}}-CH_3$$

At 20 °C, K_c for this equilibrium is 32.8 dm^3 mol^{-1}. If 100 cm^3 of 0.1M solution of acetone in ethanol is mixed with 100 cm^3 of 0.2M solution of hydrocyanic acid in ethanol, what weight of acetone cyanohydrin will be formed at equilibrium?

   (Atomic weights: H = 1 ; C = 12 ; N = 14 ; O = 16)

6 A sealed vessel containing 0.0023 mole hydrogen iodide gas was heated at 900 K until the equilibrium

$$2HI(g) \rightleftharpoons H_2(g) + I_2(g)$$

was attained. The bulb was then cooled rapidly and opened under potassium iodide solution (in which the iodine dissolves). The resulting solution was titrated with 0.1 M sodium thiosulphate solution, 6.4 cm³ being required to react completely with the iodine.

$$2S_2O_3^{2-}(aq) + I_2(aq) \longrightarrow S_4O_6^{2-}(aq) + 2I^-(aq)$$

i Calculate the number of moles of iodine present at equilibrium.
ii How many moles of hydrogen were present at equilibrium?
iii How many moles of hydrogen iodide were present at equilibrium?
iv Calculate the value of K_c for this equilibrium at 900 K.

7 For the equilibrium

$$C_2H_5OH(l) + C_2H_5CO_2H(l) \rightleftharpoons C_2H_5CO_2C_2H_5(l) + H_2O(l)$$
$$\text{propionic acid} \quad \text{ethyl propionate}$$

$K_c = 7.5$ at 50 °C. What weight of ethanol must be mixed with 60 g propionic acid at 50 °C in order to obtain 80 g ethyl propionate in the equilibrium mixture?

(Atomic weights: $H = 1$; $C = 12$; $O = 16$)

8 Sulphur dioxide and oxygen in the ratio 2 moles : 1 mole were mixed at constant temperature and a constant 9 atmospheres pressure in the presence of a catalyst. At equilibrium, one third of the sulphur dioxide had been converted into sulphur trioxide:

$$2SO_2(g) + O_2(g) \rightleftharpoons 2SO_3(g)$$

Calculate the equilibrium constant (K_p) for this reaction under these conditions.

9 $2NO_2(g) \rightleftharpoons 2NO(g) + O_2(g)$

For this equilibrium, a particular equilibrium mixture has the composition 0.96 mole $NO_2(g)$, 0.04 mole $NO(g)$, 0.02 mole $O_2(g)$ at 700 K and 0.2 atmosphere.

i Calculate the equilibrium constant K_p for this reaction under the stated conditions.
ii Calculate the average molecular weight of the mixture under the stated conditions.

10   For the following equilibrium K_p at 373 K is 3.8×10^{-5}

$$CO_2(g) + H_2(g) \rightleftharpoons CO(g) + H_2(g)$$

i   Suppose that 1 mole of carbon dioxide and 1 mole of hydrogen were put into a vessel at 2 atmospheres and 373 K. Calculate how many moles of carbon monoxide would be present when equilibrium is reached.

ii   Suppose that 1 mole of carbon dioxide and 1 mole of hydrogen were mixed at 4 atmospheres and 373 K. How many moles of carbon monoxide would be present when equilibrium is reached? Comment on the relationship between your answers to *i* and *ii*.

iii K_p for the reaction at 298 K is 10^{-5}. Would you expect the yield of carbon monoxide at 298 K to be greater than, less than, or the same as, the yield from the same mixture at 373 K and at the same pressure? Give the reasons for your decision.

iv   Deduce whether the forward reaction is exothermic or endothermic. Give the reasons for your decision.

11   0.2 mole of carbon dioxide was heated with excess carbon in a closed vessel until the following equilibrium was attained.

$$CO_2(g) + C(s) \rightleftharpoons 2CO(g)$$

It was found that the average molecular weight of the gaseous equilibrium mixture was 36.

i   Calculate the mole fraction of carbon monoxide in the equilibrium gaseous mixture.

ii   The pressure at equilibrium in the vessel was 12 atmospheres. Calculate K_p for the equilibrium at the temperature of the experiment.

iii   Calculate the mole fraction of carbon monoxide which would be present in the equilibrium mixture if the pressure were reduced to 2 atmospheres at the same temperature.

*12   A saturated solution of strontium carbonate was filtered. When 50 cm^3 of the filtrate were added to 50 cm^3 of molar sodium carbonate solution, some strontium carbonate was precipitated. Calculate the concentration of strontium ions (moles per cubic decimetre) remaining in the solution. All the work was done at 25 °C.

*13   Which of the following pairs of 0.001M solutions should form a precipitate when equal volumes are mixed at 25 °C? Explain how you arrive at your decision, and, if there is a precipitate, name it.

i   silver nitrate and sodium chloride
ii calcium hydroxide and sodium carbonate
iii silver nitrate and potassium bromate
iv magnesium sulphate and sodium hydroxide

*14   A student attempted to determine the concentration of barium ions in a solution of barium chloride by the following procedure: To 25 cm³ of the boiling solution of barium chloride he added 25 cm³ of M sulphuric acid. He filtered the precipitate of barium sulphate, washed it with 200 cm³ of water, and then ignited the filter paper and its contents in a weighed crucible. After cooling, he added two drops of concentrated sulphuric acid and heated the crucible again, until the excess of sulphuric acid had evaporated. The crucible was re-weighed, and the weight of barium sulphate, after subtracting the weight of the filter paper ash, was found to be 0.570 g.

i   Calculate the apparent concentration of barium ions (mol dm⁻³) in the original solution.

ii What percentage error would be caused by washing the precipitate with 200 cm³ of water at 25 °C?

iii Was it necessary to wash the precipitate? Give your reasons.

iv Suggest a reason for adding some concentrated sulphuric acid to the crucible towards the end of the ignition.

400

Index

References to specific salts are indexed under the name of the appropriate cation; references to substituted organic compounds are indexed under the name of the parent compound.

a

acetic acid, dimerization, 342
    model, 292
    preparation, 278
    reactions, 292–3
acetone, as solvent, 358, 364
    dipole moment and permittivity, 358
acetone–aniline system, 338
acetone–tetrachloromethane system, 335, 348
acetone–trichloromethane system, to study, 329–31, 334–5
    hydrogen bonding in, 337
acetyl chloride, hydrolysis, 298
    model, 296
actinides, 16
acyl chlorides, preparation, 295
    reactions, 305
addition reactions, 274–5
alcohols, bond energies, 182–6
    compared with phenols, 286–7
    liquid flow, 342–3
    nomenclature, 314–15
    preparation: from acyl chlorides, 305; from alkenes, 307;
        from carboxylic acids, 295
    reactions, 275–85
    solubility, 369
aldehydes, preparation: from acyl chlorides, 305; from alcohols, 284
    solubility, 369
alkalis, reaction with halogens, 114
alkanes, bond energies, 186–7
    insolubility in liquid ammonia, 371
    nomenclature, 310–12
    preparation, 304
    van der Waals' forces, 349
alkanes, halogeno-, as intermediates, 298
    nomenclature, 315
    preparation: from alcohols, 284; from alkenes, 274–5; from silver carboxylates, 295
    rates of hydrolysis, 297
    reactions, 304
alkenes, addition reactions, 274–5
    nomenclature, 313
    preparation: from alcohols, 283; from halogenoalkanes, 304

alkyl cyanides, 307
    preparation, 304
Alloprene, 132
alpha-particle scattering, 70–2
    Geiger–Marsden paper, 94–6
Alsophila pometaria, 285
aluminium, electron density map, 259
    electronic structure, 111
    ion, solvation, 364
    physical data, 21, 232
    reaction with chlorine, 23–5
aluminium chloride, analysis, 29–31
    heat of formation, 32
    preparation, 23–5
    solubility in liquid sulphur dioxide, 372
    structure, 233
aluminium oxide, heat of formation, 180–1
    structure, 233
aluminium potassium sulphate, preparation, 149–50
alums, 149–50
amethocaine, 306
amides, preparation: from acyl chlorides, 305; from ammonium salts, 295
amines, bond length and energy, 245
    preparation: from alkyl cyanides, 307; from halogenoalkanes, 304
    solubility, 369
ammonia, bond angles, 243–4
    coordinated, 368
    distribution between immiscible solvents, 378
    heat of formation, 172
    liquid: as solvent, 371–2; dipole moment and permittivity, 358; hydrogen bonding, 338, 339
    physical data, 341
ammonia–boron trifluoride addition compound, 245–6
ammonium iodide, tetramethyl-, rate of formation, 356
ammonium ion, electronic structure, 246
anaesthetics, 129, 306
analysis, atomic-emission spectroscopic, 99–104
    volumetric, of chlorides, 27–31
Ångstrom unit, definition, 218n.
anhydrite, 344–5
aniline–acetone system, 338
anisic acid, electron density map, 238
anisotropy, 230–2
antibacterials, chlorine solution, 133–4
    phenols, 290–1
antifreeze, 285

Models to show the structures of some minerals

Zinc blende, ZnS

Marcasite, FeS$_2$

Calcite, CaCO$_3$

Mica, K$_2$O. 3Al$_2$O$_3$. 6SiO$_2$. 2H$_2$O